LANCHESTER LIBRARY
D1629851

Fuzzy Set Theory—and Its Applications, Second, Revised Edition

Fuzzy Set Theory—and Its Applications

Second, Revised Edition

H.-J. Zimmermann

Kluwer Academic Publishers
Boston/Dordrecht/London

Distributors for North America:

Kluwer Academic Publishers
101 Philip Drive
Assinippi Park
Norwell, Massachusetts 02061 USA

Distributors for all other countries:

Kluwer Academic Publishers Group
Distribution Centre
Post Office Box 322
3300 AH Dordrecht, THE NETHERLANDS

Library of Congress Cataloging-in-Publication Data

Zimmermann, H.-J. (Hans-Jürgen), 1934–
Fuzzy set theory and its applications/H.-J. Zimmermann.—2nd ed.
p. cm.
Includes bibliographical references and index.
ISBN 0-7923-9075-X
1. Fuzzy sets. 2. Operations research. I. Title.
QA248.Z55 1990
511.3'22—dc20 90-38077
CIP

Printed on acid-free paper.
Printed in the United States of America

Contents

List of Figures

List of Tables

Foreword

As its name implies, the theory of fuzzy sets is, basically, a theory of graded concepts—a theory in which everything is a matter of degree or, to put it figuratively, everything has elasticity.

In the two decades since its inception, the theory has matured into a wide-ranging collection of concepts and techniques for dealing with complex phenomena which do not lend themselves to analysis by classical methods based on probability theory and bivalent logic. Nevertheless, a question that is frequently raised by the skeptics is: Are there, in fact, any significant problem-areas in which the use of the theory of fuzzy sets leads to results that could not be obtained by classical methods?

Professor Zimmermann's treatise provides an affirmative answer to this question. His comprehensive exposition of both the theory and its applications explains in clear terms the basic concepts that underlie the theory and how they relate to their classical counterparts. He shows through a wealth of examples the ways in which the theory can be applied to the solution of realistic problems, particularly in the realm of decision analysis, and motivates the theory by applications in which fuzzy sets play an essential role.

An important issue in the theory of fuzzy sets which does not have a counterpart in the theory of crisp sets relates to the combination of fuzzy sets through disjunction and conjunction or, equivalently, union and intersection. Professor Zimmermann and his associates at the Technical University of Aachen have made many important contributions to this problem and were the first to introduce the concept of a parametric family of connectives which can be chosen to fit a particular application. In recent years, this issue has given rise to an extensive literature dealing with *t*-norms and related concepts which link some aspects of the theory of fuzzy sets to the theory of probabilistic metric spaces developed by Karl Menger.

Another important issue addressed in Professor Zimmermann's treatise relates to the distinction between the concepts of probability and possibility, with the latter concept having a close connection with that of membership in a fuzzy set. The concept of possibility plays a particularly important role in the representation of meaning, in the management of uncertainty in expert systems, and in applications of the theory of fuzzy sets to decision analysis.

As one of the leading contributors to and practitioners of the use of fuzzy sets in decision analysis, Professor Zimmermann is uniquely qualified to address the complex issues arising in fuzzy optimization problems and, especially, fuzzy mathematical programming and multicriterion decision making in a fuzzy environment. His treatment of these topics is comprehensive, up-to-date, and illuminating.

In sum, Professor Zimmermann's treatise is a major contribution to the literature of fuzzy sets and decision analysis. It presents many original results and incisive analyses. And, most importantly, it succeeds in providing an excellent introduction to the theory of fuzzy sets—an introduction that makes it possible for an uninitiated reader to obtain a clear view of the theory and learn about its applications in a wide variety of fields.

The writing of this book was a difficult undertaking. Professor Zimmermann deserves to be congratulated on his outstanding accomplishment and thanked for contributing so much over the past decade to the advancement of the theory of fuzzy sets as a scientist, educator, administrator, and organizer.

L.A. Zadeh

Preface

Since its inception 20 years ago the theory of fuzzy sets has advanced in a variety of ways and in many disciplines. Applications of this theory can be found, for example, in artificial intelligence, computer science, control engineering, decision theory, expert systems, logic, management science, operations research, pattern recognition, and robotics. Theoretical advances have been made in many directions. In fact it is extremely difficult for a newcomer to the field or for somebody who wants to apply fuzzy set theory to his problems to recognize properly the present "state of the art." Therefore, many applications use fuzzy set theory on a much more elementary level than appropriate and necessary. On the other hand, theoretical publications are already so specialized and assume such a background in fuzzy set theory that they are hard to understand. The more than 4000 publications that exist in the field are widely scattered over many areas and in many journals. Existing books are edited volumes containing specialized contributions or monographs that focus only on specific areas of fuzzy sets, such as pattern recognition [Bezdek 1981], switching functions [Kandel and Lee 1979], or decision making [Kickert 1978]. Even the excellent survey book by Dubois and Prade [1980a] is primarily intended as a research compendium for insiders rather than an introduction to fuzzy set theory or a textbook. This lack of a comprehensive and modern text is particularly recognized by newcomers to the field and by those who want to teach fuzzy set theory and its applications.

The primary goal of this book is to help to close this gap—to provide a textbook for courses in fuzzy set theory and a book that can be used as an introduction.

One of the areas in which fuzzy sets have been applied most extensively is in modeling for managerial decision making. Therefore, this area has been selected for more detailed consideration. The information has been divided into two volumes. The first volume contains the basic theory of fuzzy sets and some areas of application. It is intended to provide extensive coverage of the theoretical and applicational approaches to fuzzy sets. Sophisticated formalisms have not been included. I have tried to present the basic theory and its extensions as detailed as necessary to be comprehended by those who have not been exposed to fuzzy set theory. Examples and exercises serve to illustrate the concepts even more clearly. For the interested or more advanced reader, numerous references to recent literature are included that should facilitate studies of specific areas in more detail and on a more advanced level.

The second volume is dedicated to the application of fuzzy set theory to the area of human decision making. It is self-contained in the sense that all concepts used are properly introduced and defined. Obviously this cannot be done in the same breadth as in the first volume. Also the coverage of fuzzy concepts in the second volume is restricted to those that are directly used in the models of decision making.

It is advantageous but not absolutely necessary to go through the first volume before studying the second. The material in both volumes has served as texts in teaching classes in fuzzy set theory and decision making in the United States and in Germany. Each time the material was used, refinements were made, but the author welcomes suggestions for further improvements.

The target groups were students in business administration, management science, operations research, engineering, and computer science. Even though no specific mathematical background is necessary to understand the books, it is assumed that the students have some background in calculus, set theory, operations research, and decision theory.

I would like to acknowledge the help and encouragement of all the students, particularly those at the Naval Postgraduate School in Monterey and at the Institute of Technology in Aachen (F.R.G.), who improved the manuscripts before they became textbooks. I also thank Mr. Hintz who helped to modify the different versions of the book, worked out the examples, and helped to make the text as understandable as possible. Ms. Grefen typed the manuscript several times without losing her patience. I am also indebted to Kluwer Academic Publishers for making the publication of this book possible.

H.-J. Zimmermann

Preface for the Revised Edition

Since this book was first published in 1985, Fuzzy Set Theory has had an unexpected growth. It was further developed theoretically and it was applied to new areas. A number of very good books have appeared, primarily dedicated to special areas such as Possibility Theory [Dubois and Prade 1988a], Fuzzy Control [Sugeno 1985a; Pedrycz 1989], Behavioral and Social Sciences [Smithson 1987], and others have been published. Many new edited volumes, either dedicated to special areas or with a much wider scope, have been added to the existing ones. Thousands of articles have been published on fuzzy sets in various journals. Successful real applications of fuzzy set theory have also increased in number and in quality. In particular, applications of fuzzy control, fuzzy computers, expert system shells with capabilities to process fuzzy information, and fuzzy decision support systems have become known and have partly already proven their superiority over more traditional tools.

One thing, however, does not seem to have changed since 1985: access to the area has not become easier for newcomers. I do not know of any introductory yet comprehensive book or textbook that will facilitate entering into the area of fuzzy sets or that can be used in classwork.

I am, therefore, very grateful to Kluwer Academic Publishers for having agreed to publish a revised edition of the book, which four times has already been printed without improvement. In this revised edition all typing and other errors have been eliminated. All chapters have been updated. The chapters on possibility theory (8), on fuzzy logic and approximate reasoning (9), on expert systems and fuzzy control (10), on decision making (12), and on fuzzy set models in operations research (13) have been restructured and rewritten. Exercises have been added to almost all chapters and a teacher's manual is available on request.

The intention of the book, however, has not changed: While the second volume [Zimmermann 1987] focuses on decision making and expert systems and introduces fuzzy set theory only where and to the extent that it is needed, this book tries to offer a didactically prepared text which requires hardly any special mathematical background of the reader. It tries to introduce fuzzy set theory as comprehensively as possible, without delving into very theoretical areas or presenting any mathematical proofs which do not contribute to a better understanding. It rather offers numerical examples wherever possible. I would like to thank very much Mr. C. von Altrock, Ms. B. Lelke, Mr. R. Weber, and Dr. B. Werners for their active participation in preparing this revised edition. Mr. Andrée and Mr. Lehmann kindly prepared the figures. Ms. Oed typed and retyped manuscripts over and over again and helped us to arrive at the final manuscript of the book. We are all obliged to Kluwer Academic Publishers for the opportunity to publish this volume and for the good cooperation in preparing it.

H.-J. Zimmermann

Fuzzy Set Theory—and Its Applications, Second, Revised Edition

1 INTRODUCTION TO FUZZY SETS

1.1 Crispness, Vagueness, Fuzziness, Uncertainty

Most of our traditional tools for formal modeling, reasoning, and computing are crisp, deterministic, and precise in character. By crisp we mean dichotomous, that is, yes-or-no-type rather than more-or-less type. In conventional dual logic, for instance, a statement can be true or false—and nothing in between. In set theory, an element can either belong to a set or not; and in optimization, a solution is either feasible or not. Precision assumes that the parameters of a model represent exactly either our perception of the phenomenon modeled or the features of the real system that has been modeled. Generally precision also implies that the model is unequivocal, that is, that it contains no ambiguities.

Certainty eventually indicates that we assume the structures and parameters of the model to be definitely known, and that there are no doubts about their values or their occurrence. If the model under consideration is a formal model [Zimmermann 1980, p. 127], that is, if it does not pretend to model reality adequately, then the model assumptions are in a sense arbitrary, that is, the model builder can freely decide which model characteristics he chooses. If, however, the model or theory asserts to be factual [Popper 1959; Zimmermann 1980], that is, conclusions drawn from these models have a bearing on reality and they are supposed to model reality adequately, then the modeling language has to be suited to model the characteristics of the situation under study appropriately.

The utter importance of the modeling language is recognized by Apostel, when he says:

> The relationship between formal languages and domains in which they have models must in the empirical sciences necessarily be guided by two considerations that are by no means as important in the formal sciences:
> (a) The relationship between the language and the domain must be closer because they are in a sense produced through and for each other;
> (b) extensions of formalisms and models must necessarily be considered because everything introduced is introduced to make progress in the description of the objects studied. Therefore we should say that the formalization of the concept of approximate constructive necessary satisfaction is the main task of semantic study of models in the empirical sciences. [Apostel 1961, p. 26]

Because we request that a modeling language be unequivocal and nonredundant on one hand and, at the same time, catches semantically in its terms all that is important and relevant for the model, we seem to have the following problem. Human thinking and feeling, in which ideas, pictures, images, and value systems are formed, first of all has certainly more concepts or comprehensions than our daily language has words. If one considers, in addition, that for a number of notions we use several words (synonyms) then it becomes quite obvious that the power (in a set theoretic sense) of our thinking and feeling is much higher than the power of a living language. If in turn we compare the power of a living language with the logical language, then we will find that logic is even poorer. Therefore it seems to be impossible to guarantee a one-to-one mapping of problems and systems in our imagination and a model using a mathematical or logical language.

One might object that logical symbols can arbitrarily be filled with semantic contents and that by doing so the logical language becomes much richer. It will be shown that it is very often extremely difficult to appropriately assign semantic contents to logical symbols.

The usefulness of the mathematical language for modeling purposes is undisputed. However, there are limits to the usefulness and the possibility of using classical mathematical language, based on the dichotomous character of set theory, to model particular systems and phenomena in the social sciences: "There is no idea or proposition in the field, which can not be put into mathematical language, although the utility of doing so can very well be doubted" [Brand 1961]. Schwarz [1962] brings up another argument against the unreflected use of mathematics, if he states: "An argument, which is only convincing if it is precise loses all its force if the assumptions on which it is based are slightly changed, while an argument, which is convincing but imprecise may well be stable under small perturbations of its underlying axioms." For factual models or modeling languages two major complications arise:

1. Real situations are very often not crisp and deterministic and they cannot be described precisely.
2. The complete description of a real system often would require by far more detailed data than a human being could ever recognize simultaneously, process and understand.

This situation has already been recognized by thinkers in the past. In 1923 the philosopher B. Russell [1923] referred to the first point when he wrote:

> All traditional logic habitually assumes that precise symbols are being employed. It is therefore not applicable to this terrestrial life but only to an imagined celestial existence.

L. Zadeh referred to the second point when he wrote: "As the complexity of a system increases, our ability to make precise and yet significant statements about its behaviour diminishes until a threshold is reached beyond which precision and significance (or relevance) become almost mutually exclusive characteristics." [Zadeh 1973a]

Let us consider characteristic features of real-world systems again: Real situations are very often uncertain or vague in a number of ways. Due to lack of information the future state of the system might not be known completely. This type of uncertainty (stochastic character) has long been handled appropriately by probability theory and statistics. This Kolmogoroff-type probability is essentially frequentistic and bases on set-theoretic considerations. Koopman's probability refers to the truth of statements and therefore bases on logic. On both types of probabilistic approaches it is assumed, however, that the events (elements of sets) or the statements, respectively, are well defined. We shall call this type of uncertainty or vagueness *stochastic uncertainty* by contrast to the vagueness concerning the description of the semantic meaning of the events, phenomena or statements themselves, which we shall call *fuzziness*.

Fuzziness can be found in many areas of daily life, such as in engineering [see, for instance, Blockley 1980], in medicine [see Vila and Delgado

1983], in meteorology [Cao and Chen 1983], in manufacturing [Mamdani 1981]; and others. It is particularly frequent, however, in all areas in which human judgment, evaluation, and decisions are important. These are the areas of decision making, reasoning, learning, and so on. Some reasons for this have already been mentioned. Others are that most of our daily communication uses "natural languages" and a good part of our thinking is done in it. In these natural languages the meaning of words is very often vague. The meaning of a word might even be well defined, but when using the word as a label for a set, the boundaries within which objects do or do not belong to the set become fuzzy or vague. Examples are words such as "birds" (how about penguins, bats, etc.?), "red roses," but also terms such as "tall men," "beautiful women," "creditworthy customers." In this context we can probably distinguish two kinds of fuzziness with respect to their origins: intrinsic fuzziness and informational fuzziness. The former is the fuzziness to which Russell's remark referred and it is illustrated by "tall men." This term is fuzzy because the meaning of tall is fuzzy and dependent on the context (height of observer, culture, etc.). An example of the latter is the term "creditworthy customers": A creditworthy customer can possibly be described completely and crisply if we use a large number of descriptors. These are more, however, than a human being could handle simultaneously. Therefore the term, which in psychology is called a "subjective category," becomes fuzzy. One could imagine that the subjective category creditworthiness is decomposed into two smaller subjective categories, each of which needs fewer descriptors to be completely described. This process of decomposition could be continued until the descriptions of the subjective categories generated are reasonably defined. On the other hand, the notion "creditworthiness" could be constructed by starting with the smallest subjective subcategories and aggregating them hierarchically.

For creditworthiness this concept structure, which has a symmetrical structure, was developed together with 50 credit clerks of banks.

Credit experts distinguish between the financial basis and the personality of an applicant. The financial basis comprises all realities, movables, assets, liquid funds, and others. The evaluation of the economic situation depends on the actual securities, that is, the difference between property and debts, and on the liquidity, that is, the continuous difference between income and expenses.

On the other hand, personality denotes the collection of traits by which a potent and serious person is distinguished. The achievement potential bases on the mental and physical capacity as well as on the individual's motivation. The business conduct includes economical standards. While

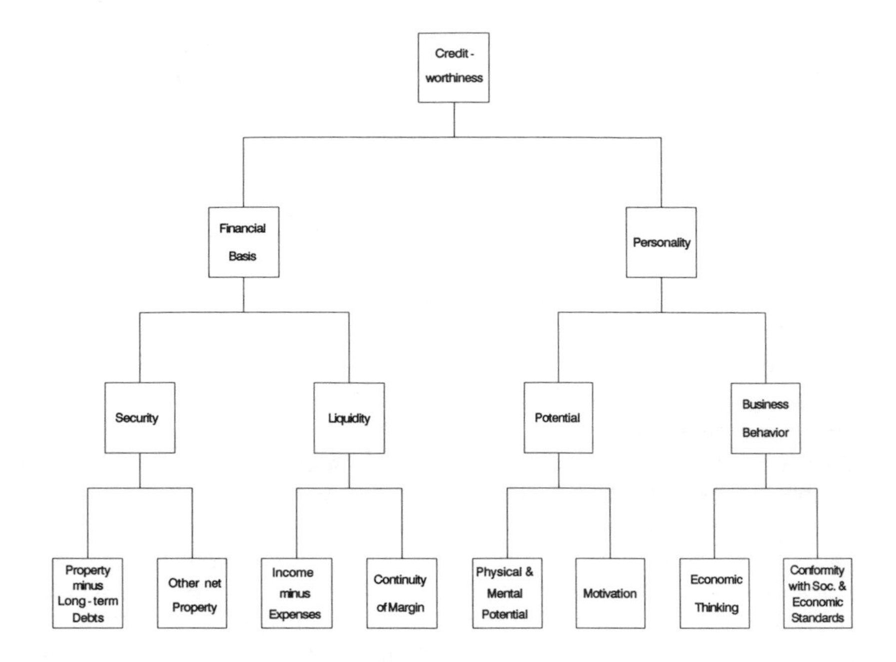

Figure 1–1. Concept hierarchy of creditworthiness.

the former means setting of realistic goals, reasonable planning, and economic criteria success, the latter is directed toward the applicant's disposition to obey business laws and mutual agreements. Hence a creditworthy person lives in secure circumstances and guarantees a successful, profit-oriented cooperation. See figure 1–1.

In chapter 14 we will return to this figure and elaborate on the type of aggregation.

1.2 Fuzzy Set Theory

The first publications in fuzzy set theory by Zadeh [1965] and Goguen [1967, 1969] show the intention of the authors to generalize the classical notion of a set and a proposition (statement) to accommodate fuzziness in the sense described in paragraph 1.1.

Zadeh [1965, p. 339] writes: "The notion of a fuzzy set provides a convenient point of departure for the construction of a conceptual framework which parallels in many respects the framework used in the case of ordinary sets, but is more general than the latter and, potentially, may prove to have a much wider scope of applicability, particularly in the fields of pattern classification and information processing. Essentially, such a framework provides a natural way of dealing with problems in which the source of imprecision is the absence of sharply defined criteria of class membership rather than the presence of random variables."

"Imprecision" here is meant in the sense of vagueness rather than the lack of knowledge about the value of a parameter as in tolerance analysis. Fuzzy set theory provides a strict mathematical framework (there is nothing fuzzy about fuzzy set theory!) in which vague conceptual phenomena can be precisely and rigorously studied. It can also be considered as a modeling language well suited for situations in which fuzzy relations, criteria, and phenomena exist.

Fuzziness has so far not been defined uniquely semantically, and probably never will. It will mean different things, depending on the application area and the way it is measured. In the meantime, numerous authors have contributed to this theory. In 1984 as many as 4000 publications may already exist. The specialization of those publications conceivably increases, making it more and more difficult for newcomers to this area to find a good entry and to understand and appreciate the philosophy, formalism, and applicational potential of this theory. Roughly speaking, fuzzy set theory in the last two decades has developed along two lines:

1. As a formal theory which, when maturing, became more sophisticated and specified and was enlarged by original ideas and concepts as well as by "embracing" classical mathematical areas such as algebra, graph theory, topology, and so on by generalizing (fuzzifying) them.
2. As a very powerful modeling language, that can cope with a large fraction of uncertainties of real-life situations. Because of its generality it can be well adapted to different circumstances and contexts. In many cases this will mean, however, the context-dependent modification and specification of the original concepts of the formal fuzzy set theory. Regrettably this adaption has not yet progressed to a satisfactory level, leaving an abundance of challenges for the ambitious researcher and practitioner.

It seems desirable that an introductionary textbook be available to help students get started and find their way around. Obviously such a textbook cannot cover the entire body of the theory in appropriate detail. This book will therefore proceed as follows:

Part I of this book, containing chapters 2 to 8, will develop the formal framework of fuzzy mathematics. Due to space limitations and for didactical reasons two restrictions will be observed:

1. Topics that are of high mathematical interest but require a very solid mathematical background and those that are not of obvious applicational relevance will not be discussed.
2. Most of the discussion will proceed along the lines of the early concepts of fuzzy set theory. At appropriate times, however, the additional potential of fuzzy set theory by using other axiomatic frameworks resulting in other operators will be indicated or described. The character of these chapters will obviously have to be formal.

The second part of the book, chapters 9 to 14, will then survey the most interesting applications of fuzzy set theory. At that stage the student should be in a position to recognize possible extensions and improvements of the applications presented. Chapter 12 on decision making in fuzzy environments might be considered as unduly brief, compared with the available literature. This area, however, has been taken up in a second volume and discussed in much more detail. This seems justified since on one hand it might not be of interest to a good number of persons being interested in fuzzy set theory from another angle and on the other hand it can be considered as the most advanced of the application areas of fuzzy set theory.

I FUZZY MATHEMATICS

This first part of the book is devoted to the formal framework of the theory of fuzzy sets. Chapter 2 provides basic definitions of fuzzy sets and algebraic operations which will then serve for further considerations. Even though we shall use one version of terminology and one set of symbols consistently throughout the book, alternative ways of denoting fuzzy sets will be mentioned because they have become common. Chapter 3 extends the basic theory of fuzzy sets by introducing additional concepts and alternative operators. Chapter 4 is devoted to fuzzy measures, measures of fuzziness, and other important measures which are needed for applications presented either in part two of this book or in the second volume on decision making in a fuzzy environment. Chapter 5 introduces the extension principle, which will be very useful for the following chapters and covers fuzzy arithmetic. Chapters 6 and 7 will then treat fuzzy relations, graphs, and functions. Chapter 8 focuses on some special topics, such as the relationship between fuzzy set theory, probability theory, and other classical areas.

2 FUZZY SETS—BASIC DEFINITIONS

2.1 Basic Definitions

A **classical** (crisp) **set** is normally defined as a collection of elements or objects $x \in X$ which can be finite, countable, or overcountable. Each single element can either belong to or not belong to a set A, $A \subseteq X$. In the former case, the statement "x belongs to A" is true, whereas in the latter case this statement is false.

Such a classical set can be described in different ways: one can either enumerate (list) the elements that belong to the set; describe the set analytically, for instance, by stating conditions for membership ($A = \{x \mid x \leq 5\}$); or define the member elements by using the characteristic function, in which 1 indicates membership and 0 nonmembership. For a fuzzy set, the characteristic function allows various degrees of membership for the elements of a given set.

Definition 2–1

If X is a collection of objects denoted generically by x then a *fuzzy set* $\widetilde{A}$ in X is a set of ordered pairs:

$$\tilde{A} = \{(x, \mu_{\tilde{A}}(x)) \mid x \in X\}$$

$\mu_{\tilde{A}}(x)$ is called the membership function or grade of membership (also degree of compatibility or degree of truth) of x in $\tilde{A}$ which maps X to the membership space M. (When M contains only the two points 0 and 1, $\tilde{A}$ is nonfuzzy and $\mu_{\tilde{A}}(x)$ is identical to the characteristic function of a nonfuzzy set.) The range of the membership function is a subset of the nonnegative real numbers whose supremum is finite. Elements with a zero degree of membership are normally not listed.

Example 2–1a

A realtor wants to classify the house he offers to his clients. One indicator of comfort of these houses is the number of bedrooms in it. Let $X = \{1, 2, 3, 4, \ldots, 10\}$ be the set of available types of houses described by x = number of bedrooms in a house. Then the fuzzy set "comfortable type of house for a 4-person family" may be described as

$$\tilde{A} = \{(1, .2), (2, .5), (3, .8), (4, 1), (5, .7), (6, .3)\}$$

In the literature one finds different ways of denoting fuzzy sets:

1. A fuzzy set is denoted by an ordered set of pairs, the first element of which denotes the element and the second the degree of membership (as in definition 2–1).

Example 2–1b

$\tilde{A}$ = "real numbers considerably larger than 10"

$$\tilde{A} = \{(x, \mu_{\tilde{A}}(x)) \mid x \in X\}$$

where

$$\mu_{\tilde{A}}(x) = \begin{cases} 0, & x \leq 10 \\ (1 + (x - 10)^{-2})^{-1}, & x > 10 \end{cases}$$

Example 2–1c

$\tilde{A}$ = "real numbers close to 10"

$$\tilde{A} = \{(x, \mu_{\tilde{A}}(x)) \mid \mu_{\tilde{A}}(x) = (1 + (x - 10)^2)^{-1}\}$$

See figure 2–1.

2. A fuzzy set is represented solely by stating its membership function [for instance Negoita and Ralescu 1975].

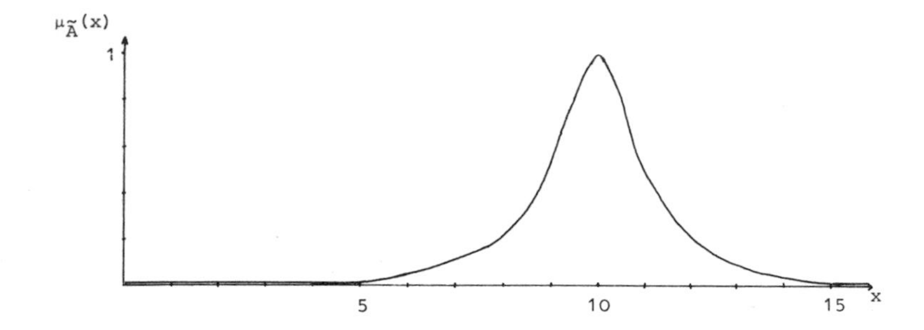

Figure 2–1. Real numbers close to 10.

3. $\widetilde{A} = \mu_{\tilde{A}}(x_1)/x_1 + \mu_{\tilde{A}}(x_2)/x_2 \ldots = \sum_{i=1}^{n} \mu_{\tilde{A}}(x_i)/x_i$

$$\text{or} \int_X \mu_{\tilde{A}}(\underline{x})/x$$

Example 2–1d

$\widetilde{A}$ = "integers close to 10"

$$\widetilde{A} = 0.1/7 + 0.5/8 + 0.8/9 + 1/10 + 0.8/11 + 0.5/12 + 0.1/13$$

Example 2–1e

$\widetilde{A}$ = "real numbers close to 10"

$$\widetilde{A} = \int_{\mathbb{R}} \frac{1}{1 + (x - 10)^2} \bigg/ x$$

It has already been mentioned that the membership function is not limited to values between 0 and 1. If $\sup_x \mu_{\tilde{A}}(x) = 1$ the fuzzy set $\widetilde{A}$ is called normal. A nonempty fuzzy set $\widetilde{A}$ can always be normalized by dividing $\mu_{\tilde{A}}(x)$ by $\sup_x \mu_{\tilde{A}}(x)$: As a matter of convenience we will generally assume that fuzzy sets are normalized. For the representation of fuzzy sets we will use the notation 1 illustrated in examples 2–1b and 2–1c, respectively.

A fuzzy set is obviously a generalization of a classical set and the membership function a generalization of the characteristic function. Since we are generally referring to a universal (crisp) set X some elements of a fuzzy set may have the degree of membership zero. Often it is appropriate to consider those elements of the universe which have a nonzero degree of membership in a fuzzy set.

Definition 2–2

The *support* of a fuzzy set $\tilde{A}$, $S(\tilde{A})$, is the crisp set of all $x \in X$ such that $\mu_{\tilde{A}}(x) > 0$.

Example 2–2

Let us consider example 2–1a again: The support of $S(\tilde{A}) = \{1, 2, 3, 4, 5, 6\}$. The elements (types of houses) $\{7, 8, 9, 10\}$ are not part of the support of $\tilde{A}$!

A more general and even more useful notion is that of an α-level-set.

Definition 2–3

The (crisp) set of elements that belong to the fuzzy set $\tilde{A}$ at least to the degree α is called the *α-level set*:

$$A_\alpha = \{x \in X \mid \mu_{\tilde{A}}(x) \geq \alpha\}$$

$A'_\alpha = \{x \in X \mid \mu_{\tilde{A}}(x) > \alpha\}$ is called "strong α-level set" or "strong α-cut."

Example 2–3

We refer again to example 2–1a and list possible α-level sets:

$$A_{.2} = \{1, 2, 3, 4, 5, 6\}$$
$$A_{.5} = \{2, 3, 4, 5\}$$
$$A_{.8} = \{3, 4\}$$
$$A_1 = \{4\}$$

The strong α-level set for $\alpha = .8$ is $A'_{.8} = \{4\}$.

Convexity also plays a role in fuzzy set theory. By contrast to classical set theory, however, convexity conditions are defined with reference to the membership function rather than the support of a fuzzy set.

Definition 2–4

A fuzzy set $\widetilde{A}$ is *convex* if

$$\mu_{\tilde{A}}(\lambda x_1 + (1-\lambda)x_2) \geq \min(\mu_{\tilde{A}}(x_1), \mu_{\tilde{A}}(x_2)), x_1, x_2 \in X, \lambda \in [0, 1]$$

Alternatively, a fuzzy set is convex if all α-level sets are convex.

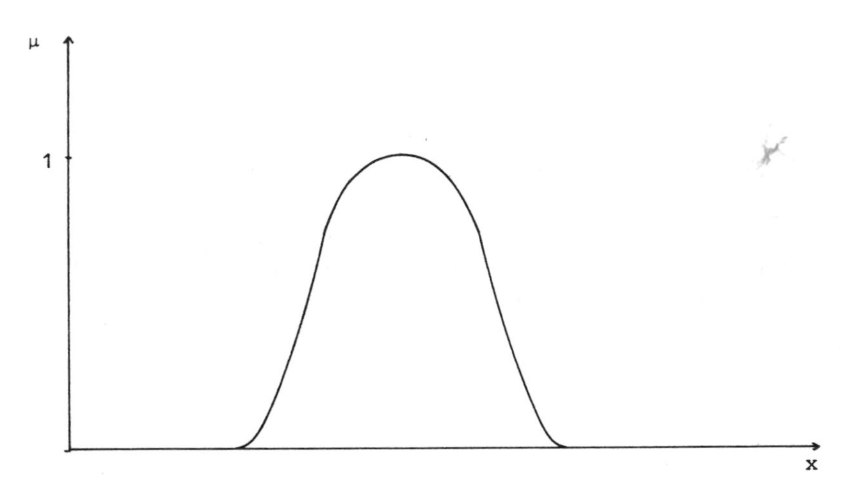

Figure 2–2a. Convex fuzzy set.

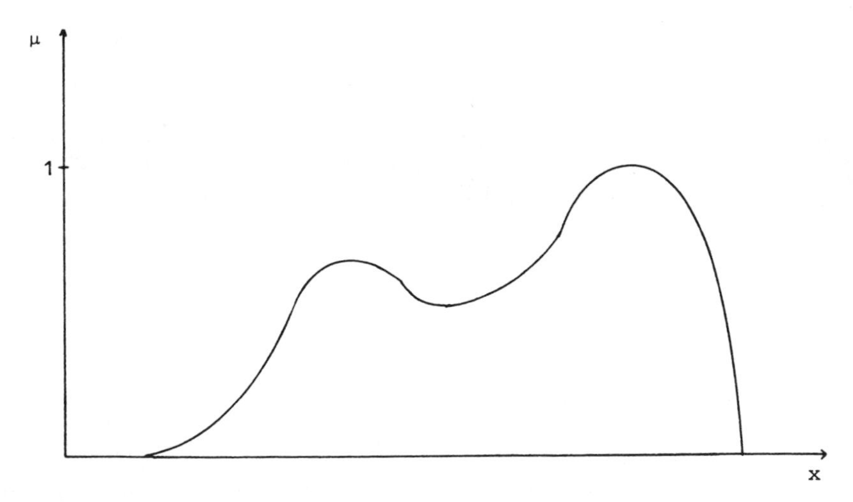

Figure 2–2b. Nonconvex fuzzy set.

Example 2–4

Figure 2–2a depicts a convex fuzzy set, whereas figure 2–2b illustrates a nonconvex fuzzy set.

One final feature of a fuzzy set which we will use frequently in later chapters is its cardinality or "power" [Zadeh 1981c].

Definition 2–5

For a finite fuzzy set $\tilde{A}$ the *cardinality* $|\tilde{A}|$ is defined as

$$|\tilde{A}| = \sum_{x \in X} \mu_{\tilde{A}}(x)$$

$\|\tilde{A}\| = \dfrac{|\tilde{A}|}{|X|}$ is called the *relative cardinality* of $\tilde{A}$.

Obviously the relative cardinality of a fuzzy set depends on the cardinality of the universe. So you have to choose the same universe if you want to compare fuzzy sets by their relative cardinality.

Example 2–5

For the fuzzy set "comfortable type of house for a 4-person family" from example 2–1a the cardinality is

$$|\tilde{A}| = .2 + .5 + .8 + 1 + .7 + .3 = 3.5$$

Its relative cardinality is

$$\|\tilde{A}\| = \frac{3.5}{10} = 0.35$$

The relative cardinality can be interpreted as the fraction of elements of X being in $\tilde{A}$, weighted by their degrees of membership in $\tilde{A}$. For infinite X the cardinality is definted by $|\tilde{A}| = \int_x \mu_{\tilde{A}}(x)\, dx$. Of course $|\tilde{A}|$ does not always exist.

2.2 Basic Set-Theoretic Operations for Fuzzy Sets

The membership function is obviously the crucial component of a fuzzy set. It is therefore not surprising that operations with fuzzy sets are defined via their membership functions. We shall first present the concepts suggested by Zadeh in 1965 [Zadeh 1965, p. 310]. They constitute a consistent framework for the theory of fuzzy sets. They are, however, not the only possible way to extend classical set theory consistently. Zadeh and other authors have suggested alternative or additional definitions for set theoretic operations which will be discussed in chapter 3.

Definition 2–6

The membership function $\mu_{\tilde{C}}(x)$ of the *intersection* $\tilde{C} = \tilde{A} \cap \tilde{B}$ is pointwise defined by

$$\mu_{\tilde{C}}(x) = \min \{\mu_{\tilde{A}}(x), \mu_{\tilde{B}}(x)\}, \qquad x \in X$$

Definition 2–7

The membership function $\mu_{\tilde{D}}(x)$ of the *union* $\tilde{D} = \tilde{A} \cup \tilde{B}$ is pointwise defined by

$$\mu_{\tilde{D}}(x) = \max \{\mu_{\tilde{A}}(x), \mu_{\tilde{B}}(x)\}, \qquad x \in X$$

Definition 2–8

The membership function of the *complement* of a fuzzy set $\tilde{A}$, $\mu_{\complement\tilde{A}}(x)$ is defined by

$$\mu_{\complement\tilde{A}}(x) = 1 - \mu_{\tilde{A}}(x), \qquad x \in X$$

Example 2–6

Let $\tilde{A}$ be the fuzzy set "comfortable type of house for a 4-person-family" from example 2–1a and $\tilde{B}$ be the fuzzy set "large type of house" defined as

$$\tilde{B} = \{(3, .2), (4, .4), (5, .6), (6, .8), (7, 1), (8, 1)\}$$

The intersection $\tilde{C} = \tilde{A} \cap \tilde{B}$ is then

$$\tilde{C} = \{(3, .2), (4, .4), (5, .6), (6, .3)\}$$

The union $\tilde{D} = \tilde{A} \cup \tilde{B}$ is:

$$\tilde{D} = \{(1, .2), (2, .5), (3, .8), (4, 1) (5, .7) (6, .8), (7, 1), (8, 1)\}$$

The complement $\complement\tilde{B}$ which might be interpreted as "not large type of house" is

$$\complement\tilde{B} = \{(1, 1), (2, 1), (3, .8), (4, .6), (5, .4), (6, .2), (9, 1), (10, 1)\}$$

Example 2–7

Let us assume that

$\tilde{A}$ = "x considerable larger than 10"
$\tilde{B}$ = "x approximately 11," characterized by

$$\tilde{A} = \{(x, \mu_{\tilde{A}}(x)) \mid x \in X\}$$

where

$$\mu_{\tilde{A}}(x) = \begin{cases} 0, & x \leq 10 \\ (1 + (x-10)^{-2})^{-1} & x > 10 \end{cases}$$

and

$$\tilde{B} = \{(x, \mu_{\tilde{B}}(x)) \mid x \in X\}$$

where

$$\mu_{\tilde{B}}(x) = (1 + (x-11)^4)^{-1}$$

Then

$$\mu_{\tilde{A}\cap\tilde{B}}(x) = \begin{cases} \min[(1 + (x-10)^{-2})^{-1}, (1 + (x-11)^4)^{-1}] & \text{for } x > 10 \\ 0 & \text{for } x \leq 10 \end{cases}$$

(x considerably larger than 10 and approximately 11)

$$\mu_{\tilde{A}\cup\tilde{B}}(x) = \max[(1 + (x-10)^{-2})^{-1}, (1 + (x-11)^4)^{-1}], \qquad x \in X$$

Figure 2–3 depicts the above.

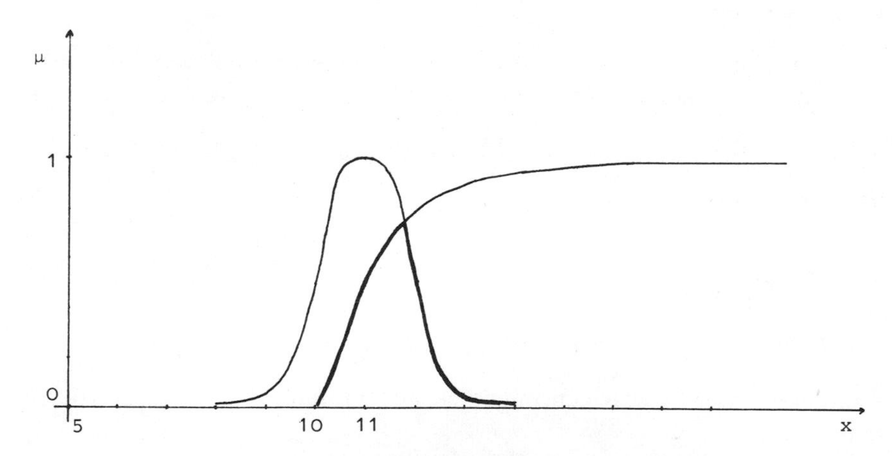

Figure 2–3. Union and intersection of fuzzy sets.

It has already been mentioned that min and max are not the only operators that could have been chosen to model the intersection or union of fuzzy sets respectively. The question arises, why those and not others? Bellman and Giertz addressed this question in 1973 axiomatically [Bellman and Giertz 1973, p. 151]. They argued from a logical point of view, interpreting the intersection as "logical and," the union as "logical or," and the fuzzy set $\widetilde{A}$ as the statement "The element x belongs to set $\widetilde{A}$" can be accepted as more or less true. It is very instructive to follow their line of argument, which is an excellent example for an axiomatic justification of specific mathematical models. We shall therefore sketch their reasoning: Consider two statements, S and T, for which the truth values are μ_S and μ_T, respectively, $\mu_S, \mu_T \in [0, 1]$. The truth value of the "and" and "or" combination of these statements, $\mu(S \text{ and } T)$ and $\mu(S \text{ or } T)$, both from the interval [0, 1] are interpreted as the values of the membership functions of the intersection and union, respectively, of S and T. We are now looking for two real-valued functions f and g such that

$$\mu_{S \text{ and } T} = f(\mu_S, \mu_T)$$
$$\mu_{S \text{ or } T} = g(\mu_S, \mu_T)$$

Bellman and Giertz feel that the following restrictions are reasonably imposed on f and g:

i. f and g are nondecreasing and continuous in μ_S and μ_T.
ii. f and g are symmetric, that is,

$$f(\mu_S, \mu_T) = f(\mu_T, \mu_S)$$
$$g(\mu_S, \mu_T) = g(\mu_T, \mu_S)$$

iii. $f(\mu_S, \mu_S)$ and $g(\mu_S, \mu_S)$ are strictly increasing in μ_S.
iv. $f(\mu_S, \mu_T) \leq \min(\mu_S, \mu_T)$ and $g(\mu_S, \mu_T) \geq \max(\mu_S, \mu_T)$. That implies that accepting the truth of the statements "S and T" requires more, and accepting the truth of the statement "S or T" less than accepting S or T alone as true.
v. $f(1, 1) = 1$ and $g(0, 0) = 0$.
vi. Logically equivalent statements must have equal truth values and fuzzy sets with the same contents must have the same membership functions, that is,

$$S_1 \text{ and } (S_2 \text{ or } S_3)$$

is equivalent to

$$(S_1 \text{ and } S_2) \text{ or } (S_1 \text{ and } S_3)$$

and therefore must be equally true.

Bellman and Giertz now formalize the above assumptions as follows: Using the symbols $\wedge$ for "and" (= intersection) and $\vee$ for "or" (= union), this amounts to the following 7 restrictions, to be imposed on the two commutative (see (ii)) and associative (see (vi)) binary compositions $\wedge$ and $\vee$ on the closed interval [0, 1] which are mutually distributive (see (vi)) with respect to one another.

1. $\mu_S \wedge \mu_T = \mu_T \wedge \mu_S$
   $\mu_S \vee \mu_T = \mu_T \vee \mu_S$
2. $(\mu_S \wedge \mu_T) \wedge \mu_U = \mu_S \wedge (\mu_T \wedge \mu_U)$
   $(\mu_S \vee \mu_T) \vee \mu_U = \mu_S \vee (\mu_T \vee \mu_U)$
3. $\mu_S \wedge (\mu_T \vee \mu_U) = (\mu_S \wedge \mu_T) \vee (\mu_S \wedge \mu_U)$
   $\mu_S \vee (\mu_T \wedge \mu_U) = (\mu_S \vee \mu_T) \wedge (\mu_S \vee \mu_U)$
4. $\mu_S \wedge \mu_T$ and $\mu_S \vee \mu_T$ are continuous and nondecreasing in each component
5. $\mu_S \wedge \mu_S$ and $\mu_S \vee \mu_S$ are strictly increasing in μ_S (see (iii))
6. $\mu_S \wedge \mu_T \leq \min(\mu_S, \mu_T)$
   $\mu_S \vee \mu_T \geq \max(\mu_S, \mu_T)$ (see (iv))
7. $1 \wedge 1 = 1$
   $0 \vee 0 = 0$ (see (v))

Bellman and Giertz then prove mathematically [see Bellman and Giertz 1973, p. 154] that

$$\mu_{S \wedge T} = \min(\mu_S, \mu_T) \quad \text{and} \quad \mu_{S \vee T} = \max(\mu_S, \mu_T)$$

For the complement it would be reasonable to assume that if statement "S" is true, its complement "non S" is false, or if $\mu_S = 1$ then $\mu_{\text{non}S} = 0$ and vice versa. The function h (as complement in analogy to f and g for intersection and union) should also be continuous and monotonicly decreasing and we would like the complement of the complement to be the original statement (in order to be in line with traditional logic and set theory). These requirements, however, are not enough to determine uniquely the mathematical form of the complement. Bellman and Giertz require in addition, that $\mu_{\bar{S}}(1/2) = 1/2$. Other assumptions are certainly possible and plausible.

Exercises

1. Model the following expressions as fuzzy sets:
   a. Large integers
   b. Very small numbers

c. Medium-sized men
d. Numbers approximately between 10 and 20
e. High speeds for racing cars

2. Determine all α-level sets and all strong α-level sets for the following fuzzy sets:
   a. $\tilde{A} = \{(3, 1), (4, .2), (5, .3), (6, .4), (7, .6), (8, .8), (10, 1), (12, .8), (14, .6)\}$
   b. $\tilde{B} = \{(x, \mu_{\tilde{B}}(x) = ((1 + (x - 10)^2)^{-1}\}$
   for $\alpha = .3, .5, .8$
   c. $\tilde{C} = \{(x, \mu_{\tilde{C}}(x) \mid x \in R\}$
   where $\mu_{\tilde{C}}(x) = 0$ for $x \leq 10$
   $\mu_{\tilde{C}}(x) = (1 + (x - 10)^{-2})^{-1}$ for $x > 10$
3. Which of the fuzzy sets of exercise 2 are convex and which are not?
4. Let $X = \{1, 2, \ldots, 10\}$. Determine the cardinalities and relative cardinalities of the following fuzzy sets:
   a. $\tilde{A}$ from exercise 2a
   b. $\tilde{B} = \{(2, .4), (3, .6), (4, .8), (5, 1), (6, .8), (7, .6), (8, .4)\}$
   c. $\tilde{C} = \{(2, .4), (4, .8), (5, 1), (7, .6)\}$
5. Determine the intersections and unions of the following fuzzy sets:
   a. The fuzzy sets $\tilde{A}$, $\tilde{B}$, and $\tilde{C}$ from exercise 4
   b. $\tilde{B}$ and $\tilde{C}$ from exercise 2
6. Determine the intersection and the union of the complements of fuzzy sets $\tilde{B}$ and $\tilde{C}$ from exercise 4

3 EXTENSIONS

3.1 Types of Fuzzy Sets

In chapter 2 the basic definition of a fuzzy set was given and the original set-theoretic operations were discussed. The membership space was assumed to be the space of real numbers, membership functions were crisp functions, and the operations corresponded essentially to the operations of dual logic or Boolean algebra.

Different extensions of the basic concept discussed in chapter 2 are possible. They may concern the definition of a fuzzy set or they may concern the operations with fuzzy sets. With respect to the definition of a fuzzy set, different structures may be imposed on the membership space and different assumptions may be made concerning the membership function. These extensions will be treated in section 3.1.

It was assumed in chapter 2 that the logical "and" corresponds to the set-theoretic intersection which in turn is modeled by the min-operator. The same type of relationship was assumed for the logical "or," the union, and the max-operator. Departing from the well-established systems of dual logic and Boolean algebra, alternative and additional definitions for terms such as intersection and union, for their interpretation as "and" and "or," and for their mathematical models can be conceived. These concepts will be discussed in section 3.2.

So far we have considered fuzzy sets with crisply defined membership functions or degrees of membership. It is doubtful whether, for instance, human beings have or can have a crisp image of membership functions in their minds. Zadeh [1973a, p. 52] therefore suggested the notion of a fuzzy set whose membership function itself is a fuzzy set. If we call fuzzy sets, such as considered so far, as type 1 fuzzy sets then a type 2 fuzzy set can be defined as follows.

Definition 3–1

A *type m fuzzy set* is a fuzzy set in X whose membership values are type $m-1$, $m > 1$ fuzzy sets on [0, 1].

The operations intersection, union, and complement defined so far are no longer adequate for type 2 fuzzy sets. We will, however, postpone, the discussions of adequate operators until section 5.2, that is, until we have presented the extension principle, which shall prove very useful for this purpose. By the same token by which we introduced type 2 fuzzy sets, it could be argued that there is no obvious reason why the membership functions of type 2 fuzzy sets should be crisp. A natural extension of these type 2 fuzzy sets is therefore the definition of type m fuzzy sets.

Definition 3–2

A *type m fuzzy set* is a fuzzy set in X whose membership values are type $m-$, $m>1$ fuzzy sets on [0, 1].

From a practical point of view, such type m fuzzy sets for large m (even for $m \geqq 3$) are hard to deal with and it will be extremely difficult or even impossible to measure them or to visualize them. We will, therefore, not even try to define the usual operations on them.

There have been other attempts to include vagueness going beyond the fuzziness of ordinary type 1 fuzzy sets. One example is the "stochastic fuzzy model" of Norwich and Turksen [1981, 1984]. Those authors were mainly concerned with the measurement and the scale level of membership functions. They view a fuzzy set as a family of random variables whose density functions are estimated by that stochasticity [Norwich and Turksen 1984, p. 21].

Hirota [1981] also considers fuzzy sets for which the "value of membership functions is a random variable."

Definition 3–3 [Hirota 1981, p. 35]

A *probabilistic set A* on X is defined by a defining function μ_A,

$$\mu_A: X \times \Omega \ni (x, \omega) \rightarrow \mu_A(x, \omega) \in \Omega_C$$

where $\mu_A(x, \cdot)$ is the (B, B_C)-measurable function for each fixed $x \in X$.

For Hirota a probabilistic set A with the defining function $\mu_A(x, \omega)$ is contained in a probabilistic set B with $\mu_B(x, \omega)$, if for each $x \in X$ there exists a $E \in B$ which satisfies $P(E) = 1$ and $\mu_A(x, \omega) \leq \mu_B(x, \omega)$ for all $\omega \in E$. $(\Omega, \mathrm{B}, \mathrm{P})$ is called the parameter space.

One of the main advantages of the notion of probabilistic sets in modeling fuzzy and stochastic features of a system is asserted to be the applicability of moment analysis, that is, the possibility of computing moments such as expectation, variance. Figure 3–1 indicates the difference between the appearance of fuzzy sets and probabilistic sets [Hirota 1981, p. 33]. Of course, the mathematical properties of probabilistic sets differ from those of fuzzy sets and so do the mathematical models for intersection, union, and so on.

A more general definition of a fuzzy set than is given in definition 2–1 is that of an L-fuzzy set [Goguen 1967; De Luca and Termini 1972]. By contrast to the above definition, the membership function of an L-fuzzy-set maps into a partially ordered set, L. Since the interval [0, 1] is a poset (partially ordered set) the fuzzy set in definition 2–1 is a special L-fuzzy set.

Further attempts at representing vague and uncertain data with different types of fuzzy sets were made by Atanassov and Stoeva [Atanassov and Stoeva 1983; Atanassov 1986], who defined a generalization of the notion of fuzzy sets—the *intuitonistic* fuzzy sets—and by Pawlak [Pawlak 1982], who developed the theory of rough sets, where grades of membership are expressed by a concept of approximation.

Definition 3–4 [Atanassov and Stoeva 1983]

Given an underlying set X of objects, an *intuitonistic fuzzy set* (IFS) A is a set of ordered triples,

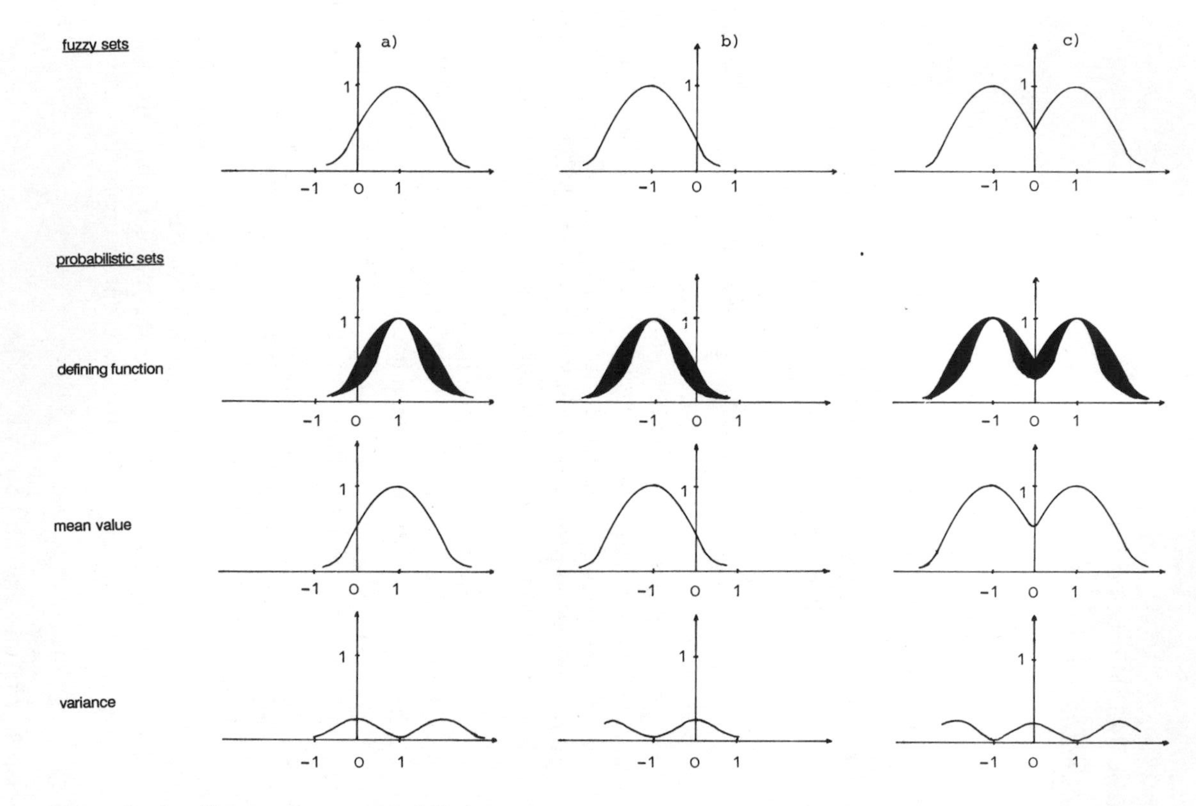

Figure 3–1. Fuzzy sets vs. probabilistic sets.

$$A = \{(x, \mu_A(x), \nu_A(x)) \mid x \in X\}$$

where $\mu_A(x)$ and $\nu_A(x)$ are functions mapping from X into $[0, 1]$. For each $x \in X$, $\mu_A(x)$ represents the degree of membership of the element x to the subset A of X, $\nu_A(x)$ gives the degree of nonmembership. For the functions $\mu_A(x)$ and $\nu_A(x)$ mapping into $[0, 1]$ the condition $0 \leq \mu_A(x) + \nu_A(x) \leq 1$ holds.

Ordinary fuzzy sets over X may be viewed as special intuitonistic fuzzy sets with the nonmembership function $\nu_A(x) = 1 - \mu_A(x)$. In the same way as fuzzy sets, intuitonistic L-fuzzy sets were defined by mapping the membership functions into a partial ordered set L [Atanassov and Stoeva 1984].

Definition 3–5 [Pawlak 1985, p. 99; Pawlak et al. 1988]

Let U denote a set of objects called universe and let $R \subset U \times U$ be an equivalence relation on U. The pair $A = (U, R)$ is called an approximation space. For $u, v \in U$ and $(u, v) \in R$, u and v belong to the same equivalence class and we say that they are *indistinguishable* in A. Therefore the relation R is called an *indiscernibility* relation. Let $[x]_R$ denote an equivalence class (elementary set of A) of R containing element x, then lower and upper approximations for a subset $X \subseteq U$ in A—denoted $\underline{A}(X)$ and $\overline{A}(X)$ respectively—are defined as follows:

$$\underline{A}(X) = \{x \in U \mid [x]_R \subset X\}$$
$$\overline{A}(X) = \{x \in U \mid [x]_R \cap X \neq \emptyset\}$$

If an object x belongs to the lower approximation space of X in A then "x surely belongs to X in A," $x \in \overline{A}(X)$ means that "x possibly belongs to X in A."

For the subset $X \subseteq U$ representing a concept of interest, the approximation space $A = (U, R)$ can be characterized with three distinct regions of X in A: the so called positive region $\underline{A}(X)$, the boundary region $\overline{A}(X) - \underline{A}(X)$, and the negative region $U - \overline{A}(X)$.

The characterization of objects in X by the indiscernibility relation R is not precise enough if the boundary region $\overline{A}(X) - \underline{A}(X)$ is not empty. For this case it may be impossible to say whether an object belongs to X or not and so the set X is said to be nondefinable in A and X is a *rough set.*

Pawlak [1985] shows that the concept of approximation given by the equivalence relation R and the approximation space may not, in general, be replaced by a membership function similar to that introduced by Zadeh.

In order to take probabilistic informations crucial to nondeterministic classification problems into account, a natural probabilistic extension of the rough-set model has been proposed [Pawlak et al. 1988].

3.2 Further Operations on Fuzzy Sets

For the time being we return to ordinary fuzzy sets (type 1 fuzzy sets) and consider additional operations on them which have been defined in the literature and which will be useful or even necessary for later chapters.

3.2.1 Algebraic Operations

Definition 3–6

The *cartesian product* of fuzzy sets is defined as follows: Let $\tilde{A}_1, \ldots, \tilde{A}_n$ be fuzzy sets in $X_1, \ldots, X_n$. The cartesian product is then a fuzzy set in the product space $X_1 \times \ldots \times X_n$ with the membership function

$$\mu_{(\tilde{A}_1 x \ldots x \tilde{A}_n)}(x) = \min_i \{\mu_{\tilde{A}_i}(x_i) \mid x = (x_1, \ldots, x_n), x_i \in X_i\}$$

Definition 3–7

The *mth power* of a fuzzy set $\tilde{A}$ is a fuzzy set with the membership function

$$\mu_{\tilde{R}^m}(x) = [\mu_{\tilde{A}}(x)]^m, \qquad x \in X$$

Additional algebraic operations are defined as follows:

Definition 3–8

The *algebraic sum* (probabilistic sum) $\tilde{C} = \tilde{A} + \tilde{B}$ is defined as

$$\tilde{C} = \{(x, \mu_{\tilde{A}+\tilde{B}}(x)) \mid x \in X\}$$

where

$$\mu_{\tilde{A}+\tilde{B}}(x) = \mu_{\tilde{A}}(x) + \mu_{\tilde{B}}(x) - \mu_{\tilde{A}}(x) \cdot \mu_{\tilde{B}}(x)$$

Definition 3–9

The *bounded sum* $\widetilde{C} = \widetilde{A} \oplus \widetilde{B}$ is defined as

$$\widetilde{C} = \{(x, \mu_{\tilde{A}\oplus\tilde{B}}(x)) \mid x \in X\}$$

where

$$\mu_{\tilde{A}\oplus\tilde{B}}(x) = \min\{1, \mu_{\tilde{A}}(x) + \mu_{\tilde{B}}(x)\}$$

Definition 3–10

The *bounded difference* $\widetilde{C} = \widetilde{A} \ominus \widetilde{B}$ is defined as

$$\widetilde{C} = \{(x, \mu_{\tilde{A}\ominus\tilde{B}}(x)) \mid x \in X\}$$

where

$$\mu_{\tilde{A}\ominus\tilde{B}}(x) = \max\{0, \mu_{\tilde{A}}(x) + \mu_{\tilde{B}}(x) - 1\}$$

Definition 3–11

The *algebraic product* of two fuzzy sets $\widetilde{C} = \widetilde{A} \cdot \widetilde{B}$ is defined as

$$\widetilde{C} = \{(x, \mu_{\tilde{A}}(x) \cdot \mu_{\tilde{B}}(x)) \mid x \in X\}$$

Example 3–1

Let $\widetilde{A}(x) = \{(3, .5), (5, 1), (7, .6)\}$
$\widetilde{B}(x) = \{(3, 1), (5, .6)\}$
The above definitions are then illustrated by the following results:

$$\begin{aligned}
\widetilde{A} \times \widetilde{B} &= \{[(3; 3), .5], [(5; 3), 1], [(7; 3), .6] \\
&\quad [(3; 5), .5], [(5; 5), .6], [(7; 5), .6]\} \\
\widetilde{A}^2 &= \{(3, .25), (5, 1), (7, .36)\} \\
\widetilde{A} + \widetilde{B} &= \{(3, 1), (5, 1), (7, .6)\} \\
\widetilde{A} \oplus \widetilde{B} &= \{(3, 1), (5, 1), (7, .6)\} \\
\widetilde{A} \ominus \widetilde{B} &= \{(3, .5), (5, .6)\} \\
\widetilde{A} \cdot \widetilde{B} &= \{(3, .5), (5, .6)\}
\end{aligned}$$

3.2.2 Set-Theoretic Operations

In chapter 2 the intersection of fuzzy sets, interpreted as the logical "and," was modeled as the min-operator and the union, interpreted as "or," as the max-operator. Other operators have also been suggested. These suggestions vary with respect to the generality or adaptibility of the operators as well as to the degree to which and how they are justified. Justification ranges from intuitive argumentation to empirical or axiomatic justification. Adaptability ranges from uniquely defined, for example, nonadaptable, concepts via parametrized "families" of operators to general classes of operators which satisfy certain properties.

We shall investigate the two basic classes of operators: operators for the intersection and union of fuzzy sets—referred to as triangular norms and conorms—and the class of averaging operators, which model connectives for fuzzy sets between *t*-norms and *t*-conorms. Each class contains parametrized as well as nonparametrized operators.

t-norms. For the intersection of fuzzy sets Zadeh [Zadeh 1965] suggested the min-operator and the algebraic product $\tilde{A} \cdot \tilde{B}$. The "bold intersection" [Giles 1976] was modeled by the "bounded sum" as defined above. The min, product, and bounded-sum operators belong to the so-called *triangular* or *t-norms*. Operators belonging to this class have drawn a lot of interest in the recent past. They can be characterized as follows:

Definition 3–12 [Dubois and Prade 1980a, p. 17]

t-norms are two-valued functions from $[0,1] \times [0,1]$ which satisfy the following conditions:

1. $t(0, 0) = 0;\ t(\mu_{\tilde{A}}(x), 1) = t(1, \mu_{\tilde{A}}(x)) = \mu_{\tilde{A}}(x), \quad x \in X$
2. $t(\mu_{\tilde{A}}(x), \mu_{\tilde{B}}(x)) \leq t(\mu_{\tilde{C}}(x), \mu_{\tilde{D}}(x))$
   if $\mu_{\tilde{A}}(x) \leq \mu_{\tilde{C}}(x)$ and $\mu_{\tilde{B}}(x) \leq \mu_{\tilde{D}}(x)$ (monotonicity)
3. $t(\mu_{\tilde{A}}(x), \mu_{\tilde{B}}(x)) = t(\mu_{\tilde{B}}(x), \mu_{\tilde{A}}(x))$ (commutativity)
4. $t(\mu_{\tilde{A}}(x), t(\mu_{\tilde{B}}(x), \mu_{\tilde{C}}(x))) = t(t(\mu_{\tilde{A}}(x), \mu_{\tilde{B}}(x)), \mu_{\tilde{C}}(x))$ (associativity)

The functions *t* define a general class of intersection operators for fuzzy sets. The operators belonging to this class of *t*-norms are, in particular, associative (see 4) and therefore it is possible to compute the membership values for the intersection of more than two fuzzy sets by recursively applying a *t*-norm operator [Bonissone and Decker 1986, p. 220].

t-conorms (or s-norms). For the union of fuzzy sets the max-operator, the algebraic sum [Zadeh 1965] and the "bold union" [Giles 1976]—modeled by the "bounded sum"—have been suggested.

Corresponding to the class of intersection operators, a general class of aggregation operators for the union of fuzzy sets called *triangular conorms* or *t-conorms* (sometimes referred to as *s-norms*) is defined analoguously [Dubois and Prade 1985, p. 90; Mizumoto 1989, p. 221]. The max-operator, algebraic sum and bounded sum considered above belong to this class.

Definition 3–13 [Dubois and Prade 1985, p. 90]

t-conorms or *s*-norms are associative, commutative, and monotonic two placed functions *s*, which map from $[0, 1] \times [0, 1]$ into $[0, 1]$. These properties are formulated with the following conditions:

1. $s(1, 1) = 1; \quad s(\mu_{\tilde{A}}(x), 0) = s(0, \mu_{\tilde{A}}(x)) = \mu_{\tilde{A}}(x), \qquad x \in X$
2. $s(\mu_{\tilde{A}}(x), \mu_{\tilde{B}}(x)) \leq s(\mu_{\tilde{C}}(x), \mu_{\tilde{D}}(x))$ if $\mu_{\tilde{A}}(x) \leq \mu_{\tilde{C}}(x)$ and $\mu_{\tilde{B}}(x) \leq \mu_{\tilde{D}}(x)$ (monotonicity)
3. $s(\mu_{\tilde{A}}(x), \mu_{\tilde{B}}(x)) = s(\mu_{\tilde{B}}(x), \mu_{\tilde{A}}(x))$ (commutativity)
4. $s(\mu_{\tilde{A}}(x), s(\mu_{\tilde{B}}(x), \mu_{\tilde{C}}(x))) = s(s(\mu_{\tilde{A}}(x), \mu_{\tilde{B}}(x)), \mu_{\tilde{C}}(x))$ (associativity)

t-norms and *t*-conorms are related in a sense of logical duality. Alsina [Alsina 1985] defined a *t*-conorm as a two-place function *s* mapping from $[0, 1] \times [0, 1]$ in $[0, 1]$ such that the function *t*, defined as

$$t(\mu_{\tilde{A}}(x), \mu_{\tilde{B}}(x)) = 1 - s(1 - \mu_{\tilde{A}}(x), 1 - \mu_{\tilde{B}}(x))$$

is a *t*-norm. So any *t*-conorm *s* can be generated from a *t*-norm *t* through this transformation. More general, Bonissone and Decker [1986] showed, that for suitable negation operators like the complement operator for fuzzy sets—defined as $n(\mu_{\tilde{A}}(x)) = 1 - \mu_{\tilde{A}}(x)$ (see chapter 2)—pairs of *t*-norms *t* and *t*-conorms *s* satisfy the following generalization of DeMorgans law [Bonissone and Decker 1986, p. 220]:

$$s(\mu_{\tilde{A}}(x), \mu_{\tilde{B}}(x)) = n(t(n(\mu_{\tilde{A}}(x)), n(\mu_{\tilde{B}}(x)))) \quad \text{and}$$
$$t(\mu_{\tilde{A}}(x), \mu_{\tilde{B}}(x)) = n(s(n(\mu_{\tilde{A}}(x)), n(\mu_{\tilde{B}}(x)))), \qquad x \in X$$

Typical dual pairs of nonparametrized *t*-norms and *t*-conorms are compiled below [Bonissone and Decker 1986, p. 221; Mizumoto 1989, p. 220]:

$$t_w(\mu_{\tilde{A}}(x), \mu_{\tilde{B}}(x)) = \begin{cases} \min\{\mu_{\tilde{A}}(x), \mu_{\tilde{B}}(x)\} & \text{if } \max\{\mu_{\tilde{A}}(x), \mu_{\tilde{B}}(x)\} = 1 \\ 0 & \text{otherwise} \end{cases} \quad \text{drastic product}$$

$$s_w(\mu_{\tilde{A}}(x), \mu_{\tilde{B}}(x)) = \begin{cases} \max\{\mu_{\tilde{A}}(x), \mu_{\tilde{B}}(x)\} & \text{if } \min\{\mu_{\tilde{A}}(x), \mu_{\tilde{B}}(x)\} = 0 \\ 1 & \text{otherwise} \end{cases} \quad \text{drastic sum}$$

$t_1(\mu_{\tilde{A}}(x), \mu_{\tilde{B}}(x)) = \max\{0, \mu_{\tilde{A}}(x) + \mu_{\tilde{B}}(x) - 1\}$ bounded difference

$s_1(\mu_{\tilde{A}}(x), \mu_{\tilde{B}}(x)) = \min\{1, \mu_{\tilde{A}}(x) + \mu_{\tilde{B}}(x)\}$ bounded sum

$t_{1.5}(\mu_{\tilde{A}}(x), \mu_{\tilde{B}}(x)) = \dfrac{\mu_{\tilde{A}}(x) \cdot \mu_{\tilde{B}}(x)}{2 - [\mu_{\tilde{A}}(x) + \mu_{\tilde{B}}(x) - \mu_{\tilde{A}}(x) \cdot \mu_{\tilde{B}}(x)]}$ Einstein product

$s_{1.5}(\mu_{\tilde{A}}(x), \mu_{\tilde{B}}(x)) = \dfrac{\mu_{\tilde{A}}(x) + \mu_{\tilde{B}}(x)}{1 + \mu_{\tilde{A}}(x) \cdot \mu_{\tilde{B}}(x)}$ Einstein sum

$t_2(\mu_{\tilde{A}}(x), \mu_{\tilde{B}}(x)) = \mu_{\tilde{A}}(x) \cdot \mu_{\tilde{B}}(x)$ algebraic product

$s_2(\mu_{\tilde{A}}(x), \mu_{\tilde{B}}(x)) = \mu_{\tilde{A}}(x) + \mu_{\tilde{B}}(x) - \mu_{\tilde{A}}(x) \cdot \mu_{\tilde{B}}(x)$ algebraic sum

$t_{2.5}(\mu_{\tilde{A}}(x), \mu_{\tilde{B}}(x)) = \dfrac{\mu_{\tilde{A}}(x) \cdot \mu_{\tilde{B}}(x)}{\mu_{\tilde{A}}(x) + \mu_{\tilde{B}}(x) - \mu_{\tilde{A}}(x) \cdot \mu_{\tilde{B}}(x)}$ Hamacher product

$s_{2.5}(\mu_{\tilde{A}}(x), \mu_{\tilde{B}}(x)) = \dfrac{\mu_{\tilde{A}}(x) + \mu_{\tilde{B}}(x) - 2\mu_{\tilde{A}}(x) \cdot \mu_{\tilde{B}}(x)}{1 - \mu_{\tilde{A}}(x) \cdot \mu_{\tilde{B}}(x)}$ Hamacher sum

$t_3(\mu_{\tilde{A}}(x), \mu_{\tilde{B}}(x)) = \min\{\mu_{\tilde{A}}(x), \mu_{\tilde{B}}(x)\}$ minimum

$s_3(\mu_{\tilde{A}}(x), \mu_{\tilde{B}}(x)) = \max\{\mu_{\tilde{A}}(x), \mu_{\tilde{B}}(x)\}$ maximum

These operators are ordered as follows:

$$t_w \le t_1 \le t_{1.5} \le t_2 \le t_{2.5} \le t_3$$
$$s_3 \le s_{2.5} \le s_2 \le s_{1.5} \le s_1 \le s_w$$

We notice, that this order implies that for any fuzzy sets $\tilde{A}$ and $\tilde{B}$ in X with membership values between 0 and 1 any intersection operator that is a t-norm is bounded by the min-operator and the operator t_w. A t-conorm is bounded by the max-operator and the operator s_w, respectively, [Dubois and Prade 1982a, p. 42]:

$$t_w(\mu_{\tilde{A}}(x), \mu_{\tilde{B}}(x)) \le t(\mu_{\tilde{A}}(x), \mu_{\tilde{B}}(x)) \le \min\{\mu_{\tilde{A}}(x), \mu_{\tilde{B}}(x)\}$$
$$\max\{\mu_{\tilde{A}}(x), \mu_{\tilde{B}}(x)\} \le s(\mu_{\tilde{A}}(x), \mu_{\tilde{B}}(x)) \le s_w(\mu_{\tilde{A}}(x), \mu_{\tilde{B}}(x)), \qquad x \in X$$

It may be desirable to extend the range of the previously described operators in order to adapt them to the context in which they are used. To this end different authors suggested parametrized families of t-norms and t-conorms, often maintaining the associativity property.

For illustration purposes we review some interesting parametrized operators. Some of these operators and their equivalence to the logical "and" and "or" respectively have been justified axiomatically. We shall sketch the axioms on which the Hamacher-operator rests in order to give the reader the opportunity to compare the axiomatic system of Bellman and Giertz (min/max) on one hand with that of the Hamacher-operator (which is essentially a family of product operators) on the other.

Definition 3–14 [Hamacher 1978]

The *intersection* of two fuzzy sets $\tilde{A}$ and $\tilde{B}$ is defined as

$$\tilde{A} \cap \tilde{B} = \{(x, \mu_{\tilde{A}\cap\tilde{B}}(x)) \mid x \in X\}$$

where

$$\mu_{\tilde{A}\cap\tilde{B}}(x) = \frac{\mu_{\tilde{A}}(x)\mu_{\tilde{B}}(x)}{\gamma + (1-\gamma)(\mu_{\tilde{A}}(x) + \mu_{\tilde{B}}(x) - \mu_{\tilde{A}}(x)\mu_{\tilde{B}}(x))}, \quad \gamma \geq 0$$

Hamacher wants to derive a mathematical model for the "and" operator. His basic axioms are:

A1. The operator $\wedge$ is associative, that is, $\tilde{A} \wedge (\tilde{B} \wedge \tilde{C}) = (\tilde{A} \wedge \tilde{B}) \wedge \tilde{C}$.
A2. The operator $\wedge$ is continuous.
A3. The operator $\wedge$ is injective in each argument, that is,

$$(\tilde{A} \wedge \tilde{B}) = (\tilde{A} \wedge \tilde{C}) \Rightarrow \tilde{B} = \tilde{C}$$
$$(\tilde{A} \wedge \tilde{B}) = (\tilde{C} \wedge \tilde{B}) \Rightarrow \tilde{A} = \tilde{C}$$

(this is the essential difference between the Hamacher-operator and the Bellman-Giertz axioms).

A4. $\mu_{\tilde{A}}(x) = 1 \Rightarrow \mu_{\tilde{A}\wedge\tilde{A}}(x) = 1$

He then proves that a function $f: R \rightarrow [0, 1]$ exists with

$$\mu_{\tilde{A}\wedge\tilde{B}}(x) = f(f^{-1}(\mu_{\tilde{A}}(x)) + f^{-1}(\mu_{\tilde{B}}(x)))$$

If f is a rational function in $\mu_{\tilde{A}}(x)$ and $\mu_{\tilde{B}}(x)$ then the only possible operator is that shown in definition 3–14. (For $\gamma = 1$, this reduces to the algebraic product!).

Notice that the Hamacher-operator is the only H strict t-norm that can be expressed as a rational function [Mizumoto 1989, p. 223].

Definition 3–15 [Hamacher 1978]

The *union* of two fuzzy sets $\tilde{A}$ and $\tilde{B}$ is defined as

$$\tilde{A} \cup \tilde{B} = \{(x, \mu_{\tilde{A}\cup\tilde{B}}(x)) \mid x \in X\}$$

where

$$\mu_{\tilde{A}\cup\tilde{B}}(x) = \frac{(\gamma' - 1)\mu_{\tilde{A}}(x)\mu_{\tilde{B}}(x) + \mu_{\tilde{A}}(x) + \mu_{\tilde{B}}(x)}{1 + \gamma'\mu_{\tilde{A}}(x)\mu_{\tilde{B}}(x)}, \quad \gamma' \geqslant -1$$

For $\gamma' = 0$ the Hamacher-union-operator reduces to the algebraic sum.
Yager [1980] defined another triangular family of operators.

Definition 3–16 [Yager 1980]

The *intersection* of fuzzy sets $\tilde{A}$ and $\tilde{B}$ is defined as

$$\tilde{A} \cap \tilde{B} = \{(x, \mu_{\tilde{A}\cap\tilde{B}}(x)) \mid x \in X\}$$

where

$$\mu_{\tilde{A}\cap\tilde{B}}(x) = 1 - \min\{1, ((1 - \mu_{\tilde{A}}(x))^p + (1 - \mu_{\tilde{B}}(x))^p)^{1/p}\}, \qquad p \geq 1$$

The *union* of fuzzy sets is defined as

$$\tilde{A} \cup \tilde{B} = \{(x, \mu_{\tilde{A}\cup\tilde{B}}(x)) \mid x \in X\}$$

where

$$\mu_{\tilde{A}\cup\tilde{B}}(x) = \min\{1, (\mu_{\tilde{A}}(x)^p + \mu_{\tilde{B}}(x)^p)^{1/p}\}, \qquad p \geq 1$$

His intersection-operator converges to the min-operator for $p \to \infty$ and his union operator to the max-operator for $p \to \infty$.

For $p = 1$ the Yager-intersection becomes the "bold-intersection" of definition 3–10. The union operator converges to the maximum-operator for $p \to \infty$ and to the bold union for $p = 1$. Both operators satisfy the DeMorgan laws, and are commutative, associative for all p, monotonically nondecreasing in $\mu(x)$, and include the classical cases of dual logic. They are, however, not distributive.

Dubois and Prade [1980c, 1982a] also proposed a commutative and associative parametrized family of aggregation operators:

Definition 3–17 [Dubois and Prade 1980c, 1982a]

The *intersection* of two fuzzy sets $\tilde{A}$ and $\tilde{B}$ is defined as

$$\tilde{A} \cap \tilde{B} = \{(x, \mu_{\tilde{A}\cap\tilde{B}}(x)) \mid x \in X\}$$

where

$$\mu_{\tilde{A}\cap\tilde{B}}(x) = \frac{\mu_{\tilde{A}}(x) \cdot \mu_{\tilde{B}}(x)}{\max\{\mu_{\tilde{A}}(x), \mu_{\tilde{B}}(x), \alpha\}}, \qquad \alpha \in [0, 1]$$

This intersection-operator is decreasing with respect to α and lies between $\max\{\mu_{\tilde{A}}(x), \mu_{\tilde{B}}(x)\}$ (which is the resulting operation for $\alpha = 0$) and the algebraic product $\mu_{\tilde{A}}(x) \cdot \mu_{\tilde{B}}(x)$ (for $\alpha = 1$). The parameter α is a kind of threshold since the following relationships hold for the defined intersection operation [Dubois and Prade 1982a, p. 47]:

$$\mu_{\tilde{A}\cap\tilde{B}}(x) = \min\{\mu_{\tilde{A}}(x), \mu_{\tilde{B}}(x)\} \qquad \text{for } \mu_{\tilde{A}}(x), \mu_{\tilde{B}}(x) \in [\alpha, 1]$$

$$\mu_{\tilde{A}\cap\tilde{B}}(x) = \frac{\mu_{\tilde{A}}(x) \cdot \mu_{\tilde{B}}(x)}{\alpha} \qquad \text{for } \mu_{\tilde{A}}(x), \mu_{\tilde{B}}(x) \in [0, \alpha]$$

Definition 3–18 [Dubois and Prade 1980c, 1982a]

For the *union* of two fuzzy sets $\tilde{A}$ and $\tilde{B}$, defined as

$$\tilde{A} \cup \tilde{B} = \{(x, \mu_{\tilde{A}\cup\tilde{B}}(x)) \mid x \in X\}$$

Dubois and Prade suggested the following operation, where $\alpha \in [0, 1]$,

$$\mu_{\tilde{A}\cup\tilde{B}}(x) = \frac{\mu_{\tilde{A}}(x) + \mu_{\tilde{B}}(x) - \mu_{\tilde{A}}(x) \cdot \mu_{\tilde{B}}(x) - \min\{\mu_{\tilde{A}}(x), \mu_{\tilde{B}}(x), (1-\alpha)\}}{\max\{(1-\mu_{\tilde{A}}(x)), (1-\mu_{\tilde{B}}(x)), \alpha\}}$$

All the operators mentioned so far include the case of dual logic as special case. The question may arise: Why are there unique definitions for intersection (= and) and union (= or) in dual logic and traditional set theory and so many suggested definitions in fuzzy set theory? The answer is simply that many operators (for instance product and min-operator) perform in exactly the same way if the degrees of membership are restricted to the values 0 or 1. If this is not longer requested they lead to different results.

This triggers yet another question: Are the only ways to "combine" or aggregate fuzzy sets the intersection or union—or the logical "and" or "or,"—respectively, or are there other possibilities of aggregation? The answer to this question is definitely yes. There are other ways of combining fuzzy sets to fuzzy statements; "and" and "or" are only limiting special cases. Generalized models for the logical "and" and "or" are given by the "fuzzy and" and "fuzzy or" [Werners 1984]. Furthermore, a number of authors have suggested general connectives, which are (so far) of particular importance for decision analysis and for other applications of fuzzy set theory. These operators are general in the sense that they do not distinguish between the intersection and union of fuzzy sets.

Here we shall only mention some of these general connectives. A detailed discussion of them and the description of still others can be found in volume 2 in the context of decision making in fuzzy environments.

Averaging Operators. A straightforward approach for aggregating fuzzy sets, for instance in the context of decision making, would be to use the aggregating procedures frequently used in utility theory or multi-criteria decision theory. They realize the idea of trade-offs between conflicting goals when compensation is allowed, and the resulting trade-offs lie between the most optimistic lower bound and the most pessimistic upper bound, that is, they map between the minimum and the maximum degree of membership of the aggregated sets. Therefore they are called averaging operators. Operators such as the weighted and unweighted arithmetic or geometric mean are examples of nonparametric averaging operators. In fact, they are adequate models for human aggregation procedures in decision environments and have empirically performed quite well [Thole, Zimmermann, and Zysno 1979]. Procedures and results of empirical research done in the context of human decision making are investigated in section 14.3.

The fuzzy aggregation operators "fuzzy and" and "fuzzy or" suggested by Werners [1984] combine the minimum and maximum operator, respectively, with the arithmetic mean. The combination of these operators leads to very good results with respect to empirical data [Zimmermann and Zysno 1983] and allows compensation between the membership values of the aggregated sets.

Definition 3–19 [Werners 1988, p. 297]

The "fuzzy and" operator is defined as

$$\mu_{\widetilde{\text{and}}}(\mu_{\tilde{A}}(x), \mu_{\tilde{B}}(x)) = \gamma \cdot \min\{\mu_{\tilde{A}}(x), \mu_{\tilde{B}}(x)\} + \frac{(1-\gamma)(\mu_{\tilde{A}}(x) + \mu_{\tilde{B}}(x))}{2}$$

$$x \in X, \gamma \in [0, 1]$$

The "fuzzy or" operator is defined as

$$\mu_{\widetilde{\text{or}}}(\mu_{\tilde{A}}(x), \mu_{\tilde{B}}(x)) = \gamma \cdot \max\{\mu_{\tilde{A}}(x), \mu_{\tilde{B}}(x)\} + \frac{(1-\gamma)(\mu_{\tilde{A}}(x) + \mu_{\tilde{B}}(x))}{2}$$

$$x \in X, \gamma \in [0, 1]$$

The parameter γ indicates the degree of nearness to the strict logical meaning of "and" and "or," respectively. For $\gamma = 1$ the "fuzzy and" becomes the minimum operator, the "fuzzy or" reduces to the maximum operator. $\gamma = 0$ yields for both the arithmetic mean.

Additional averaging aggregation procedures are symmetric summation operators, which as well as the arithmetic or geometric mean operators indicate some degree of compensation but in contrast to the latter are not associative. Examples of symmetric summation operators are the operators M_1, M_2, and N_1, N_2, known as symmetric summations and symmetric differences, respectively. Here the aggregation of two fuzzy sets $\tilde{A}$ and $\tilde{B}$ is pointwise defined as follows:

$$M_1(\mu_{\tilde{A}}(x), \mu_{\tilde{B}}(x)) = \frac{\mu_{\tilde{A}}(x) + \mu_{\tilde{B}}(x) - \mu_{\tilde{A}}(x) \cdot \mu_{\tilde{B}}(x)}{1 + \mu_{\tilde{A}}(x) + \mu_{\tilde{B}}(x) - 2\mu_{\tilde{A}}(x) \cdot \mu_{\tilde{B}}(x)}$$

$$M_2(\mu_{\tilde{A}}(x), \mu_{\tilde{B}}(x)) = \frac{\mu_{\tilde{A}}(x) \cdot \mu_{\tilde{B}}(x)}{1 - \mu_{\tilde{A}}(x) - \mu_{\tilde{B}}(x) + 2\mu_{\tilde{A}}(x) \cdot \mu_{\tilde{B}}(x)}$$

$$N_1(\mu_{\tilde{A}}(x), \mu_{\tilde{B}}(x)) = \frac{\max\{\mu_{\tilde{A}}(x), \mu_{\tilde{B}}(x)\}}{1 + |\mu_{\tilde{A}}(x) - \mu_{\tilde{B}}(x)|}$$

$$N_2(\mu_{\tilde{A}}(x), \mu_{\tilde{B}}(x)) = \frac{\min\{\mu_{\tilde{A}}(x), \mu_{\tilde{B}}(x)\}}{1 - |\mu_{\tilde{A}}(x) - \mu_{\tilde{B}}(x)|}$$

A detailed description of the properties of nonparametric averaging operators is reported in [Dubois and Prade 1984]. For further details of symmetric summation operators the reader is referred to [Silvert 1979].

The above-mentioned averaging operators indicate a "fix" compensation between the logical "and" and the logical "or." In order to describe a variety of phenomena in decision situations, several operators with different compensations are necessary. An operator that is more general in the sense that the compensation between intersection and union is expressed by a parameter γ was suggested and empirically tested by Zimmermann and Zysno [1980] under the name "compensatory and."

Definition 3–20 [Zimmermann and Zysno 1980]

The *"compensatory and" operator* is defined as follows:

$$\mu_{\tilde{A}i,\text{comp}}(x) = \left(\prod_{i=1}^{m} \mu_i(x)\right)^{(1-\gamma)} \left(1 - \prod_{i=1}^{m}(1 - \mu_i(x))\right)^{\gamma}, \qquad x \in X, 0 \leq \gamma \leq 1$$

This "γ-operator" is obviously a combination of the algebraic product (modeling the logical "and") and the algebraic sum (modeling the "or"). It is pointwise injective (except at zero and one), continuous, monotonous, and commutative. It also satisfies the DeMorgan laws and is in accordance with the truth tables of dual logic. The parameter indicates where the actual operator is located between the logical "and" and "or."

Other operators following the idea of parametrized compensation are defined by taking linear convex combinations of noncompensatory operators modeling the logical "and" and "or." The aggregation of two fuzzy sets $\widetilde{A}$ and $\widetilde{B}$ by the convex combination between the min- and max-operator is defined as

$$\mu_1(\mu_{\tilde{A}}(x), \mu_{\tilde{B}}(x)) = \gamma \cdot \min\{\mu_{\tilde{A}}(x), \mu_{\tilde{B}}(x)\} + (1-\gamma) \cdot \max\{\mu_{\tilde{A}}(x), \mu_{\tilde{B}}(x)\} \qquad \gamma \in [0, 1]$$

Combining the algebraic product and algebraic sum we obtain the following operation

$$\mu_2(\mu_{\tilde{A}}(x), \mu_{\tilde{B}}(x)) = \gamma\mu_{\tilde{A}}(x) \cdot \mu_{\tilde{B}}(x) + (1-\gamma) \cdot [\mu_{\tilde{A}}(x) + \mu_{\tilde{B}}(x) - \mu_{\tilde{A}}(x) \cdot \mu_{\tilde{B}}(x)] \qquad \gamma \in [0, 1]$$

This class of operators is again in accordance with the dual logic truth tables. But Zimmermann and Zysno showed that the "compensatory and" operator is more adequate in human decision making than are these operators [Zimmermann and Zysno 1980, p. 50].

The relationships between different aggregation operators for aggregating two fuzzy sets $\widetilde{A}$ and $\widetilde{B}$ with respect to the three classes of *t*-norms, *t*-conorms, and averaging operators are represented in figure 3–2.

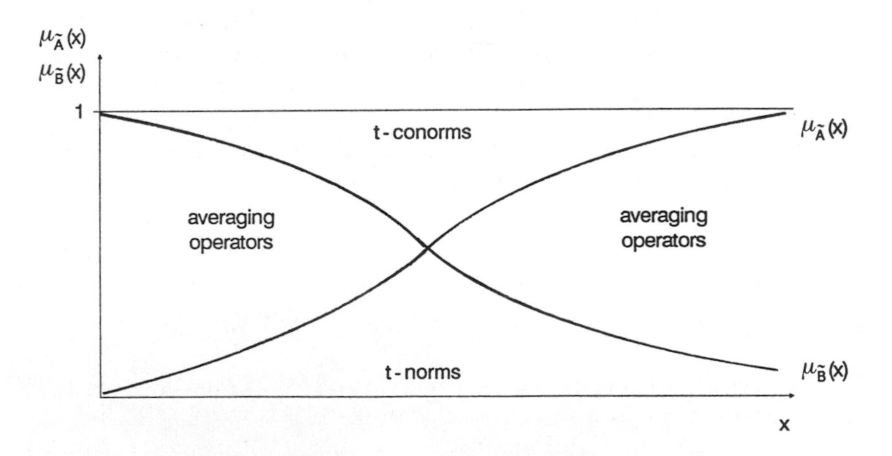

Figure 3–2. Mapping of *t*-norms, *t*-conorms, and averaging operators.

Table 3–1. Classification of compensatory and noncompensatory operators.

	Distinguishing operators	*General operators*
Compensatory	fuzzy and fuzzy or	compensatory and convex combinations of min and max symmetric summations arithmetic mean geometric mean
Non compensatory	*t*-norms *t*-conorms min max	

A taxonomie with respect to the compensatory property of distinguishing operators, which differentiate between the intersection and union of fuzzy sets, and general operators is presented in table 3–1. Table 3–2 summarizes the classes of aggregation operators for fuzzy sets reported in this chapter and compiles some references. Table 3–3 represents the relationship between parametrized families of operators and the presented *t*-norms and *t*-conorms with respect to special values of their parameters.

3.2.3 Criteria for Selecting Appropriate Aggregation Operators

The variety of operators for the aggregation of fuzzy sets might be confusing and might make it difficult to decide which one to use in a specific model or situation. Which rules can be used for such a decision?

The following eight important criteria according to which operators can be classified are not quite disjunct; hopefully they may be helpful in selecting the appropriate connective.

1. Axiomatic Strength. We have listed the axioms that Bellman-Giertz and Hamacher, respectively, wanted their operators to satisfy. Obviously, everything else being equal, an operator is the better the less limiting are the axioms it satisfies.

Table 3–2. Classification of aggregation operators.

References	*Intersection operators t-norms*	*Averaging operators*	*Union operators t-conorms*
		non-parametrized	
Zadeh 1965	minimum		maximum
	algebraic product		algebraic sum
Giles 1976	bounded sum		bounded difference
Hamacher 1978	Hâmacher product		Hamacher sum
Mizumoto 1982	Einstein product		Einstein sum
Dubois and Prade 1980, 1982	drastic product		drastic sum
Dubois and Prade 1984		arithmetic mean geometric mean	
Silvert 1979		symmetric summation and differences	
		parametrized families	
Hamacher 1978	Hamacher-intersection-operators		Hamacher-union-operators
Yager 1980	Yager-intersection-operators		Yager-union-operators
Dubois and Prade 1980a, 1982a, 1984	Dubois-intersection-operators		Dubois-union-operators
Werners 1984		"fuzzy and", "fuzzy or"	
Zimmermann and Zysno 1980		"compensatory and", convex comb. of maximum and minimum, or algebraic product and algebraic sum	

Table 3–3. Relationship between parameterized operators and their parameters.

Parametrized operators	*Types of t-norms and t-conorms*											
	drastic		*bounded*		*Einstein*		*algebraic*		*Hamacher*			
	prod.	*sum*	*sum*	*diff.*	*prod.*	*sum*	*prod.*	*sum*	*prod.*	*sum*	*min*	*max*
Hamacher												
intersection	$\gamma \to \infty$				$\gamma = 2$		$\gamma = 1$		$\gamma = 0$			
union		$\gamma' \to \infty$				$\gamma' = 1$		$\gamma' = 0$		$\gamma' = -1$		
Yager												
intersection	$p \to 0$		$p = 1$								$p \to \infty$	
union		$p \to 0$		$p = 1$								$p \to \infty$
Dubois												
intersection							$\alpha = 1$				$\alpha = 0$	
union								$\alpha = 1$				$\alpha = 0$

2. Empirical Fit. If fuzzy set theory is used as a modeling language for real situations or systems, it is not only important that the operators satisfy certain axioms or have certain formal qualities (such as associativity, commutativity), which are certainly of importance from a mathematical point of view, but the operators must also be appropriate models of real-system behavior; and this can normally be proven only by empirical testing.

3. Adaptability. It is rather unlikely that the type of aggregation is independent of the context and semantic interpretation, that is, whether the aggregation of fuzzy sets models a human decision, a fuzzy controller, a medical diagnostic system, or a specific inference rule in fuzzy logic. If one wants to use a very small number of operators to model many situations, then these operators have to be adaptable to the specific context. This can, for instance, be achieved by parametrization. Thus min- and max-operators cannot be adapted at all. They are acceptable in situations in which they fit and under no other circumstances. (Of course, they have other advantages, such as numerical efficiency). By contrast, Yager's operators or the γ-operator can be adapted to certain contexts by setting the p's or γ's appropriately.

4. Numerical Efficiency. Comparing the min-operator with, for instance, Yager's intersection operator or the γ-operator it becomes quite obvious, that the latter two require considerably more computational effort than the former. In practice, this might be quite important, in particular when large problems have to be solved.

5. Compensation. The logical "and" does not allow for compensation at all, that is, an element of the intersection of two sets cannot compensate a low degree of belonging to one of the intersected sets by a higher degree of belonging to another of them; in (dual) logic one can not compensate by higher truth of one statement for lower truth of another statement when combining them by "and." By compensation, in the context of aggregation operators for fuzzy sets, we mean the following: Given that the degree of membership to the aggregated fuzzy set is

$$\mu_{\mathrm{Agg}}(x_k) = f(\mu_{\tilde{A}}(x_k), \mu_{\tilde{B}}(x_k)) = k$$

f is compensatory if $\mu_{\mathrm{Agg}}(x_k) = k$ is obtainable for a different $\mu_{\tilde{A}}(x_k)$ by a change in $\mu_{\tilde{B}}(x_k)$. Thus the min-operator is not compensatory while the product operator, the γ-operator, and so forth are.

6. Range of Compensation. If one would use a convex combination of min- and max-operator, a compensation could obviously occur in the range between min and max. The product operator allows compensation in the open interval (0, 1). In general, the larger the range of compensation the better the compensatory operator.

7. Aggregating Behavior. Considering normal or subnormal fuzzy sets, the degree of membership in the aggregated set depends very frequently on the number of sets combined. Combining fuzzy sets by the product operator, for instance, each additional fuzzy set "added" will normally decrease the resulting aggregate degrees of membership. This might be a desirable feature, it might, however, also be not adequate. Goguen, for instance, argues that for formal reasons the resulting degree of membership should be nonincreasing [Goguen 1967].

8. Required Scale Level of Membership Functions. The scale level (nominal, interval, ratio, or absolute) on which membership information can be obtained depends on a number of factors. Different operators may require different scale levels of membership information to be admissible. (For instance, the min-operator is still admissible for ordinal information while the product operator, strictly speaking, is not!). In general, again all else being equal, the operator that requires the lowest scale level is the most preferable from the point of view of information gathering.

Exercises

1. The product and the bounded difference have both been suggested as models for the intersection. Compute the intersection of fuzzy sets $\tilde{B}$ and $\tilde{C}$ from exercise 4 of chapter 2 and compare the three alternative models for the intersection: Minimum, product, and bounded difference.
2. The bounded sum and the algebraic sum have been suggested as alternative models for the union of fuzzy sets. Compute the union of the fuzzy sets $\tilde{B}$ and $\tilde{C}$ of exercise 4 of chapter 2 using the above-mentioned models, and compare the result with the result of exercise 5 chapter 2.
3. Determine the intersection of $\tilde{B}$ and $\tilde{C}$ in exercise 4 chapter 2 by using the
   a. Hamacher operator with $\gamma = .25$; $.5$; $.75$
   b. Yager operator with $p = 1, 5, 10$.

4. Which of the intersection operators mentioned in chapter 3 are compensatory and which not? Are the "compensatory" operators compensatory for the entire range [0, 1] and for the entire domain of their parameters (γ, p, etc.). If not, what are the limits of compensation?
5. Prove, that the following properties are satisfied by Yager's union operator:
   a. $\mu_{\tilde{A} \cup \tilde{B}}(x) = \mu_{\tilde{A}}(x)$ for $\mu_{\tilde{B}}(x) = 0$
   b. $\mu_{\tilde{A} \cup \tilde{B}}(x) = 1$ for $\mu_{\tilde{B}}(x) = 1$
   c. $\mu_{\tilde{A} \cup \tilde{B}}(x) \geq \mu_{\tilde{A}}(x)$ for $\mu_{\tilde{A}}(x) = \mu_{\tilde{B}}(x)$
   d. For $p \to 0$ the Yager union operator reduces to s_w (drastic sum).
6. Show for the parametrized families of fuzzy union defined by Hamacher, Yager, and Dubois that the defining functions of these operators decrease with any increase in the parameter.

4 FUZZY MEASURES AND MEASURES OF FUZZINESS

4.1 Fuzzy Measures

In order to prevent confusion about fuzzy measures and measures of fuzziness, we shall first briefly describe the meaning and features of fuzzy measures. In the late 1970s, Sugeno defined a fuzzy measure as follows Sugeno [1977]: $\mathscr{B}$ is a Borel field of the arbitrary set (universe) X.

Definition 4–1

A set function g defined on $\mathscr{B}$ which has the following properties is called a *fuzzy measure*:

1. $g(\emptyset) = 0$, $g(X) = 1$.
2. If $A, B \in \mathscr{B}$ and $A \subseteq B$, then $g(A) \leq g(B)$.
3. If $A_n \in \mathscr{B}$, $A_1 \subseteq A_2 \subseteq \ldots$ then $\lim_{n \to \infty} g(A_n) = g(\lim_{n \to \infty} A_n)$.

Sugeno's measure differs from the classical measure essentially by relaxing the additivity property [Murofushi and Sugeno 1989, p. 201]. A different approach, however, is used by Klement and Schwyhla [1982]. The interested reader is referred to their article.

Banon [1981] shows that very many measures with finite universe, such as probability measures, belief functions, plausibility measures, and so on, are fuzzy measures in the sense of Sugeno. For this book one measure—possibility—is of particular interest [see Dubois and Prade 1988a, p. 7].

In the framework of fuzzy set theory Zadeh introduced the notion of a possibility distribution and the concept of a possibility measure, which is a special type of the fuzzy measure proposed by Sugeno. A possibility measure is defined as follows: [Zadeh 1978; Higashi and Klir 1982]:

Definition 4–2

Let $P(X)$ be the power set of a set X.
A *possibility measure* is a function Π: $P(X) \rightarrow [0, 1]$ with the properties

1. $\Pi(\emptyset) = 0, \Pi(X) = 1$
2. $A \subseteq B \Rightarrow \Pi(A) \leq \Pi(B)$
3. $\Pi(\bigcup_{i \in I} A_i) = \sup_{i \in I} \Pi(A_i)$ with an index set I

It can be uniquely determined by a possibility distribution function f: $X \rightarrow [0, 1]$ by $\Pi(A) = \sup_{x \in A} f(x)$, $A \subset X$. It follows directly that f is defined by $f(x) = \Pi(\{x\})\ \forall x \in X$ [Klir and Folger 1988, p. 122].

A possibility is not always a fuzzy measure [Puri and Ralescu 1982]. It is, however, a fuzzy measure, if X is finite and if the possibility distribution is normal—that is, a mapping into [0, 1].

Example 4–1

Let $X = \{0, 1, \ldots, 10\}$.
$\Pi(\{x\}) := $ Possibility that x is close to 8.

x	0	1	2	3	4	5	6	7	8	9	10
$\Pi(\{x\})$	.0	.0	.0	.0	.0	.1	.5	.8	1	.8	.5

$\Pi(A)$: = Possibility that A contains an integer close to 8.
$A \subset X \Rightarrow \Pi(A) = \sup_{x \in A} \Pi(\{x\})$

For $A = \{2, 5, 9\}$ we compute: $\Pi(A) = \sup_{x \in A} \Pi(\{x\})$

$$= \sup \{\Pi(\{2\})\ \Pi(\{5\}), \Pi(\{9\})\}$$

$$= \sup\{0, .1, .8\}$$
$$= .8$$

4.2 Measures of Fuzziness

Measures of fuzziness by contrast to fuzzy measures try to indicate the degree of fuzziness of a fuzzy set. A number of approaches to this end have become known. Some authors, strongly influenced by the Shannon entropy as a measure of information, and following de Luca and Termini [1972], consider a measure of fuzziness as a mapping d from the power set $P(X)$ to $[0, +\infty]$ satisfying a number of conditions. Others [Kaufmann 1975] suggested an index of fuzziness as a normalized distance, and others [Yager 1979; Higashi and Klir 1982] base their concept of a measure of fuzziness on the degree of distinction between the fuzzy set and its complement.

We shall, as an illustration, discuss two of those measures. Suppose for both cases that the support of A is finite.

The first is: Let $\mu_{\tilde{A}}(x)$ be the membership function of the fuzzy set $\tilde{A}$ for $x \in X$, X finite. It seems plausible that the measure of fuzziness $d(\tilde{A})$ should then have the following properties [de Luca and Termini 1972]:

1. $d(\tilde{A}) = 0$ if $\tilde{A}$ is a crisp set in X.
2. $d(\tilde{A})$ assumes a unique maximum if $\mu_{\tilde{A}}(x) = \frac{1}{2}\ \forall x \in X$.
3. $d(\tilde{A}) \geq d(\tilde{A}')$ if $\tilde{A}'$ is "crisper" than $\tilde{A}$, i.e., if $\mu_{\tilde{A}'}(x) \leq \mu_{\tilde{A}}(x)$ for $\mu_{\tilde{A}}(x) \leq \frac{1}{2}$ and $\mu_{\tilde{A}'}(x) \leq \mu_{\tilde{A}}(x)$ for $\mu_{\tilde{A}}(x) \geq \frac{1}{2}$.
4. $d(\complement\tilde{A}) = d(\tilde{A})$ where $\complement\tilde{A}$ is the complement of $\tilde{A}$.

De Luca and Termini suggested as a measure of fuzziness the "entropy"[1] of a fuzzy set [de Luca and Termini 1972, p. 305], which they defined as follows:

Definition 4–3a

The entropy as a *measure of a fuzzy set* $\tilde{A} = \{(x, \mu_{\tilde{A}}(x)\}$ is defined as

$$d(\tilde{A}) = H(\tilde{A}) + H(\complement\tilde{A}), \qquad x \in X$$
$$H(\tilde{A}) = -K \sum_{i=1}^{n} \mu_{\tilde{A}}(x_i) \ln(\mu_{\tilde{A}}(x_i))$$

[1] Also employed in thermodynamics, information theory, and statistics [Capocelli and de Luca 1973].

where n is the number of elements in the support of $\tilde{A}$ and K is a positive constant.

Using Shannon's function $S(x) = -x \ln x - (1-x) \ln (1-x)$, they simplify the expression in definition 4–3a to arrive at:

Definition 4–3b

The entropy d as a *measure of fuzziness* of a fuzzy set $\tilde{A} = \{x, \mu_{\tilde{A}}(x)\}$ is defined as

$$d(\tilde{A}) = K \sum_{i=1}^{n} S(\mu_{\tilde{A}}(x_i))$$

Example 4–2

Let $\tilde{A}$ = "integers close to 10" (see example 2–1d)

$$\tilde{A} = \{(7, .1), (8, .5), (9, .8), (10, 1), (11, .8), (12, .5), (13, .1)\}$$

Let $K = 1$, so

$$d(\tilde{A}) = .325 + .693 + .501 + 0 + .501 + .693 + .325 = 3.038$$

Let furthermore $\tilde{B}$ = "integers quite close to 10"

$$\tilde{B} = \{(6, .1), (7, .3), (8, .4), (9, .7), (10, 1), (11, .8), (12, .5), (13, .3), (14, .1)\}$$

$$d(\tilde{B}) = .325 + .611 + .673 + .611 + 0 + .501 + .693 + .611 + .325 = 4.35$$

The second measure is: Knopfmacher [1975], Loo [1977], Gottwald [1979b], and others based their contributions on de Luca's and Termini's suggestion in some respects.

If $\tilde{A}$ is a fuzzy set in X and $\complement\tilde{A}$ its complement, then by contrast to crisp sets, it is not necessarily true that

$$\tilde{A} \cup \complement\tilde{A} = X$$
$$\tilde{A} \cap \complement\tilde{A} = \emptyset$$

This means that fuzzy sets do not always satisfy the law of the excluded middle, which is one of their major distinctions from traditional crisp sets.

Some authors [Yager 1979; Higashi and Klir 1982] consider the relationship between $\tilde{A}$ and $\complement\tilde{A}$ to be the essence of fuzziness.

Yager [1979] notes that the requirement of distinction between $\tilde{A}$ and $\complement\tilde{A}$ is not satisfied by fuzzy sets. He therefore suggests that any measure of fuzziness should be a measure of the lack of distinction between $\tilde{A}$ and $\complement\tilde{A}$ or $\mu_{\tilde{A}}(x)$ and $\mu_{\complement\tilde{A}}(x)$.

As a possible metric to measure the *distance* between a fuzzy set and its complement Yager suggests:

Definition 4–4

$$D_p(\tilde{A}, \complement\tilde{A}) = \left[\sum_{i=1}^{n} |\mu_{\tilde{A}}(x_i) - \mu_{\complement\tilde{A}}(x_i)|^p\right]^{1/p} \qquad p = 1, 2, 3, \ldots$$

Let $S = \text{supp}(\tilde{A})$: $D_p(S, \complement S) = \|S\|^{1/p}$

Definition 4–5 [Yager 1979]

A *measure of the fuzziness* of $\tilde{A}$ can be defined as

$$f_p(\tilde{A}) = 1 - \frac{D_p(\tilde{A}, \complement\tilde{A})}{\|\text{supp}(\tilde{A})\|}$$

So $f_p(\tilde{A}) \in [0, 1]$. This measure also satisfies properties 1 to 4 required by de Luca and Termini (see above).

For $p = 1$, $D_p(\tilde{A}, \complement\tilde{A})$ yields the Hamming metric

$$D_1(\tilde{A}, \complement\tilde{A}) = \sum_{i=1}^{n} |\mu_{\tilde{A}}(x_i) - \mu_{\complement\tilde{A}}(x_i)|$$

Because $\mu_{\complement\tilde{A}}(x) = 1 - \mu_{\tilde{A}}(x)$, this becomes

$$D_1(\tilde{A}, \complement\tilde{A}) = \sum_{i=1}^{n} |2\mu_{\tilde{A}}(x_i) - 1|$$

For $p = 2$, we arrive at the Euclidean metric

$$D_2(\tilde{A}, \complement\tilde{A}) = \left(\sum_{i=1}^{n} (\mu_{\tilde{A}}(x_i) - \mu_{\complement\tilde{A}}(x_i))^2\right)^{1/2}$$

and for $\mu_{\complement\tilde{A}}(x) = 1 - \mu_{\tilde{A}}(x)$, we have

$$D_2(\tilde{A}, \complement A) = \left(\sum_{i=1}^{n} (2\mu_{\tilde{A}}(x_i) - 1)^2\right)^{1/2}$$

Example 4–3

Let $\tilde{A}$ = "integers close to 10" and
$\tilde{B}$ = "integers quite close to 10" be defined as in example 4–2.
Applying the above derived formula we compute for $p = 1$,

$$D_1(\tilde{A}, \complement\tilde{A}) = .8 + 0 + .6 + 1 + .6 + 0 + .8$$
$$= 3.8$$
$$\| \text{supp}\,(\tilde{A}) \| = 7$$

so $f_1(\tilde{A}) = 1 - \frac{3.8}{7} = 0.457$

Analogously:

$$D_1(\tilde{B}, \complement\tilde{B}) = 4.6$$
$$\| \text{supp}\,(\tilde{B}) \| = 9$$

so $f_1(\tilde{B}) = 1 - \frac{4.6}{9} = .489$

Similarly, for $p = 2$, we obtain

$$D_2(\tilde{A}, \complement\tilde{A}) = 1.73$$
$$\| \text{supp}\,(\tilde{A}) \|^{1/2} = 2.65$$

so $f_2(\tilde{A}) = 1 - \frac{1.73}{2.65} = 0.347$

and

$$D_2(\tilde{B}, \complement\tilde{B}) = 1.78$$
$$\| \text{supp}\,(\tilde{B}) \|^{1/2} = 3$$

so $f_2(\tilde{B}) = 1 - \frac{1.78}{3} = 0.407$

The reader should realize that the complement of a fuzzy set is not uniquely defined [see, Bellman and Giertz 1973; Dubois and Prade 1982a; Lowen 1978]. It is therefore not surprising that for other definitions of the complement and for other measures of distance, other measures of fuzziness will result, even though they all focus on the distinction between a fuzzy set and its complement [see for example, Klir 1987, p. 141]. Those variations, as well as extension of measures of fuzziness to nonfinite supports, will not be considered here; neither will the approaches that define fuzzy measures of fuzzy sets [Yager 1979].

Exercises

1. Let $\widetilde{A}$ be defined as in example 4–2.
   $\widetilde{B}' = \{(8, .5), (9, .9), (10, 1), (11, .8), (12, .5)\}$
   $\widetilde{C}' = \{(6, .1), (7, .1), (8, .5), (9, .8), (10, 1), (11, .8), (12, .5), (13, .1), (14, .1)\}$
   Is $\widetilde{A}$ crisper than $\widetilde{B}$ (or $\widetilde{C}$)?
   Compute as measures of fuzziness:
   a. the entropy (with $K = 1$)
   b. f_1
   c. f_2 for all three sets.
   Compare the results.
2. Determine the maximum of the entropy of $d(\widetilde{A})$ in dependence of the cardinality of the support of $\widetilde{A}$.
3. Consider $\widetilde{A}$ as in exercise 1. Determine $\widetilde{A} \cap ¢\widetilde{A}$ and $\widetilde{A} \cup ¢\widetilde{A}$. For which (special) fuzzy sets does the equality hold?
4. Consider Example 4–1. Compute the possibilities of the following sets:

$$A_1 = \{1, 2, 3, 4, 5, 6\}, A_2 = \{1, 5, 8, 9\}, A_3 = \{7, 9\}$$

5 THE EXTENSION PRINCIPLE AND APPLICATIONS

5.1 The Extension Principle

One of the most basic concepts of fuzzy set theory which can be used to generalize crisp mathematical concepts to fuzzy sets is the extension principle. In its elementary form it was already implied in Zadeh's first contribution [1965]. In the meantime, modifications have been suggested [Zadeh 1973a; Zadeh et al. 1975; Jain 1976]. Following Zadeh [1973a] and Dubois and Prade [1980a] we define the extension principle as follows:

Definition 5–1

Let X be a cartesian product of universes $X = X_1 \ldots X_r$, and $\widetilde{A}_1, \ldots, \widetilde{A}_r$ be r fuzzy sets in $X_1, \ldots, X_r$, respectively. f is a mapping from X to a universe Y, $y = f(x_1, \ldots, x_r)$. Then the extension principle allows us to define a *fuzzy set $\widetilde{B}$ in Y* by

$$\widetilde{B} = \{(y, \mu_{\widetilde{B}}(y)) \mid y = f(x_1, \ldots, x_r), (x_1, \ldots, x_r) \in X\}$$

where

$$\mu_{\widetilde{B}}(y) = \begin{cases} \sup\limits_{(x_1, \ldots, x_r) \in f^{-1}(y)} \min\{\mu_{\widetilde{A}_1}(x_1), \ldots, \mu_{\widetilde{A}_r}(x_r)\} & \text{if } f^{-1}(y) \neq \emptyset \\ 0 & \text{otherwise} \end{cases}$$

where f^{-1} is the inverse of f.

For $r = 1$, the extension principle, of course, reduces to

$$\widetilde{B} = f(\widetilde{A}) = \{(y, \mu_{\widetilde{B}}(y)) \mid y = f(x), x \in X\}$$

where

$$\mu_{\widetilde{B}}(y) = \begin{cases} \sup\limits_{x \in f^{-1}(y)} \mu_{\widetilde{A}}(x), & \text{if } f^{-1}(y) \neq \emptyset \\ 0 & \text{otherwise} \end{cases}$$

Example 5–1

Let $\widetilde{A} = \{(-1, .5), (0, .8), (1, 1), (2, .4)\}$

$f(x) = x^2$

Then by applying the extension principle we obtain

$$\widetilde{B} = f(\widetilde{A}) = \{(0, .8), (1, 1), (4, .4)\}$$

Figure 5–1 illustrates the relationship:

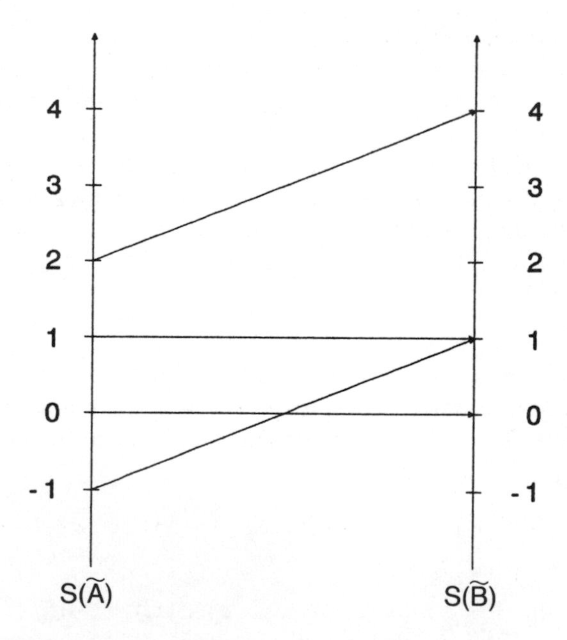

Figure 5–1. The extension principle.

The extension principle as stated in definition 5–1 can and has been modified by using the algebraic sum (definition 3–8) rather than sup, and the product rather than min [Dubois and Prade 1980a]. Since, however, it is generally used as defined in definition 5–1 we will restrict our considerations to this "classical" version.

5.2 Operations for Type 2 Fuzzy Sets

The extension principle can be used to define set theoretic operations for type 2 fuzzy sets such as defined in definition 3–1.

We shall consider only fuzzy sets of type 2 with discrete domains. Let two fuzzy sets of type 2 be defined by

$$\widetilde{A}(x) = \{(x, \mu_{\tilde{A}}(x))\} \quad \text{and} \quad \widetilde{B}(x) = \{(x, \mu_{\tilde{B}}(x))\}$$

where

$$\mu_{\tilde{A}}(x) = \{(u_i, \mu_{ui}(x)) \mid x \in X, u_i, \mu_{ui}(x) \in [0, 1]\}$$
$$\mu_{\tilde{B}}(x) = \{(v_j, \mu_{vj}(x)) \mid x \in X, v_j, \mu_{vj}(x) \in [0, 1]\}$$

The μ_i and v_j are degrees of membership of type-1 fuzzy sets and the $\mu_{ui}(x)$ and $\mu_{vj}(x)$, respectively, their membership functions. Using the extension principle the set theoretic operations can be defined as follows [Mizumoto and Tanaka 1976]:

Definition 5–2

Let two *fuzzy sets of type 2* be defined as above. The membership function of their *union* is then defined by

$$\begin{aligned}\mu_{\tilde{A} \cup \tilde{B}}(x) &= \mu_{\tilde{A}}(x) \cup \mu_{\tilde{B}}(x) \\ &= \{(w, \mu_{\tilde{A} \cup \tilde{B}}(w)) \mid w = \max\{u_i, v_j\}, u_i, v_j \in [0, 1]\}\end{aligned}$$

where

$$\mu_{\tilde{A} \cup \tilde{B}}(w) = \sup_{w = \max\{u_i, v_j\}} \min\{\mu_{u_i}(x), \mu_{v_j}(x)\}$$

Their *intersection* is defined by

$$\begin{aligned}\mu_{\tilde{A} \cap \tilde{B}}(x) &= \mu_{\tilde{A}}(x) \cap \mu_{\tilde{B}}(x) \\ &= \{(w, \mu_{\tilde{A} \cap \tilde{B}}(w)) \mid w = \min\{u_i, v_j\}, u_i, v_j \in [0, 1]\}\end{aligned}$$

where

$$\mu_{\tilde{A} \cap \tilde{B}}(w) = \sup_{w=\min\{u_i, v_j\}} \min\{\mu_{u_i}(x), \mu_{v_j}(x)\}$$

and the *complement* of $\tilde{A}$ by

$$\tilde{\mu}_{\complement \tilde{A}}(x) = \{[(1 - u_i), \mu_{\tilde{A}}(u_i)]\}$$

Example 5–2

Let $X = 1, \ldots, 10$, $\tilde{A}$ = small integers
$\tilde{B}$ = integers close to 4

defined by

$$\tilde{A} = \{(x, \mu_{\tilde{A}}(x))\}$$
$$\tilde{B} = \{(x, \mu_{\tilde{B}}(x))\}$$

where, for $x = 3$,

$$\begin{aligned}\mu_{\tilde{A}}(3) &= \{(u_i, \mu_{u_i}(3)) \mid i = 1, \ldots, 3\} \\ &= \{(.8, 1), (.7, .5), (.6, .4)\} \\ \mu_{\tilde{B}}(3) &= \{(v_j, \mu_{v_j}(3)) \mid j = 1, \ldots, 3\} \\ &= \{(1, 1), (.8, .5), (.7, .3)\}\end{aligned}$$

Compute $\mu_{\tilde{A} \cap \tilde{B}}$:

u_i	v_j	$w = \min\{u_i, v_j\}$	$\mu_{u_i}(3)$	$\mu_{v_j}(3)$	$\min\{\mu_{u_i}(3), \mu_{v_j}(3)\}$
.8	1	.8	1	1	1
.8	.8	.8	1	.5	.5
.8	.7	.7	1	.3	.3
.7	1	.7	.5	1	.5
.7	.8	.7	.5	.5	.5
.7	.7	.7	.5	.3	.3
.6	1	.6	.4	1	.4
.6	.8	.6	.4	.5	.4
.6	.7	.6	.4	.3	.3

Next you have to compute the supremum of the degrees of membership of all pairs (u_i, v_j) which yield w as minimum:

$$\sup_{.8=\min\{u_i, v_j\}} \{1, .5\} = 1$$

$$\sup_{.7=\min\{u_i, v_j\}} \{.3, .5, .5, .3\} = .5$$

$$\sup_{.6=\min\{u_i,v_j\}} \{.4, .4, .3\} = .4$$

So you obtain the membership function of $x = 3$ as the fuzzy set:

$$\mu_{\tilde{A}\cap\tilde{B}}(3) = \{(.8, 1), (.7, .5), (.6, .4)\}$$

Mizumoto and Tanaka [1976, p. 318] show that type 2 fuzzy sets such as defined above are idempotent, commutative, and associative and satisfy the DeMorgan laws. They are, however, not distributive and do not satisfy the absorbtion laws, the identity laws, or the complement laws.

Example 5–2 is a good indication of the computational effort involved in operations with type 2 fuzzy sets. The reader should realize that in this example the degrees of membership of only *one* element of the type 2 fuzzy set is computed. For all other elements such as $x = 4$, $x = 5$, ... etc. of the sets $\widetilde{A} * \widetilde{B}$ the corresponding calculations would be necessary. Here "*" can be any set-theoretic operation mentioned so far.

5.3 Algebraic Operations with Fuzzy Numbers

Definition 5–3

A *fuzzy number* $\widetilde{M}$ is a convex normalized fuzzy set $\widetilde{M}$ of the real line $\mathbb{R}$ such that

1. It exists exactly one $x_0 \in \mathbb{R}$ with $\mu_{\widetilde{M}}(x_0) = 1$ (x_0 is called the mean value of $\widetilde{M}$).
2. $\mu_{\widetilde{M}}(x)$ is piecewise continuous.

Nowadays definition 5–3 is very often modified. For the sake of computational efficiency and ease of data acquisition, trapezoidal membership functions are often used. Figure 5–2 shows such a fuzzy set, which could be called "approximately 5" and which would normally be defined as the quadrupel $\{3, 4, 6, 7\}$. Strictly speaking, it is a fuzzy interval (see section 5.3.2). A triagular fuzzy number is, of course, a special case of this.

Definition 5–4

A fuzzy number $\widetilde{M}$ is called *positive* (negative) if its membership function is such that $\mu_{\widetilde{M}}(x) = 0$, $\forall\, x < 0$ ($\forall\, x > 0$).

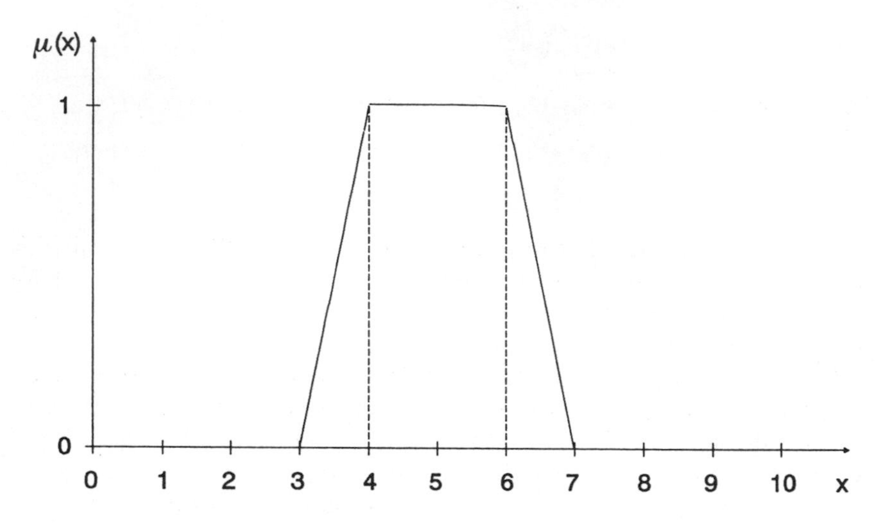

Figure 5–2. Trapezoidal "fuzzy number."

Example 5–3

The following fuzzy sets are fuzzy numbers:

approximately 5 = {(3, .2), (4, .6), (5, 1), (6, .7), (7, .1)}
approximately 10 = {(8, .3), (9, .7), (10, 1), (11, .7), (12, .3)}

But {(3, .8), (4, 1), (5, 1), (6, .7)} is not a fuzzy number because $\mu(4)$ and also $\mu(5) = 1$.

We are all familiar with algebraic operations with crisp numbers. If we want to use fuzzy sets in applications we will have to deal with fuzzy numbers and the extension principle is one way to extend algebraic operations from crisp to fuzzy numbers.

We need a few more definitions: Let $F(\mathbb{R})$ be the set of real fuzzy numbers and $X = X_1 \otimes X_2$. We can define the following properties of binary operations:

Definition 5–5

A binary operation $*$ in $\mathbb{R}$ is called *increasing* (decreasing) if

for $x_1 > y_1$ and $x_2 > y_2$

$$x_1 * x_2 > y_1 * y_2 \qquad (x_1 * x_2 < y_1 * y_2)$$

Example 5–4

$f(x, y) = x + y$	is an increasing operation.
$f(x, y) = x \cdot y$	is an increasing operation on $\mathbb{R}^+$.
$f(x, y) = -(x + y)$	is a decreasing operation.

If the normal algebraic operations $+, -, \cdot, :$ are extended to operations on fuzzy numbers they shall be denoted by $\oplus, \ominus, \odot, \oslash$.

Theorem 5–1 [See Dubois and Prade 1980a, p. 44]

If $\tilde{M}$ and $\tilde{N}$ are fuzzy numbers whose membership functions are continuous and surjective from $\mathbb{R}$ to [0, 1] and $*$ is a continuous increasing (decreasing) binary operation, then $\tilde{M} \circledast \tilde{N}$ is a fuzzy number whose membership function is continuous and surjective from $\mathbb{R}$ to [0, 1].

Dubois and Prade [1980a] present procedures to determine the membership functions $\mu_{\tilde{M} \circledast \tilde{N}}$ on the basis of $\mu_{\tilde{M}}$ and $\mu_{\tilde{N}}$.

Theorem 5–2

If $\tilde{M}, \tilde{N} \in F(\mathbb{R})$ with $\mu_{\tilde{N}}(x)$ and $\mu_{\tilde{M}}(x)$ continuous membership functions, then by application of the extension principle for the binary operation $*: \mathbb{R} \otimes \mathbb{R} \rightarrow \mathbb{R}$ the membership function of the fuzzy number $\tilde{M} \circledast \tilde{N}$ is given by

$$\mu_{\tilde{M} \circledast \tilde{N}}(z) = \sup_{z = x \times y} \min \{\mu_{\tilde{M}}(x), \mu_{\tilde{N}}(y)\}$$

Properties of the Extended Operation $\circledast$

Remark 5–1 [Dubois and Prade 1980a, p. 45]

1. For any commutative operation $*$ the extended operation $\circledast$ is also commutative.
2. For any associative operation $*$ the extended operation $\circledast$ is also associative.

5.3.1 Special Extended Operations

For unary operations $f: X \rightarrow Y$, $X = X_1$ (see definitions 5–1) the extension principle reduces for all $\tilde{M} \in F(\mathbb{R})$ to

$$\mu_{f(\widetilde{M})}(z) = \sup_{x \in f^{-1}(z)} \mu_{\widetilde{M}}(x)$$

Example 5–5

1. For $f(x) = -x$ the opposite of a fuzzy number $\widetilde{M}$ is given with $-\widetilde{M} = \{(x, \mu_{-\widetilde{M}}(x)) \mid x \in X\}$, where $\mu_{-\widetilde{M}}(x) = \mu_{\widetilde{M}}(-x)$.
2. If $f(x) = \frac{1}{x}$ then the inverse of a fuzzy number $\widetilde{M}$ is given with $\widetilde{M}^{-1} = \{(x, \mu_{\widetilde{M}^{-1}}(x)) \mid x \in X\}$, where $\mu_{\widetilde{M}^{-1}}(x) = \mu_{\widetilde{M}}(\frac{1}{x})$.
3. For $\lambda \in \mathbb{R} \setminus \{0\}$ and $f(x) = \lambda \cdot x$ then the scalar multiplication of a fuzzy number is given by $\lambda\widetilde{M} = \{(x, \mu_{\lambda\widetilde{M}}(x)) \mid x \in X\}$, where $\mu_{\lambda\widetilde{M}}(x) = \mu_{\widetilde{M}}(\lambda \cdot x)$

In the following we shall apply the extension principle to binary operations. A generalization to n-ary operations is straightforward.

Extended addition. Since addition is an increasing operation according to theorem 5–1, we get for the extended addition $\oplus$ of fuzzy numbers that $f(\widetilde{N}, \widetilde{M}) = \widetilde{N} \oplus \widetilde{M}$, $\widetilde{N}, \widetilde{M} \in F(\mathbb{R})$ is a fuzzy number—that is, $\widetilde{N} \oplus \widetilde{M} \in F(\mathbb{R})$.

Properties of $\oplus$.

1. $\ominus(\widetilde{M} \oplus \widetilde{N}) = (\ominus\widetilde{M}) \oplus (\ominus\widetilde{N})$.
2. $\oplus$ is commutative.
3. $\oplus$ is associative.
4. $0 \in \mathbb{R} \subseteq F(\mathbb{R})$ is the neutral element for $\oplus$, that is, $\widetilde{M} \oplus 0 = \widetilde{M}$, $\forall \widetilde{M} \in F(\mathbb{R})$.
5. For $\oplus$ there does not exist an inverse element, that is, $\forall \widetilde{M} \in F(\mathbb{R}) \setminus \mathbb{R}: \widetilde{M} \oplus (\ominus\widetilde{M}) \neq 0 \in \mathbb{R}$.

One of the consequences is [Yager 1980] that fuzzy equations are very difficult to solve because the variables cannot be eliminated as usual.

Extended Product. Multiplication is an increasing operation on $\mathbb{R}^+$ and a decreasing operation on $\mathbb{R}^-$. Hence, according to theorem 5–1, the product of positive fuzzy numbers or of negative fuzzy numbers results in a positive fuzzy number. Let $\widetilde{M}$ be a positive and $\widetilde{N}$ a negative fuzzy number. Then $\ominus\widetilde{M}$ is also negative and $\widetilde{M} \odot \widetilde{N} = \ominus(\ominus\widetilde{M} \odot \widetilde{N})$ results in a negative fuzzy number.

Properties of $\odot$

1. $(\ominus\tilde{M})\odot\tilde{N}=\ominus(\tilde{M}\odot\tilde{N})$.
2. $\odot$ is commutative.
3. $\odot$ is associative.
4. $\tilde{M}\odot 1=\tilde{M}$, $1\in\mathbb{R}\subseteq F(\mathbb{R})$ is the neutral element for $\odot$, that is $\tilde{M}\odot 1=\tilde{M}$, $\forall\tilde{M}\in F(\mathbb{R}))$.
5. For $\odot$ there does not exist an inverse element, that is, $\forall\tilde{M}\in F(\mathbb{R})\setminus\mathbb{R}$: $\tilde{M}\odot\tilde{M}^{-1}\neq 1$.

Theorem 5–3 [for proof see Dubois and Prade 1980a, p. 51]

If $\tilde{M}$ is either a positive or a negative fuzzy number and $\tilde{N}$ and $\tilde{P}$ are both either positive or negative fuzzy numbers then

$$\tilde{M}\odot(\tilde{N}\oplus\tilde{P})=(\tilde{M}\odot\tilde{N})\oplus(\tilde{M}\odot\tilde{P})$$

Extended Subtraction. Subtraction is neither an increasing nor a decreasing operation. Therefore theorem 5–1 is not immediately applicable. The operation $\tilde{M}\ominus\tilde{N}$ can, however, always be written as $\tilde{M}\ominus\tilde{N}=\tilde{M}\oplus(\ominus\tilde{N})$.

Applying the extension principle [Dubois and Prade 1979] yields

$$\begin{aligned}\mu_{\tilde{M}\ominus\tilde{N}}(z)&=\sup_{z=x-y}\min(\mu_{\tilde{M}}(x),\mu_{\tilde{N}}(y))\\&=\sup_{z=x+y}\min(\mu_{\tilde{M}}(x),\mu_{\tilde{N}}(-y))\\&=\sup_{z=x+y}\min(\mu_{\tilde{M}}(x),\mu_{-\tilde{N}}(y))\end{aligned}$$

Thus $\tilde{M}\ominus\tilde{N}$ is a fuzzy number whenever $\tilde{M}$ and $\tilde{N}$ are.

Extended Division. Division is also neither an increasing nor a decreasing operation. If $\tilde{M}$ and $\tilde{N}$ are strictly positive fuzzy numbers, that is, $\mu_{\tilde{M}}(x)=0$ and $\mu_{\tilde{N}}(x)=0$ $\forall x\leq 0$, however, we obtain in analogy to the extended subtraction:

$$\begin{aligned}\mu_{\tilde{M}\oslash\tilde{N}}(z)&=\sup_{z=x/y}\min(\mu_{\tilde{M}}(x),\mu_{\tilde{N}}(y))\\&=\sup_{z=xy}\min(\mu_{\tilde{M}}(x),\mu_{\tilde{N}}(\tfrac{1}{y}))\\&=\sup_{z=xy}\min(\mu_{\tilde{M}}(x),\mu_{\tilde{N}^{-1}}(y))\end{aligned}$$

$\widetilde{N}^{-1}$ is a positive fuzzy number. Hence theorem 5–1 can now be applied. The same is true if $\widetilde{M}$ and $\widetilde{N}$ are both strictly negative fuzzy numbers.

Similar results can be obtained by using other than the min-max operations, for instance those of definitions 3–7, through 3–11.

Extended operations with fuzzy numbers involve rather extensive computations as long as no restrictions are put on the type of membership functions allowed. Dubois and Prade [1979] propose a general algorithm for performing extended operations. For practical purposes, however, it will generally be more appropriate to resort to specific kinds of fuzzy numbers, as they are described in the next section. The generality is not limited considerably by limiting extended operations to fuzzy numbers in *LR*-representation or even to triangular fuzzy numbers [van Laarhoven and Pedrycz 1983] and the computational effort is very much decreased. The reader should also realize that extended operations on the basis of min-max cannot be directly applied to "fuzzy numbers" with discrete supports. As illustrated by example 5–6, the resulting fuzzy sets may no longer be convex and therefore no longer considered as fuzzy numbers.

Example 5–6

Let $\widetilde{M} = \{(1, .3), (2, 1), (3, .4)\}$
$\widetilde{N} = \{(2, .7), (3, 1), (4, .2)\}$

Then

$$\widetilde{M} \odot \widetilde{N} = \{(2, .3), (3, .3), (4, .7), (6, 1), (8, .2), \underline{(9, .4)}, (12, .2)\}$$

5.3.2 Extended Operations for LR-Representation of Fuzzy Sets

Computational efficiency is of particular importance when using fuzzy set theory to solve real problems, that is, problems of realistic size. We shall therefore consider in the following in detail the LR-representation of fuzzy sets, which increases computational efficiency without limiting the generality beyond acceptable limits:

Dubois and Prade [1979] suggested a special type of representation for fuzzy numbers of the following type: They call L (and R), which map $\mathbb{R}^+ \to [0, 1]$, and are decreasing, *shape functions* if $L(0) = 1$, $L(x) < 1$ for $\forall x > 0$; $L(x) > 0$ for $\forall x < 1$; $L(1) = 0$ or $[L(x) > 0, \forall x$ and $L(+\infty) = 0]$.

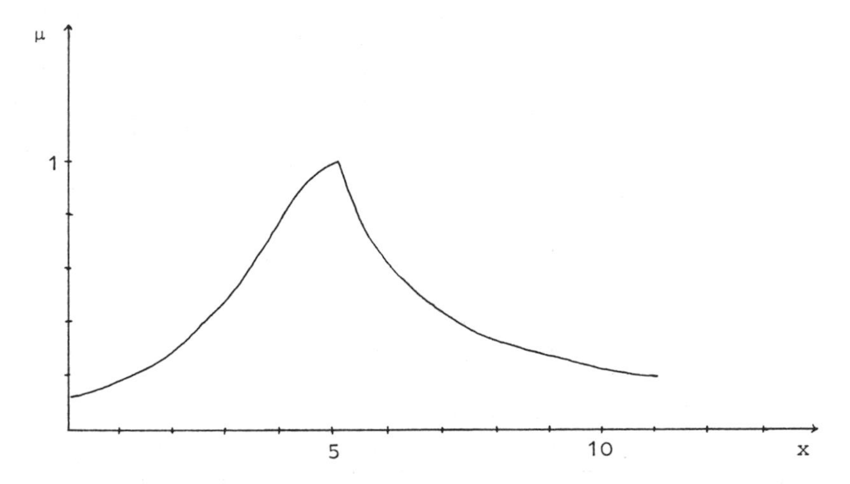

Figure 5–3. LR-representation of fuzzy numbers.

Definition 5–6

A *fuzzy number* $\widetilde{M}$ is of *LR-type* if there exist reference functions L (for left), R (for right), and scalars $\alpha > 0$, $\beta > 0$ with

$$\mu_{\widetilde{M}}(x) = \begin{cases} L\left(\dfrac{m-x}{\alpha}\right) & \text{for } x \leq m \\ R\left(\dfrac{x-m}{\beta}\right) & \text{for } x \geq m \end{cases}$$

m, called the mean value of $\widetilde{M}$, is a real number and α and β are called the left and right spreads, respectively. Symbolically $\widetilde{M}$ is denoted by $(m, \alpha, \beta)_{LR}$. (See figure 5–3.)

For $L(z)$, different functions can be chosen. Dubois and Prade [1988a, p. 50] mention, for instance, $L(x) = \max(0, \ 1-x)^p$, $L(x) = \max(0, 1-x^p)$, with $p > 0$, $L(x) = e^{-x}$ or $L(x) = e^{-x^2}$. These examples already give an impression of the wide scope of $L(z)$. One problem, of course, is to find the appropriate function in a specific context.

Example 5–7

Let

$$L(x) = \frac{1}{1+x^2}$$

$$R(x) = \frac{1}{1 + 2|x|}$$

$$\alpha = 2, \ \beta = 3, \ m = 5$$

Then

$$\mu_5(x) = \begin{cases} L\left(\frac{5-x}{2}\right) = \dfrac{1}{1 + \left(\frac{5-x}{2}\right)^2} & \text{for} \quad x \leq 5 \\ R\left(\frac{x-5}{3}\right) = \dfrac{1}{1 + \left|\frac{2(x-5)}{3}\right|} & \text{for} \quad x \geq 5 \end{cases}$$

If the m is not a real number but an interval $[\underline{m}, \bar{m}]$ then the fuzzy set $\widetilde{M}$ is not a fuzzy number but a fuzzy interval. Accordingly, a fuzzy interval in LR-representation can be defined as follows:

Definition 5–6a [Dubois and Prade 1988a, p. 48]

A *fuzzy interval* $\widetilde{M}$ is of *LR*-type if there exist shape functions L and R and four parameters $(\underline{m}, \bar{m}) \in \mathbb{R}^2 \cup \{-\infty, +\infty\}$, α, β and the membership function of $\widetilde{M}$ is

$$\mu_{\widetilde{M}}(x) = \begin{cases} L\left(\frac{\underline{m} - x}{\alpha}\right) & \text{for} \quad x \leq \underline{m} \\ 1 & \text{for} \quad \underline{m} \leq x \leq \bar{m} \\ R\left(\frac{x - \bar{m}}{\beta}\right) & \text{for} \quad x \geq \bar{m} \end{cases}$$

The fuzzy interval is then denoted by

$$\widetilde{M} = (\underline{m}, \bar{m}, \alpha, \beta)_{\mathrm{LR}}$$

This definition is very general and allows quantification of quite different types of information; for instance, if $\widetilde{M}$ is supposed to be a real crisp number for $m \in \mathbb{R}$,

$$\widetilde{M} = (m, m, 0, 0)_{LR}, \ \forall L, \ \forall R$$

If $\widetilde{M}$ is a crisp interval,

$$\widetilde{M} = (a, b, 0, 0)_{LR}, \ \forall L, \ \forall R$$

and if $\widetilde{M}$ is a "trapezoidal fuzzy number" (see definition 5–3), $L(x) = R(x) = \max(0, 1 - x)$ is implied.

For LR fuzzy numbers the computations necessary for the above-mentioned operations are considerably simplified: Dubois and Prade [1979] showed that exact formulas can be given for $\oplus$ and $\ominus$. They also suggested approximate expressions for $\odot$ and $\oslash$ [Dubois and Prade 1979] which approximate better when the spreads are smaller compared to the mean values.

Theorem 5–4

Let $\widetilde{M}$, $\widetilde{N}$ be two fuzzy numbers of LR-type:

$$\widetilde{M} = (m, \alpha, \beta)_{LR}, \quad \widetilde{N} = (n, \gamma, \delta)_{LR}$$

Then

1. $(m, \alpha, \beta)_{LR} \oplus (n, \gamma, \delta)_{LR} = (m + n, \alpha + \gamma, \beta + \delta)_{LR}$.
2. $-(m, \alpha, \beta)_{LR} = (-m, \beta, \alpha)_{LR}$.
3. $(m, \alpha, \beta)_{LR} \ominus (n, \gamma, \delta)_{RL} = (m - n, \alpha + \delta, \beta + \gamma)_{LR}$.

Example 5–8

$$L(x) = R(x) = \frac{1}{1 + x^2}$$
$$\widetilde{M} = (1, .5, .8)_{LR}$$
$$\widetilde{N} = (2, .6, .2)_{LR}$$
$$\widetilde{M} \oplus \widetilde{N} = (3, 1.1, 1)_{LR}$$
$$\widetilde{O} = (2, .6, .2)_{LR}$$
$$\ominus \widetilde{O} = (-2, .2, .6)_{LR}$$
$$\widetilde{M} \ominus \widetilde{O} = (-1, .7, 1.4)_{LR}$$

Theorem 5–5 [Dubois and Prade 1980a, p. 55]

Let $\widetilde{M}$, $\widetilde{N}$ be fuzzy numbers as in definition 5–3, then

$$(m, \alpha, \beta)_{LR} \odot (n, \gamma, \delta)_{LR} \approx (mn, m\gamma + n\alpha, m\delta + n\beta)_{LR}$$

for $\widetilde{M}$, $\widetilde{N}$ positive;

$$(m, \alpha, \beta)_{RL} \odot (n, \gamma, \delta)_{LR} \approx (mn, n\alpha - m\delta, n\beta - m\gamma)_{RL}$$

for $\widetilde{N}$ positive, $\widetilde{M}$ negative and

$$(m, \alpha, \beta)_{LR} \odot (n, \gamma, \delta)_{LR} \approx (mn, -n\beta - m\delta, -n\alpha - m\gamma)_{RL}$$

for $\tilde{M}$, $\tilde{N}$ negative.

The following example shows an application of theorem 5–5.

Example 5–9

Let $\tilde{M} = (2, .2, .1)_{LR}$
$\tilde{N} = (3, .1, .3)_{LR}$

be fuzzy numbers of *LR*-type with reference functions

$$L(z) = R(z) = \begin{cases} 1 & -1 \le z \le 1 \\ 0 & \text{else} \end{cases}$$

If we were interested in the *LR*-representation of $\tilde{M} \odot \tilde{N}$ we prove the conditions of theorem 5–5 and apply it. Thus, with

$$\mu_{\tilde{M}}(x) = \begin{cases} L\left(\frac{2-x}{.2}\right) & x \le 2 \\ R\left(\frac{x-2}{.1}\right) & x \ge 2 \end{cases}$$

$$= \begin{cases} 1 & -1 \le \frac{2-x}{.2} \le 1 \quad \text{and} \quad -1 \le \frac{x-2}{.1} \le 1 \\ 0 & \text{else} \end{cases}$$

$$= \begin{cases} 1 & 1.9 \le x \le 2.1 \\ 0 & \text{else} \end{cases}$$

it follows that $\tilde{M}$ is positive.

$$\mu_{\tilde{N}}(x) = \begin{cases} L\left(\frac{3-x}{.1}\right) & x \le 3 \\ R\left(\frac{x-3}{.3}\right) & x \ge 3 \end{cases}$$

$$= \begin{cases} 1 & 2.9 \le x \le 3.1 \\ 0 & \text{else} \end{cases}$$

shows that $\tilde{N}$ is positive.

Following theorem 5–5 for the case in which $\tilde{M}$ and $\tilde{N}$ are positive, we obtain

$$\tilde{M} \odot \tilde{N} \approx (2 \cdot 3,\ 2 \cdot .1 + 3 \cdot .2,\ 2 \cdot .3 + 3 \cdot .1)_{LR} = (6, .8, .9)_{LR}$$

Exercises

1. Let $X = \mathbb{N} \times \mathbb{N}$
   $\widetilde{A}_1 = \{(1, .6), (2, .8), (3, 1), (4, .6)\}$
   $\widetilde{A}_2 = \{(0, .5), (1, .7), (2, .9), (3, 1), (4, .4)\}$
   $f: \mathbb{N} \times \mathbb{N} \to \mathbb{N}$ be defined by

$$f(x, y) = z, \; x \in \widetilde{A}_1, \; y \in \widetilde{A}_2$$

   Determine the image $f(\widetilde{A}_1 \times \widetilde{A}_2)$ by the extension principle.
2. Compute $\mu_{\tilde{A} \cup \tilde{B}}$ and $\mu_{\complement \tilde{A}}$ for $\widetilde{A}$, $\widetilde{B}$ as in example 5–2.
3. Which of the following fuzzy sets are fuzzy numbers:
   a. $\widetilde{A} = \{(x, \mu_{\tilde{A}}(x)) \mid x \in \mathbb{R}\}$
   where

$$\mu_{\tilde{A}}(x) = \begin{cases} \left(\left(1 + \dfrac{5-x}{2}\right)^2\right)^{-1} & x \le 5 \\ \left(1 + \left|\dfrac{2(x-5)}{3}\right|\right)^{-1} & x \ge 5 \end{cases}$$

   b. $\widetilde{B} = \{(x, \mu_{\tilde{B}}(x)) \mid x \in \mathbb{R}^+\}$
   where

$$\mu_{\tilde{B}}(x) = \begin{cases} x & x \in [0, 1] \\ 1 & x \in [1, 2] \\ 3 - x & x \in [2, 3] \end{cases}$$

   c. $\widetilde{C} = \{(0, .4), (1, 1), (2, .7)\}$
4. Which of the following functions are reference functions for $x \in \mathbb{R}$
   a. $f_1(x) = |x + 1|$
   b. $f_2(x) = \dfrac{1}{1 + x^2}$
   c. $f_3(x) = \begin{cases} \frac{1}{2}x + 1 & x \in [-2, 0] \\ -2x + 1 & x \in [0, \frac{1}{2}] \\ 0 & \text{else} \end{cases}$
   d. $f_4(x) = \dfrac{1}{1 + a|x|^p} \qquad p \ge 1$
5. Let $\widetilde{M}$, $L(x)$, $R(x)$ be defined as in example 5–8. $\widetilde{N} = (-4, .1, .6)_{LR}$. Compute $\widetilde{M} \ominus \widetilde{N}$.
6. Let $\widetilde{M}$, $\widetilde{N}$ be defined as in example 5–8. Compute $\widetilde{M} \odot \widetilde{N}$.
7. Develop an approximate formula to compute $\widetilde{M} \odot \widetilde{N}$, $\widetilde{M} = (m, \alpha, \beta)_{LR}$, $\widetilde{N} = (n, \gamma, \sigma)_{LR}$. (Remember how the formula was derived for the general extended division.)

6 FUZZY RELATIONS AND FUZZY GRAPHS

6.1 Fuzzy Relations on Sets and Fuzzy Sets

Fuzzy relations are fuzzy subsets of $X \times Y$, that is, mappings from $X \rightarrow Y$. They have been studied by a number of authors, in particular by Zadeh [1965, 1971], Kaufmann [1975], and Rosenfeld [1975]. Applications of fuzzy relations are widespread and important. We shall consider some of them and point to more possible uses at the end of this chapter. We shall exemplarily consider only binary relations. A generalization to n-ary relations is straight forward.

Definition 6–1

Let $X, Y \subseteq \mathbb{R}$ be universal sets, then

$$\widetilde{R} = \{((x, y), \mu_{\widetilde{R}}(x, y)) | (x, y) \in X \times Y\}$$

is called a *fuzzy relation* on $X \times Y$.

Example 6–1

Let $X = Y = \mathbb{R}$ and $\widetilde{R}: =$ "considerably larger than." The membership function of the fuzzy relation, which is, of course, a fuzzy set on $X \times Y$ can then be

$$\mu_{\tilde{R}}(x, y) = \begin{cases} 0 & \text{for } x \leq y \\ \dfrac{(x-y)}{10y} & \text{for } y < x \leq 11y \\ 1 & \text{for } x > 11y \end{cases}$$

A different membership function for this relation could be

$$\mu_{\tilde{R}}(x, y) = \begin{cases} 0 & \text{for } x \leq y \\ (1 + (y-x)^{-2})^{-1} & \text{for } x > y \end{cases}$$

For discrete supports; fuzzy relations can also be defined by matrixes.

Example 6–2

Let $X = \{x_1, x_2, x_3\}$ and $Y = \{y_1, y_2, y_3, y_4\}$

$\tilde{R}$ = "x considerably larger than y":

	y_1	y_2	y_3	y_4
x_1	.8	1	.1	.7
x_2	0	.8	0	0
x_3	.9	1	.7	.8

and

$\tilde{Z}$ = "y very close to x":

	y_1	y_2	y_3	y_4
x_1	.4	0	.9	.6
x_2	.9	.4	.5	.7
x_3	.3	0	.8	.5

In definition 6–1 it was assumed that $\mu_{\tilde{R}}$ was a mapping from $X \times Y$ to [0, 1], that is, it assigns to each pair (x, y) a degree of membership in the unit interval. In some instances, such as in graph theory, it is useful to consider fuzzy relations that map from fuzzy sets contained in the universal sets into the unit interval. Then definition 6–1 has to be generalized [Rosenfeld 1975].

Definition 6–2

Let $X, Y \subseteq \mathbb{R}$ and

$$\tilde{A} = \{(x, \mu_{\tilde{A}}(x)) \mid x \in X\},$$
$$\tilde{B} = \{(y, \mu_{\tilde{B}}(y)) \mid y \in Y\} \quad \text{two fuzzy sets.}$$

Then $\tilde{R} = \{[(x, y), \mu_{\tilde{R}}(x, y)] \mid (x, y) \in X \times Y\}$ is a *fuzzy relation* on $\tilde{A}$ and $\tilde{B}$ if

$$\mu_{\tilde{R}}(x, y) \leq \mu_{\tilde{A}}(x), \ \forall (x, y) \in X \times Y$$

and

$$\mu_{\tilde{R}}(x, y) \leq \mu_{\tilde{B}}(y), \ \forall (x, y) \in X \times Y.$$

This definition will be particularly useful when defining fuzzy graphs: let the elements of the fuzzy relation of definition 6–2 be the nodes of a fuzzy graph which is represented by this fuzzy relation. The degrees of membership of the elements of the related fuzzy sets define the "strength" of or the flow in the respective edges of the graph, while the degrees of membership of the corresponding pairs in the relation are the "flows" or "capacities" of the graph. The additional requirement of definition 6–2 ($\mu_{\tilde{R}}(x, y) \leq \min\{\mu_{\tilde{A}}(x), \mu_{\tilde{B}}(y)\}$) then ensures that the "flows" in the edges of the graph can never exceed the flows in the respective nodes (= pair of edges).

Fuzzy relations are obviously fuzzy sets in product spaces. Therefore set-theoretic and algebraic operations can be defined for them in analogy to the definitions in chapters 2 and 3 by utilizing the extension principle.

Definition 6–3

Let $\tilde{R}$ and $\tilde{Z}$ be two fuzzy relations in the same product space; The *union/intersection* of $\tilde{R}$ with $\tilde{Z}$ is then defined by

$$\mu_{\tilde{R} \cup \tilde{Z}}(x, y) = \max\{\mu_{\tilde{R}}(x, y), \mu_{\tilde{Z}}(x, y)\}, \qquad (x, y) \in X \times Y$$
$$\mu_{\tilde{R} \cap \tilde{Z}}(x, y) = \min\{\mu_{\tilde{R}}(x, y), \mu_{\tilde{Z}}(x, y)\}, \qquad (x, y) \in X \times Y$$

Example 6–3

Let $\tilde{R}$ and $\tilde{Z}$ be the two fuzzy relations defined in example 6–2. The union of $\tilde{R}$ and $\tilde{Z}$, which can be interpreted as "x considerably larger or very close to y" is then given by

$\tilde{R} \cup \tilde{Z}$:

	y_1	y_2	y_3	y_4
x_1	.8	1	.9	.7
x_2	.9	.8	.5	.7
x_3	.9	1	.8	.8

The intersection of $\tilde{R}$ and $\tilde{Z}$ is represented by

$\tilde{R} \cap \tilde{Z}$:

	y_1	y_2	y_3	y_4
x_1	.4	0	.1	.6
x_2	0	.4	0	0
x_3	.3	0	.7	.5

So far "min" and "max" have been used to define intersection and union. Since fuzzy relations are fuzzy sets, operations can also be defined using the alternative definitions in section 3.2. Some additional concepts, such as the projection and the cylindrical extension of fuzzy relations, have been shown to be useful.

Definition 6–4

Let $\tilde{R} = \{[(x, y), \mu_{\tilde{R}}(x, y)] \mid (x, y) \in X \times Y\}$ be a fuzzy binary relation. The *first projection* of $\tilde{R}$ is then defined as

$$\tilde{R}^{(1)} = \{(x, \max_y \mu_{\tilde{R}}(x, y)) \mid (x, y) \in X \times Y\}$$

The *second projection* is defined as

$$\tilde{R}^{(2)} = \{(y, \max_x \mu_{\tilde{R}}(x, y)) \mid (x, y) \in X \times Y\}$$

and the *total projection* as

$$\tilde{R}^{(T)} = \max_x \max_y \{\mu_{\tilde{R}}(x, y) \mid (x, y) \in X \times Y\}$$

Example 6–4

Let $\widetilde{R}$ be a fuzzy relation defined by the following relational matrix. The first, second, and total projections are then shown at the appropriate places below.

	y_1	y_2	y_3	y_4	y_5	y_6	1st projection $[\mu_{\widetilde{R}^{(1)}}(x)]$
x_1	.1	.2	.4	.8	1	.8	1
$\widetilde{R}$: x_2	.2	.4	.8	1	.8	.6	1
x_3	.4	.8	1	.8	.4	.2	1
2nd projection $[\mu_{\widetilde{R}^{(2)}}(y)]$	.4	.8	1	1	1	.8	1 Total projection

The relation resulting from applying an operation of projection to another relation is also called a "shadow" [Zadeh 1973a]. Let us now consider a more general space: $X = X_1 \times \ldots \times X_n$; and let $\widetilde{R}_q$ be a projection on $X_{i_1} \times \ldots \times X_{i_k}$, where $(i_1, \ldots, i_k)$ is a subsequence of $(1, \ldots, n)$. It is obvious that distinct fuzzy relations in the same universe can have the same projection. There must, however, be a uniquely defined largest relation $\widetilde{R}_{qL}(X_1, \ldots, X_n)$ with $\mu_{\widetilde{R}_{qL}}(x_{i_1}, \ldots, x_{i_k})$ for each projection. This largest relation is called the *cylindrical extension of the projection relation.*

Definition 6–5

$\widetilde{R}_{qL} \subseteq X$ is the largest relation in X the projection of which is $\widetilde{R}_q$, $\widetilde{R}_{qL}$ is then called the *cylindrical extension* of $\widetilde{R}_q$ and $\widetilde{R}_q$ is the base of $\widetilde{R}_{qL}$.

Example 6–5

The cylindrical extension of $R^{(2)}$ (example 6–4) is

$\tilde{R}_2$:

	y_1	y_2	y_3	y_4	y_5	y_6
x_1	.4	.8	1	1	1	.8
x_2	.4	.8	1	1	1	.8
x_3	.4	.8	1	1	1	.8

Definition 6–6

Let $\tilde{R}$ be a fuzzy relation on $X = X_1 \times \cdots \times X_n$ and $\tilde{R}_1$ and $\tilde{R}_2$ be two fuzzy projections on $X_1 \times \cdots \times X_r$ and $X_s \times \cdots \times X_n$, respectively, with $s \leq r+1$ and $\tilde{R}_{1L}$, $\tilde{R}_{2L}$ their respective cylindrical extensions.

The *join* of $\tilde{R}_1$ and $\tilde{R}_2$ is then defined as $\tilde{R}_{1L} \cap \tilde{R}_{2L}$ and their *meet* as $\tilde{R}_{1L} \cup \tilde{R}_{2L}$.

6.1.1 Compositions of Fuzzy Relations

Fuzzy relations in different product spaces can be combined with each other by the operation "composition." Different versions of "composition" have been suggested which differ in their results and also with respect to their mathematical properties. The max-min composition has become the best known and the most frequently used one. However, often the so-called max-product or max-average compositions lead to results that are more appealing.

Definition 6–7

Max-min composition: Let $\tilde{R}_1(x, y)$, $(x, y) \in X \times Y$ and $\tilde{R}_2(y, z)$, $(y, z) \in Y \times Z$ be two fuzzy relations. The max-min composition $\tilde{R}_1$ max-min $\tilde{R}_2$ is then the fuzzy set

$$\tilde{R}_1 \circ \tilde{R}_2 = \{[(x, z), \max_y \{\min \{\mu_{\tilde{R}_1}(x, y), \mu_{\tilde{R}_2}(y, z)\}\}] \mid x \in X, y \in Y, z \in Z\}$$

$\mu_{\tilde{R}_1 \circ \tilde{R}_2}$ is again the membership function of a fuzzy relation on fuzzy sets (definition 6–2).

A more general definition of composition is the "max-* composition."

Definition 6–8

Let $\tilde{R}_1$ and $\tilde{R}_2$ be defined as in definition 6–7. The *max-* composition* of $\tilde{R}_1$ and $\tilde{R}_2$ is then defined as

$$\tilde{R}_1 \overset{\circ}{*} \tilde{R}_2 = \{[x, z), \max_y (\mu_{\tilde{R}_1}(x, y) * \mu_{\tilde{R}_2}(y, z))] | x \in X, y \in Y, z \in Z\}$$

If $*$ is an associative operation which is monotonically nondecreasing in each argument, then the max-$*$ composition corresponds essentially to the max-min composition. Two special cases of the max-$*$ composition are proposed in the next definition.

Definition 6–9

[Rosenfeld 1975]: Let $\tilde{R}_1$ and $\tilde{R}_2$, respectively, be defined as in definition 6–7. The *max-prod composition* $\tilde{R}_1 \overset{\circ}{\cdot} \tilde{R}_2$ and the *max-av composition* $\tilde{R}_1 \underset{\text{av}}{\circ} \tilde{R}_2$ are then defined as follows:

$$\tilde{R}_1 \overset{\circ}{\cdot} \tilde{R}_2(x, z) = \{[(x, z), \max_y \{\mu_{\tilde{R}_1}(x, y) \cdot \mu_{\tilde{R}_2}(y, z)\}] | x \in X, y \in Y, z \in Z\}$$

$$\tilde{R}_1 \underset{\text{av}}{\circ} \tilde{R}_2(x, z) = \{[(x, z), \tfrac{1}{2} \cdot \max\{\mu_{\tilde{R}_1}(x, y) + \mu_{\tilde{R}_2}(y, z)\}] | x \in X, y \in Y, z \in Z\}$$

Example 6–6

Let $\tilde{R}_1(x, y)$ and $\tilde{R}_2(y, z)$ be defined by the following relational matrixes [Kaufmann 1975, p. 62]:

$\tilde{R}_1$:

	y_1	y_2	y_3	y_4	y_5
x_1	.1	.2	0	1	.7
x_2	.3	.5	0	.2	1
x_3	.8	0	1	.4	.3

$\tilde{R}_2$:

	z_1	z_2	z_3	z_4
y_1	.9	0	.3	.4
y_2	.2	1	.8	0
y_3	.8	0	.7	1
y_4	.4	.2	.3	0
y_5	0	1	0	.8

We shall first compute the min-max-composition $\widetilde{R}_1 \circ \widetilde{R}_2(x, z)$. We shall show in detail the determination for $x = x_1$, $z = z_1$ and leave it to the reader to verify the total results shown in the matrix at the end of the detailed computations. We first perform the min operation in the minor brackets of definition 6–7:

Let $x = x_1, z = z_1$ and $y = y_i, i = 1, \ldots, 5$:

$$
\begin{aligned}
\min\{\mu_{\tilde{R}_1}(x_1, y_1), \mu_{\tilde{R}_2}(y_1, z_1)\} &= \min\{.1, .9\} = .1\\
\min\{\mu_{\tilde{R}_1}(x_1, y_2), \mu_{\tilde{R}_2}(y_2, z_1)\} &= \min\{.2, .2\} = .2\\
\min\{\mu_{\tilde{R}_1}(x_1, y_3), \mu_{\tilde{R}_2}(y_3, z_1)\} &= \min\{0, .8\} = 0\\
\min\{\mu_{\tilde{R}_1}(x_1, y_4), \mu_{\tilde{R}_2}(y_4, z_1)\} &= \min\{1, .4\} = .4\\
\min\{\mu_{\tilde{R}_1}(x_1, y_5), \mu_{\tilde{R}_2}(y_5, z_1)\} &= \min\{.7, 0\} = 0
\end{aligned}
$$

$$
\begin{aligned}
\widetilde{R}_1 \circ \widetilde{R}_2(x_1, z_1) &= ((x_1, z_1), \mu_{\tilde{R}_1 \circ \tilde{R}_2}(x_1, z_1))\\
&= ((x_1, z_1), \max\{.1, .2, 0, .4, 0\}) = ((x_1, z_1), .4))
\end{aligned}
$$

In analogy to the above computation we now determine the grades of membership for all pairs (x_i, z_j), $i = 1, \ldots, 3$, $j = 1, \ldots, 4$ and arrive at

$\widetilde{R}_1 \circ \widetilde{R}_2$:

	z_1	z_2	z_3	z_4
x_1	.4	.7	.3	.7
x_2	.3	1	.5	.8
x_3	.8	.3	.7	1

For the max-prod we obtain

$$x = x_1, z = z_1, y = y_i, i = 1, \ldots, 5:$$

$$
\begin{aligned}
\mu_{\tilde{R}_1}(x_1, y_1) \cdot \mu_{\tilde{R}_2}(y_1, z_1) &= .1 \cdot .9 = .09\\
\mu_{\tilde{R}_1}(x_1, y_2) \cdot \mu_{\tilde{R}_2}(y_2, z_1) &= .2 \cdot .2 = .04\\
\mu_{\tilde{R}_1}(x_1, y_3) \cdot \mu_{\tilde{R}_2}(y_3, z_1) &= 0 \cdot .8 = 0\\
\mu_{\tilde{R}_1}(x_1, y_4) \cdot \mu_{\tilde{R}_2}(y_4, z_1) &= 1 \cdot .4 = .4\\
\mu_{\tilde{R}_1}(x_1, y_5) \cdot \mu_{\tilde{R}_2}(y_5, z_1) &= .7 \cdot 0 = 0
\end{aligned}
$$

Hence

$$
\begin{aligned}
\widetilde{R}_1 \stackrel{\cdot}{\circ} \widetilde{R}_2(x_1, z_1) &= ((x_1, z_1), (\mu_{\tilde{R}_1 \circ \tilde{R}_2}(x_1, z_1)))\\
&= ((x_1, z_1), \max\{.09, .04, 0, .4, 0\})\\
&= ((x_1, z_1), .4)
\end{aligned}
$$

After performing the remaining computations we obtain

$\widetilde{R}_1 \overset{\circ}{\cdot} \widetilde{R}_2$:

	z_1	z_2	z_3	z_4
x_1	.4	.7	.3	.56
x_2	.27	1	.4	.8
x_3	.8	.3	.7	1

The max-av composition finally yields

i	$\mu(x_1, y_i) + \mu(y_i, z_1)$
1	1
2	.4
3	.8
4	1.4
5	.7

Hence

$$\tfrac{1}{2} \cdot \max_{y} \{\mu_{\widetilde{R}_1}(x_1, y_i) + \mu_{\widetilde{R}_2}(y_i, z_1)\} = \tfrac{1}{2} \cdot (1.4) = .7$$

$\widetilde{R}_1 \underset{av}{\circ} \widetilde{R}_2$:

	z_1	z_2	z_3	z_4
x_1	.7	.85	.65	.75
x_2	.6	1	.65	.9
x_3	.9	.65	.85	1

6.1.2 Properties of the Min-Max Composition

(For proofs and more details see, for instance, Rosenfeld 1975.)

Associativity

The max-min composition is *associative*, that is,

$$(\widetilde{R}_3 \circ \widetilde{R}_2) \circ \widetilde{R}_1 = \widetilde{R}_3 \circ (\widetilde{R}_2 \circ \widetilde{R}_1).$$

Hence $\widetilde{R}_1 \circ \widetilde{R}_1 \circ \widetilde{R}_1 = \widetilde{R}_1^3$ the 3rd power of a fuzzy relation is defined.

Reflexitivity

Definition 6–10

Let $\tilde{R}$ be a fuzzy relation in $X \times X$.

1. $\tilde{R}$ is called *reflexive* [Zadeh 1971] if

$$\mu_{\tilde{R}}(x, x) = 1 \; \forall x \in X$$

2. $\tilde{R}$ is called *ε-reflective* [Yeh 1975] if

$$\mu_{\tilde{R}}(x, x) \geqslant \varepsilon \; \forall x \in X$$

3. $\tilde{R}$ is called *weakly reflexive* [Yeh 1975] if

$$\left.\begin{array}{l} \mu_{\tilde{R}}(x, y) \leq \mu_{\tilde{R}}(x, x) \\ \mu_{\tilde{R}}(y, x) \leq \mu_{\tilde{R}}(x, x) \end{array}\right\} \forall x, y \in X.$$

Example 6–7

Let $X = \{x_1, x_2, x_3, x_4\}$ and $Y = \{y_1, y_2, y_3, y_4\}$.
The following relation "y is close to x" is reflexive:

$\tilde{R}$:	y_1	y_2	y_3	y_4
x_1	1	0	.2	.3
x_2	0	1	.1	1
x_3	.2	.7	1	.4
x_3	0	1	.4	1

If $\tilde{R}_1$ and $\tilde{R}_2$ are reflexive fuzzy relations then the max-min composition $\tilde{R}_1 \circ \tilde{R}_2$ is also reflexive.

Symmetry

Definition 6–11

A fuzzy relation $\tilde{R}$ is called *symmetric* if $\tilde{R}(x, y) = \tilde{R}(y, x)$.

Definition 6–12

A relation is called *antisymmetric* if for

$$x \neq y \quad \begin{array}{ll} \text{either} & \mu_{\tilde{R}}(x, y) \neq \mu_{\tilde{R}}(y, x) \\ \text{or} & \mu_{\tilde{R}}(x, y) = \mu_{\tilde{R}}(y, x) = 0 \end{array} \Big\} \; \forall x, y \in X$$

[Kaufmann 1975, p. 105].
A relation is called *perfectly antisymmetric* if for $x \neq y$ whenever

$$\mu_{\tilde{R}}(x, y) > 0 \quad \text{then} \quad \mu_{\tilde{R}}(y, x) = 0 \; \forall x, y \in X$$

[Zadeh 1971].

Example 6–8

$\tilde{R}_1$:

	x_1	x_2	x_3	x_4
x_1	.4	0	.1	.8
x_2	.8	1	0	0
x_3	0	.6	.7	0
x_4	0	.2	0	0

$\tilde{R}_2$:

	x_1	x_2	x_3	x_4
x_1	.4	0	.7	0
x_2	0	1	.9	.6
x_3	.8	.4	.7	.4
x_4	0	.1	0	0

$\tilde{R}_3$:

	x_1	x_2	x_3	x_4
x_1	.4	.8	.1	.8
x_2	.8	1	.0	.2
x_3	.1	.6	.7	.1
x_4	0	.2	0	0

$\tilde{R}_1$ is a perfectly antisymmetric relation while $\tilde{R}_2$ is an antisymmetric, but not perfectly antisymmetric relation. $\tilde{R}_3$ is a nonsymmetric relation, that is, there exist $x, y \in X$ with $\mu_{\tilde{R}}(x, y) \neq \mu_{\tilde{R}}(y, x)$, which is not antisymmetric and therefore also not perfectly antisymmetric.

One could certainly define other concepts, such as an α-antisymmetry $(1\mu_{\tilde{R}}(x, y) - \mu_{\tilde{R}}(y, x)1 \geq \alpha \; \forall x, y \in X)$. These concepts would probably be more in line with the basic ideas of fuzzy set theory. Since we shall not need this type of definition for our further considerations, we will abstain from any further definition in this direction.

Example 6–9

Let X and Y be defined as in example 6–8. The following relation is then a symmetric relation:

$\tilde{R}(x, y)$:

	y_1	y_2	y_3	y_4
x_1	0	.1	0	.1
x_2	.1	1	.2	.3
x_3	0	.2	.8	.8
x_4	.1	.3	.8	1

Remark 6–1

For max-min compositions the following properties hold:

1. If $\tilde{R}_1$ is reflexive and $\tilde{R}_2$ is an arbitrary fuzzy relation then $\tilde{R}_1 \circ \tilde{R}_2 \supseteq \tilde{R}_2$ and $\tilde{R}_2 \circ \tilde{R}_1 \supseteq \tilde{R}_2$.
2. If $\tilde{R}$ is reflexive then $\tilde{R} \subseteq \tilde{R} \circ \tilde{R}$.
3. If $\tilde{R}_1$ and $\tilde{R}_2$ are reflexive relations, so is $\tilde{R}_1 \circ \tilde{R}_2$.
4. If $\tilde{R}_1$ and $\tilde{R}_2$ are symmetric, then $\tilde{R}_1 \circ \tilde{R}_2$ is symmetric if $\tilde{R}_1 \circ \tilde{R}_2 = \tilde{R}_2 \circ \tilde{R}_1$.
5. If $\tilde{R}$ is symmetric, so is each power of $\tilde{R}$.

Transitivity

Definition 6–13

A fuzzy relation $\tilde{R}$ is called (max-min) *transitive* if

$$\tilde{R} \circ \tilde{R} \subseteq \tilde{R}$$

Example 6–10

Let the fuzzy relation $\tilde{R}$ be defined as

$\tilde{R}$:	x_1	x_2	x_3	x_4
x_1	.2	1	.4	.4
x_2	0	.6	.3	0
x_3	0	1	.3	0
x_4	.1	1	1	.1

Then $\tilde{R} \circ \tilde{R}$ is

	x_1	x_2	x_3	x_4
x_1	.2	.6	.4	.2
x_2	0	.6	.3	0
x_3	0	.6	.3	0
x_4	.1	1	.3	.1

Now one can easily see that $\mu_{\tilde{R} \circ \tilde{R}}(x, y) \leq \mu_{\tilde{R}}(x, y)$ holds for all $x, y \in X$.

Remark 6–2

Combinations of the above properties give some interesting results for max-min compositions:

1. If $\tilde{R}$ is symmetric and transitive, then $\mu_{\tilde{R}}(x, y) \leq \mu_{\tilde{R}}(x, x)$ for all x, $y \in X$.

2. If $\tilde{R}$ is reflexive and transitive, then $\tilde{R} \circ \tilde{R} = \tilde{R}$.
3. If $\tilde{R}_1$ and $\tilde{R}_2$ are transitive and $\tilde{R}_1 \circ \tilde{R}_2 = \tilde{R}_2 \circ \tilde{R}_1$, then $\tilde{R}_1 \circ \tilde{R}_2$ is transitive.

The properties mentioned in remarks 6–1 and 6–2 hold for the *max-min composition*. For the *max-prod composition* property 3 of remark 6–2 is also true but not properties 1 and 3 of remark 6–1 or property 1 of remark 6–2. For the *max-av composition* properties 1 and 3 of remark 6–1 hold as well as properties 1 and 3 of remark 6–2. Property 5 of remark 6–1 is true for any commutative operator.

6.2 Fuzzy Graphs

It was already mentioned that definitions 6–1 and 6–2 of a fuzzy relation can also be interpreted as defining a fuzzy graph. In order to stay in line with the terminology of traditional graph theory we shall use the following definition of a fuzzy graph.

Definition 6–14

Let E be the (crisp) set of nodes. A *fuzzy graph* is then defined by

$$\tilde{G}(x_i, x_j) = \{((x_i, x_j), \mu_{\tilde{G}}(x_i, x_j)) \mid (x_i, x_j) \in E \times E\}$$

If $\tilde{E}$ is a fuzzy set a fuzzy graph would have to be defined in analogy to definition 6–2.

Example 6–11

a. Let $E = \{A, B, C\}$.
   Considering only three possible degrees of membership, a graph could be described as follows. (See figure 6–1.)
b. Let $E = \{x_1, x_2, x_3, x_4\}$ then a fuzzy graph could be described as

$$\tilde{G}(x_i, x_j) = \{[(x_1, x_2), .3], [(x_1, x_3), .6], [(x_1, x_1), 1], [(x_2, x_1), .4], [(x_3, x_1), .2], [(x_3, x_2), .5], [(x_4, x_3), .8]\}$$

Example 6–11a shows directed fuzzy binary graphs. Graphs can of course, also be defined in higher-dimensioned product spaces. We shall, however,

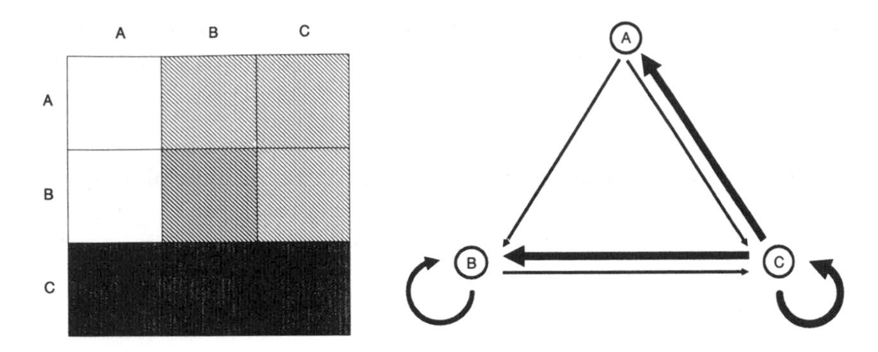

Figure 6–1. Fuzzy graphs.

focus our attention on finite undirected binary graphs, that is, we shall assume in the following that the fuzzy relation representing a graph is symmetric. The arcs can then be considered as unordered pairs of nodes. In analogy to traditional graph theory, fuzzy graph theoretic concepts can be defined.

Definition 6–15

$\widetilde{H}(x_i, x_j)$ is a *fuzzy subgraph* of $\widetilde{G}(x_i, x_j)$ if

$$\mu_{\widetilde{H}}(x_i, x_j) \leq \mu_{\widetilde{G}}(x_i, x_j) \ \forall (x_i, x_j) \in N \times N$$

$\widetilde{H}(x_i, x_j)$ *spans* graph $\widetilde{G}(x_i, x_j)$ if the node sets of $\widetilde{H}(x_i, x_j)$ and $\widetilde{G}(x_i, x_j)$ are equal, that is, if they differ only in their arc weights.

Example 6–12

Let $\widetilde{G}(x_i, x_j)$ be defined as in example 6–11b. A spanning subgraph of $\widetilde{G}(x_i, x_j)$ is then

$$\widetilde{H}(x_i, x_j) = \{[(x_1, x_2), .2], [(x_1, x_3), .4], [(x_3, x_2), .4], [(x_4, x_3), .7]\}$$

Definition 6–16

A *path* in a fuzzy graph $\widetilde{G}(x_i, x_j)$ is a sequence of distinct nodes, $x_0, x_1, \ldots, x_n$, such that for all (x_i, x_{i+1}) $\mu_{\widetilde{G}}(x_i, x_{i+1}) > 0$. The *strength* of the path is $\min\{\mu_{\widetilde{G}}(x_i, x_{i+1})\}$ for all nodes contained in the path. The *length* of a path

$n > 0$ is the number of nodes contained in the path. Each pair of nodes (x_i, x_{i+1}), $\mu(x_i, x_{i+1}) > 0$ is called an *edge* (arc) of the graph. A path is called a *cycle* if $x_0 = x_n$ and $n \geq 3$.

It would be straightforward to call the length of the shortest path between two nodes of the graph the distance between these nodes. This definition, however, has some disadvantages. It is therefore more reasonable to define the distance between two nodes as follows [Rosenfeld 1975, p. 58]:

Definition 6–17

The μ-*length* of a path $p = x_0, \ldots, x_n$ is equal to

$$L(p) = \sum_{i=1}^{n} \frac{1}{\mu(x_i, x_{i+1})}$$

The μ-*distance* $d(x_i, x_j)$ between two nodes x_i, x_j is the smallest μ-length of any path from x_i to x_j, $x_i, x_j \in \tilde{G}$.

It can then be shown [see Rosenfeld 1975, p. 88] that $d(x_i, x_j)$ is a metric (in undirected graphs!).

Definition 6–18

Two nodes that are joined by a path are called *connected nodes*.

Connectedness is a relation which is also transitive.

Definition 6–19

A fuzzy graph is a *forest* if it has no cycles, that is, it is an acyclic fuzzy graph. If the fuzzy forest is connected it is called a *tree*. (A fuzzy graph that

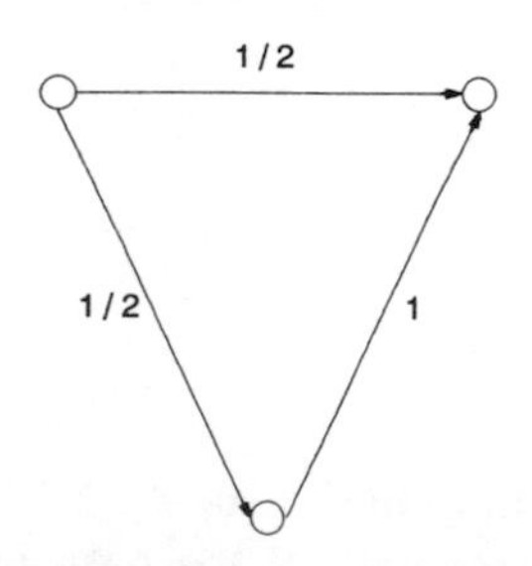

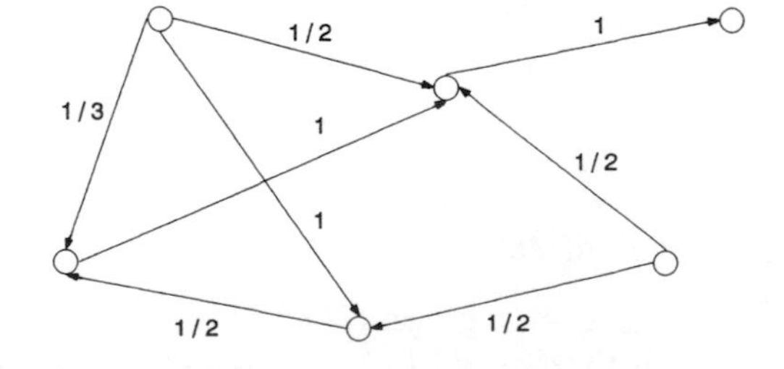

Figure 6–2. Fuzzy forests.

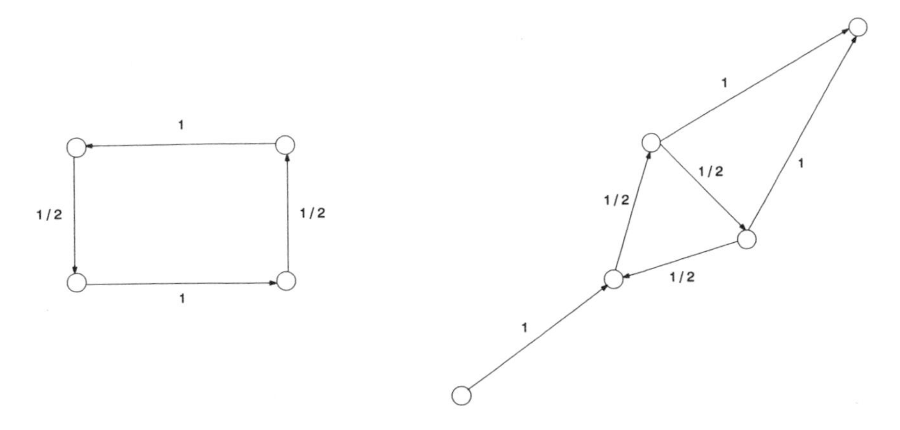

Figure 6–3. Graphs that are not forests.

is a forest has to be distinguished from a fuzzy graph that is a fuzzy forest. The latter shall not be discussed here [see Rosenfeld 1975, p. 92].)

Example 6–13

The fuzzy graphs shown in figure 6–2 are forests. The graphs shown in figure 6–3 are not.

6.3 Special Fuzzy Relations

Relations that are of particular interest to us are fuzzy relations that pertain to the similarity of fuzzy sets and those that order fuzzy sets. All of the relations discussed below are reflexive, that is, $\mu_{\tilde{R}}(x, x) = 1 \; \forall \; x \in X$ [Zadeh 1971], and they are max-min transitive, that is, $\widetilde{R} \circ \widetilde{R} \subseteq \widetilde{R}$ or $\mu_{\tilde{R}}(x, z) \geq \min \{\mu_{\tilde{R}}(x, y), \mu_{\tilde{R}}(y, z)\} \; \forall x, y, z \in X$. It should be noted that other kinds of transitivities have been defined [see Bezdek and Harris 1978]. These will, however, not be discussed here. The main difference between similarity relations and order relations is the property of symmetry or antisymmetry, respectively.

Definition 6–20

A *similarity relation* is a fuzzy relation $\mu_s(\cdot)$ which is reflexive, symmetrical, and max-min transitive.

Example 6–14

The following relation is a similarity relation [Zadeh 1971]:

	x_1	x_2	x_3	x_4	x_5	x_6
x_1	1	.2	1	.6	.2	.6
x_2	.2	1	.2	.2	.8	.2
$\tilde{R}_s$: x_3	1	.2	1	.6	.2	.6
x_4	.6	.2	.6	1	.2	.8
x_5	.2	.8	.2	.2	1	.2
x_6	.6	.2	.6	.8	.2	1

A similarity relation of a finite number of elements can also be represented by a *similarity-tree*, similar to a dendogram. In this tree each level represents an α-cut (α-level set) of the similarity relation. For the above similarity relation the similarity tree is shown below. The sets of elements on specific α-levels can be considered as similarity classes of α-level.

$\{x_1, x_2, x_3, x_4, x_5, x_6\}$ — $\tilde{R}_{0.2}$

$\{x_1, x_3, x_4, x_6\}$ $\{x_2, x_5\}$ — $\tilde{R}_{0.6}$

$\{x_1, x_3\}$ $\{x_4, x_6\}$ $\{x_2, x_5\}$ — $\tilde{R}_{0.8}$

$\{x_1, x_3\}$ $\{x_4\}$ $\{x_6\}$ $\{x_2\}$ $\{x_5\}$ — $\tilde{R}_1$

The properties of a similarity relation such as defined in definition 6–20 are rather restrictive and not quite in accordance with fuzzy-set thinking: Reflexitivity could be considered as being too restrictive and hence weakened by substituting these requirements by ε-reflexitivity or weak reflexitivity (cf. definition 6–10). The max-min transitivity can be replaced by any max-∗ transitivity listed in definition 6–10 or in remark 6–1.

We shall now turn to fuzzy order relations: As already mentioned, similarity relations and order relations are primarily distinguished by their degree of symmetry. Roughly speaking, similarity relations are fuzzy relations that are reflexive, (max-min) transitive, and symmetrical; order relations, however, are not symmetrical. To be more precise, even different kinds of fuzzy order relations differ by their degree of symmetry.

Definition 6–21

A fuzzy relation which is (max-min) transitive and reflexive is called a *fuzzy preorder relation*.

Definition 6–22

A fuzzy relation that is (min-max) transitive, reflexive, and anti-symetric is called a *fuzzy order relation*. If the relation is perfectly antisymmetrical it is called a *perfect fuzzy order relation* [Kaufmann 1975, p. 113]. It is also called a *fuzzy partial order relation* [Zadeh 1971].

Definition 6–23

A *total fuzzy order relation* [Kaufmann 1975, p. 112] or a *fuzzy linear ordering* [Dubois and Prade 1980a, p. 82; Zadeh 1971] is a fuzzy order relation such that $\forall x, y \in X; x \neq y$ either $\mu_{\tilde{R}}(x, y) > 0$ or $\mu_{\tilde{R}}(y, x) > 0$.

Any α-cut of a fuzzy linear order is a *crisp linear order*.

Example 6–15

$\tilde{R}$:	y_1	y_2	y_3	y_4
x_1	.7	.4	.8	.8
x_2	0	1	0	.2
x_3	0	.6	0	.4
x_4	0	0	0	.7

$\tilde{R}$ is a total fuzzy order relation.

Fuzzy order relations play a very important role in models for decision making in fuzzy environment. We will therefore elaborate on some particularly interesting properties in the second volume and we shall also

Table 6–1. Properties of fuzzy relations.

	Relexitivity	*Transitivity*	*Anti-Symmetry*	*Perfect Anti-Symmetry*	*Linearity*	*Symmetry*
Fuzzy preorder	X	X				
Similarity relation	X	X				X
Fuzzy order relation	X	X	X			
Perfect fuzzy order relation	X	X		X		
Total (linear) fuzzy order relation	X	X		X	X	

discuss some additional concepts in this context. Some of the properties of the special fuzzy relations defined in this chapter are summarized in table 6–1.

Exercises

1. Given an example for the membership function of the fuzzy relation $\tilde{R}$: = "considerably smaller than" in $R \times R$. Restrict $\tilde{R}$ to the first ten natural numbers and define the resulting matrix.
2. Let the two fuzzy sets $\tilde{A}$ and $\tilde{B}$ be defined as

$$\tilde{A} = \{(0, .2), (1, .3), (2, .4), (3, .5)\}$$
$$\tilde{B} = \{(0, .5), (1, .4), (2, .3), (3, .0)\}.$$

   Is the following set a fuzzy relation on $\tilde{A}$ and $\tilde{B}$?

$$\{((0, 0), .2), ((0, 2), .2), ((2, 0), .2)\}$$

   Give an example of a fuzzy relation on $\tilde{A}$ and $\tilde{B}$.
3. Consider the following matrix defining a fuzzy relation $\tilde{R}$ on $\tilde{A} \times \tilde{B}$.

$\tilde{R}$	y_1	y_2	y_3	y_4	y_5
x_1	.5	0	1	.9	.9
x_2	1	.4	.5	.3	.1
x_3	.7	.8	0	.2	.6
x_4	.1	.3	.7	1	0

Give the first and the second projection with $\mu_{\tilde{R}^{(1)}}(x)$ and $\mu_{\tilde{R}^{(2)}}(y)$ and the cylindrical extensions of the projection relations with $\mu_{\tilde{R}^{(1)}L}$ and $\mu_{\tilde{R}^{(2)}L}$.

4. Compose the following two fuzzy relations $\tilde{R}_1$ and $\tilde{R}_2$ by using the
   = max-min composition,
   = max-prod. composition and
   = max-av. composition.

R_1	y_1	y_2	y_3	y_4
x_1	.3	0	.7	.3
x_2	0	1	.2	0

$\tilde{R}_2$	z_1	z_2	z_3
y_1	1	0	1
y_2	0	.5	.4
y_3	.7	.9	.6
y_4	0	0	0

5. Discuss the reflexivity properties of the following fuzzy relation:

$\tilde{R}$	x_1	x_2	x_3
x_1	1	.7	.3
x_2	.4	.5	.8
x_3	.7	.5	1

6. Give an example for a reflexive transitive relation and verify remark 6–2.2.

7. Consider the following fuzzy graph $\tilde{G}$:

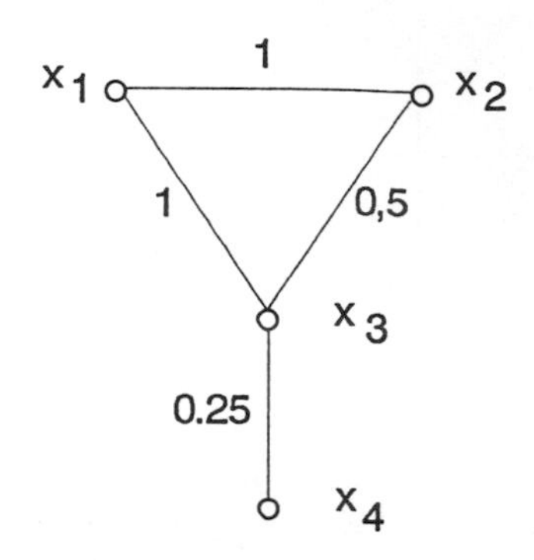

Give an example for a spanning subgraph of $\tilde{G}$!
Give all paths from x_1 to x_4 and determine their strengths and their μ lengths.
Is the above graph a forest or a tree?

8. In example 6–2, two relations are defined without specifying for which numerical values of $\{x_i\}$, $\{y_i\}$ the relations are good interpretations of the verbal relations. Give examples of numerical vectors for $\{x_i\}$ and $\{y_i\}$ such that the relations $\tilde{R}$ and $\tilde{Z}$, respectively, (in the matrixes) would express the verbal description.

7 FUZZY ANALYSIS

7.1 Fuzzy Functions on Fuzzy Sets

A fuzzy function is a generalization of the concept of a classical function. A classical function f is a mapping (correspondence) from the domain D of definition of the function into a space S; $f(D) \subseteq S$ is called the range of f. Different features of the classical concept of a function can be considered to be fuzzy rather than crisp. Therefore different "degrees" of fuzzification of the classical notion of a function are conceivable.

1. There can be a crisp mapping from a fuzzy set which carries along the fuzziness of the domain and therefore generates a fuzzy set. The image of a crisp argument would again be crisp.
2. The mapping itself can be fuzzy, thus blurring the image of a crisp argument. This we shall call a *fuzzy function*. These are called "fuzzifying functions" by Dubois and Prade [1980a, p. 106].
3. Ordinary functions can have fuzzy properties or be constrained by fuzzy constraints.

Naturally, hybrid types can be considered. We shall focus our considerations, however, only on frequently used pure cases.

Definition 7–1 [Dubois and Prade 1980a; Negoita and Ralescu 1975]

A *classical function* $f: X \to Y$ maps from a *fuzzy domain* $\tilde{A}$ in X into a fuzzy range $\tilde{B}$ in Y iff

$$\forall x \in X, \ \mu_{\tilde{B}}(f(x)) \geq \mu_{\tilde{A}}(x)$$

Given a classical function $f: X \to Y$ and a fuzzy domain $\tilde{A}$ in X the extension principle (chapter 5.1) yields the fuzzy range $\tilde{B}$ with the membership function

$$\mu_{\tilde{B}}(y) = \sup_{x \in f^{-1}(y)} \mu_{\tilde{A}}(x)$$

Hence f is a function according to definition 7–1.

Example 7–1

Let X be the set of temperatures and Y the possible demands for energy of households. $\tilde{A}$ be the fuzzy set "low temperatures" and $\tilde{B}$ the fuzzy set "high energy demands." The assignment "low temperatures" → "high energy demands" is then a fuzzy function, and the additional constraint in definition 7–1 means "the lower the temperatures, the higher the energy demands."

The correspondence between a fuzzy function and a fuzzy relation becomes even more obvious when looking at the following definition.

Definition 7–2

Let X and Y be universes and $\tilde{P}(Y)$ the set of all fuzzy sets in Y (power set).

$$\tilde{f}: X \to \tilde{P}(Y) \quad \text{is a mapping}$$

$$\tilde{f} \text{ is a } \textit{fuzzy function} \text{ iff}$$

$$\mu_{\tilde{f}(x)}(y) = \mu_{\tilde{R}}(x, y), \ \forall (x, y) \in X \times Y$$

where $\mu_{\tilde{R}}(x, y)$ is the membership function of a fuzzy relation.

Example 7–2

a. Let X be the set of all workers of a plant, $\tilde{f}$ the daily output, and y the number of processed work pieces. A fuzzy function could then be

$$\tilde{f}(x) = y$$

b. $\tilde{a}, \tilde{b} \in \mathcal{L}(\mathbb{R})$
   $X = \mathbb{R}$
   $\tilde{f}: x \rightarrow \tilde{a}x \oplus \tilde{b}$ is a fuzzy function.
c. X = set of all 1-mile runners.
   $\tilde{f}$ = possible record times.
   $\tilde{f}(x) = \{y \mid y\text{: achieved record times}\}$.

7.2 Extrema of Fuzzy Functions

Traditionally, an extremum (maximum or minimum) of a crisp function f over a given domain D is attained at a precise point x_0. If the function f happens to be the objective function of a decision model, possibly constrained by a set of other functions, then the point x_0 at which the function attains the optimum is generally called the optimal decision, that is, in classical theory there is an almost unique relationship between the extremum of the objective function and the notion of the optimal decision of a decision model.

In models in which fuzziness is involved, this unique relationship no longer exists. Extremum of a function or optimum of a decision model can be interpreted in a number of ways: In decision models the "optimal decision" is often considered to be the crisp set, D_m, which contains those elements of the fuzzy set "decision" attaining the maximum degree of membership [Bellman and Zadeh 1970, p. 150]. We shall discuss this concept in more detail in chapter 12.

The notion of an "optimal decision" as mentioned above corresponds to the concept of a "maximizing set" when considering functions in general.

Definition 7–3 [Zadeh 1972]

Let f be a real-valued function in X. Let f be bounded from below by $\inf(f)$ and from above by $\sup(f)$. The fuzzy set $\tilde{M} = \{(x, \mu_{\tilde{M}}(x))\}, x \in X$ with

$$\mu_{\tilde{M}}(x) = \frac{f(x) - \inf(f)}{\sup(f) - \inf(f)}$$

is then called the *maximizing set*.

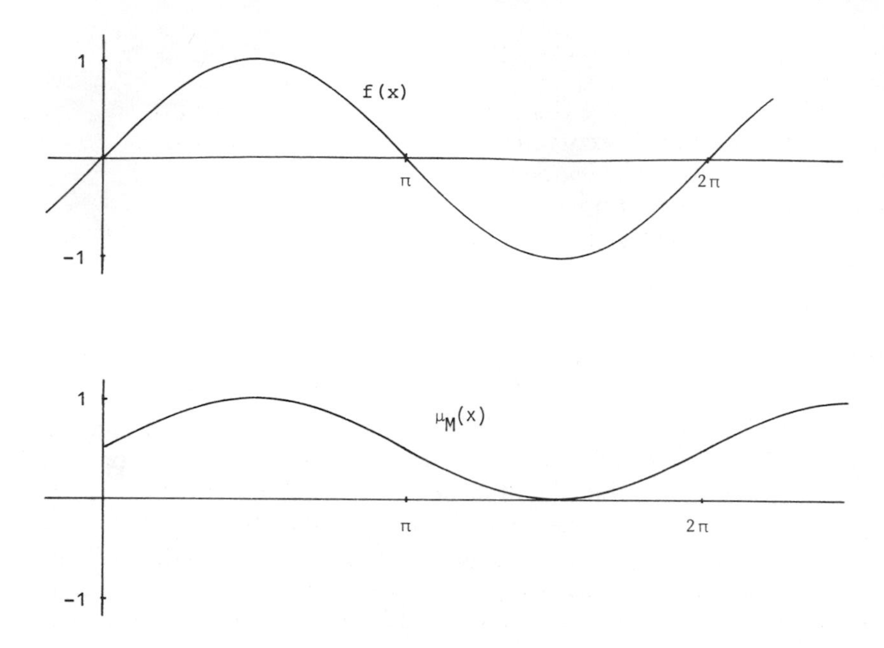

Figure 7–1. Maximizing set.

Example 7–3

$$f(x) = \sin x$$

$$\mu_{\tilde{M}}(x) = \frac{\sin x - \inf(\sin)}{\sup(\sin) - \inf(\sin)} = \frac{\sin x - (-1)}{1 - (-1)}$$
$$= \frac{\sin x + 1}{2} = \frac{1}{2}\sin x + \frac{1}{2}$$

In definition 7–3 f is a crisp real-valued function, similar to the membership function of the fuzzy set "decision," and the maximizing set provides information about the neighbourhood of the extremum of the function f, the domain of which is also crisp. The case in which the domain of f is also fuzzy will be considered in chapter 12.

Let us now consider the extrema of fuzzy functions according to definition 7–2, which are defined over a crisp domain: Since a fuzzy function $f(x)$ is a fuzzy set, say in $\mathbb{R}$, the maximum will generally not be a point in $\mathbb{R}$ but also a fuzzy set, which we shall call the "fuzzy maximum of $f(x)$." A straightforward approach is to define an extended max operation in analogy to the other extended operations defined in chapter 5. Max and min are increasing operations in $\mathbb{R}$. The maximum or minimum, respectively, of n fuzzy numbers, denoted by $\widetilde{\max}(\widetilde{M}_1, \ldots, \widetilde{M}_n)$ and $\widetilde{\min}(\widetilde{M}_1, \ldots, \widetilde{M}_n)$ is again a fuzzy number. Dubois and Prade [1980a, p. 58] present rules for computing $\widetilde{\max}$ and $\widetilde{\min}$ and also comment on the properties of $\widetilde{\max}$ and $\widetilde{\min}$. The reader is referred to the above reference for further details.

Definition 7–4

Let $\tilde{f}(x)$ be a fuzzy function from X to $\mathbb{R}$, defined over a crisp and finite domain D. The *fuzzy maximum* of $f(x)$ is then defined as

$$\widetilde{M} = \widetilde{\max_{x \in D}} \tilde{f}(x) = \{(\sup \tilde{f}(x), \mu_{\widetilde{M}}(x)) \mid x \in D\}$$

For $|D| = n$ the membership function of $\widetilde{\max} \tilde{f}(x)$ is given by

$$\mu_{\widetilde{M}}(x) = \min_{j=1,\ldots,n} \mu_{\tilde{f}(x_j)}(f(x_j)), \qquad f(x) \in D$$

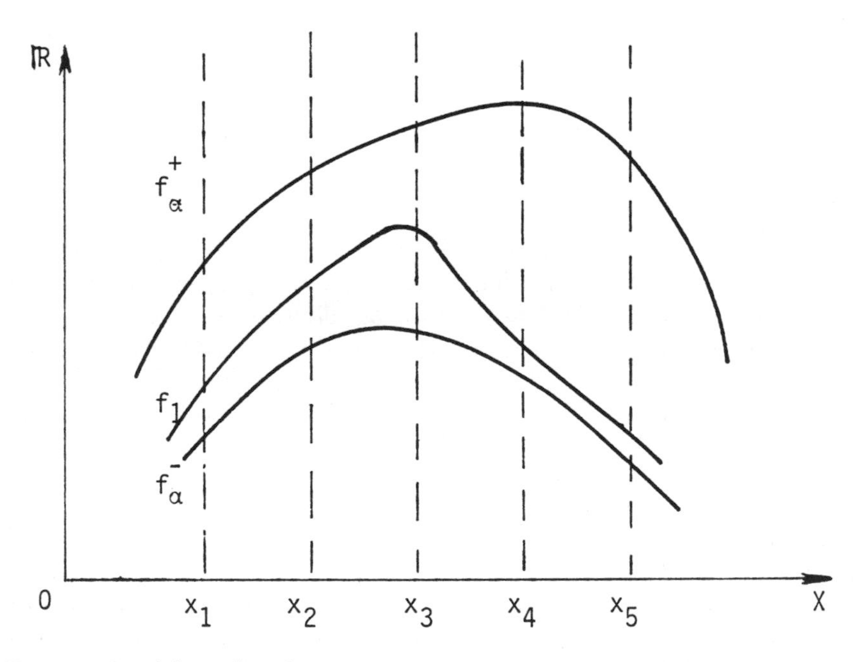

Figure 7–2. A fuzzy function.

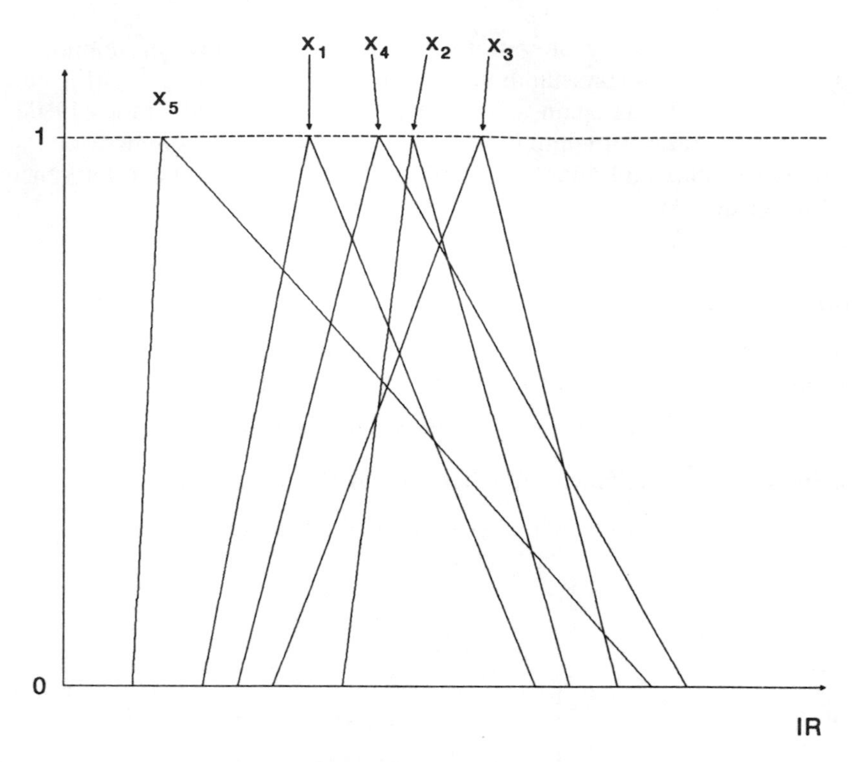

Figure 7–3. Triangular fuzzy numbers representing a fuzzy function.

Example 7–4 [Dubois and Prade 1980a, p. 105]

Let $\tilde{f}(x)$ be a fuzzy function from $\mathbb{R}$ to $\mathbb{R}$ such that, for any x $\tilde{f}(x)$ is a triangular fuzzy number. The domain $D = \{x_1, x_2, x_3, x_4, x_5\}$. Figure 7–2 sketches such a function by showing for the domain D "level-curves" of $\tilde{f}(x)$: f_1 is the curve for which $\mu_{\tilde{f}(x)}(f_1(x)) = 1$, and for f_α^+ and f_α^-, respectively

$$\mu_{\tilde{f}(x)}(f_\alpha^-(x)) = \mu_{\tilde{f}(x)}(f_\alpha^+(x)) = \alpha$$

The triangular fuzzy numbers representing the function $\tilde{f}(x)$ at $x = x_1, x_2, x_3, x_4$, and x_5 are shown in figure 7–3.

We can make the following observation: Since the level-curves in figure 7–2 are not parallel to each other their maxima are attained at different x_i: $\max f_\alpha^+ = f_\alpha^+(x_4)$, $\max f_1(x) = f_1(x_3)$ and $\max f_\alpha^-(x) = f_\alpha^-(x_2)$. Thus x_1 and x_5 do certainly not "belong" to the maximum of $f(x)$. We can easily determine the fuzzy set "maximum of $\tilde{f}(x)$" as defined in definition 7–4 by looking at figure 7–4 and observing that, for

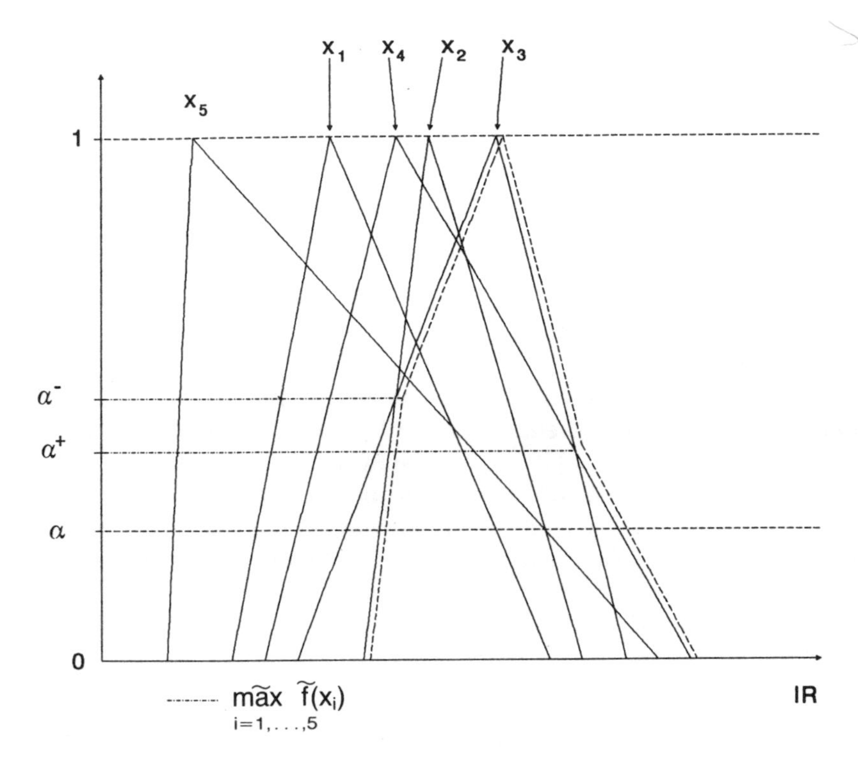

Figure 7–4. The maximum of a fuzzy function.

$$\alpha \in [0, \alpha^-]: f^-(x_2) \geq f_\alpha^-(x_i) \ \forall i$$
$$\alpha \in [\alpha^-, 1]: f^-(x_3) \geq f_\alpha^-(x_i) \ \forall i$$
$$\alpha \in [\alpha^+, 1]: f^+(x_3) \geq f_\alpha^+(x_i) \ \forall i$$
$$\alpha \in [0, \alpha^+]: f^+(x_4) \geq f_\alpha^+(x_i) \ \forall i$$

with α^- and α^+ such that $f_\alpha^-(x_2) = f_\alpha^-(x_3)$ and $f_\alpha^+(x_4) = f_\alpha^+(x_3)$, respectively.

The maximum of $\widetilde{f}(x)$ is therefore

$$\widetilde{M} = \{(x_2, \alpha^-), (x_3, 1), (x_4, \alpha^+)\}$$

This set is indicated in figure 7–4 by the dashed line.

Dubois and Prade [1980a, p. 101] suggest additional possible interpretations of fuzzy extrema, which might be very appropriate in certain situations. We shall, however, not discuss them here and rather proceed to consider possible notions of the integral of a fuzzy set or a fuzzy function.

7.3 Integration of Fuzzy Functions

Quite different suggestions have been made to define fuzzy integrals, integrals of fuzzy functions, and integrals of crisp functions over fuzzy domains or with fuzzy ranges.

One of the first concepts of a fuzzy integral was put forward by Sugeno [1972, 1977] who considered fuzzy measures and suggested a definition of a fuzzy integral which is a generalization of Lebesque integrals: "From the viewpoint of functionals, fuzzy integrals are merely a kind of nonlinear functionals (precisely speaking, monotonous functionals), while Lebesque integrals are linear ones" [Sugeno 1977, p. 92].

We shall focus our attention on approaches along the line of Riemann integrals. The main references for the following are Dubois and Prade [1980a, 1982b], Aumann [1965], and Nguyen [1978].

The classical concept of integration of a real-valued function over a closed interval can be generalized in four ways: The function can be a fuzzy function that is to be integrated over a crisp interval, or it can be integrated over a fuzzy interval (that is, an interval with fuzzy foundations). Alternatively we may consider integrating a fuzzy function as defined in definitions 7–1 or 7–2 over a crisp or a fuzzy interval.

7.3.1 Integration of a Fuzzy Function over a Crisp Interval

We shall now consider a fuzzy function $\tilde{f}$, according to definition 7–2 which shall be integrated over the crisp interval $[a, b]$. The fuzzy function $\tilde{f}(x)$ is supposed to be a fuzzy number, that is, a piecewise continuous convex normalized fuzzy set on $\mathbb{R}$.

We shall further assume, that the α-level-curves (see definition 2.3) $\mu_{\tilde{f}(x)}(y) = \alpha$ for all $\alpha \in [0, 1]$ and α and x as parameters have exactly two continuous solutions, $y = f_\alpha^+(x)$ and $y = f_\alpha^-(x)$ for $\alpha \neq 1$ and only one for $\alpha = 1$. f_α^+ and f_α^- are defined such that

$$f_{\alpha'}^+(x) \geq f_\alpha^+(x) \geq f(x) \geq f_\alpha^- \geq f_{\alpha'}^-$$

for all $\alpha' \geq \alpha$.

The integral of any continuous α-level curve of $\tilde{f}$ over $[a, b]$ always exists.

One may now define the integral $\tilde{I}(a, b)$ of $\tilde{f}(x)$ over $[a, b]$ as a fuzzy set in which the degree of membership α is assigned to the integral of any α-level curve of $\tilde{f}(x)$ over $[a, b]$.

Definition 7–5

Let $f(x)$ be a fuzzy function from $[a, b] \subseteq \mathbb{R}$ to $\mathbb{R}$ such that $\forall x \in [a, b]\ \tilde{f}(x)$ is a fuzzy number and $f_\alpha^-(x)$ and $f_\alpha^+(x)$ are α-level curves as defined above. The *integral* of $\tilde{f}(x)$ over $[a, b]$ is then defined to be the fuzzy set

$$\tilde{I}(a, b) = \left\{\left(\int_a^b f_\alpha^-(x)\,dx + \int_a^b f_\alpha^+(x)\,dx,\ \alpha\right)\right\}$$

This definition is consistent with the extension principle according to which

$$\mu_{\int_a^b f}(y) = \sup_{\substack{g \in y \\ y = \int_a^b g}} \inf_{x \in [a,b]} \mu_{\tilde{f}(x)}(g(x)), \qquad y \in \mathbb{R}$$

where $y = \{g\colon [a, b] \to \mathbb{R} \mid g \text{ integrable}\}$ (see Dubois and Prade 1980a, p. 107; 1982 p. 5.).

The determination of the integral $\tilde{I}(a, b)$ becomes somewhat easier if the fuzzy function is assumed to be of the LR type (see definition 5–6). We shall therefore assume that $\tilde{f}(x) = (f(x), s(x), t(x))_{LR}$ is a fuzzy number in LR representation for all $x \in [a, b]$. f, s, and t are assumed to be positive integrable functions on $[a, b]$. Dubois and Prade [1980a, p. 109] have shown that under these conditions

$$\tilde{I}(a, b) = \left(\int_a^b f(x)\,dx,\ \int_a^b s(x)\,dx,\ \int_a^b t(x)\,dx\right)_{LR}$$

It is then sufficient to integrate the mean value and the spread functions of $\tilde{f}(x)$ over $[a, b]$ and the result will again be an LR fuzzy number.

Example 7–5

Consider the fuzzy function $\tilde{f}(x) = (f(x), s(x), t(x))_{LR}$ with the mean function $f(x) = x^2$, the spread functions $s(x) = x/4$ and

$$t(x) = \frac{x}{2}$$

$$L(x) = \frac{1}{1 + x^2}$$

$$R(x) = \frac{1}{1 + 2|x|}$$

Determine the integral from $a = 1$ to $b = 4$, that is, compute $\int_1^4 \tilde{f}$.

According to the above formula we compute

$$\int_a^b f(x)\,dx = \int_1^4 x^2\,dx = 21$$

$$\int_a^b s(x)\,dx = \int_1^4 \frac{x}{4}dx = 1.875$$

$$\int_a^b t(x)\,dx = \int_1^4 \frac{x}{2}dx = 3.75$$

This yields the fuzzy number $\tilde{I}(a, b) = (21, 1.875, 3.75)_{LR}$ as the value of the fuzzy integral.

Some Properties of Integrals of Fuzzy Functions. Let A_α be the α-level set of the fuzzy set $\tilde{A}$. The support $S(\tilde{A})$ of $\tilde{A}$ is then $S(\tilde{A}) = \bigcup_{\alpha\in[0,1]} A_\alpha$. The fuzzy set $\tilde{A}$ can now be written as

$$\tilde{A} = \bigcup_{\alpha\in[0,1]} \alpha A_\alpha = \bigcup_{\alpha\in[0,1]} \{(x, \mu_{\alpha A_\alpha}(x) \mid x \in A_\alpha)\}$$

where

$$\mu_{\alpha A_\alpha}(x) = \begin{cases} \alpha & \text{for } x \in A_\alpha \\ 0 & \text{for } x \notin A_\alpha \end{cases}$$

(see Nguyen [1978, p. 369]).

Let $\tilde{A}$ represent a fuzzy integral, that is,

$$\tilde{A} = \int_I \tilde{f}$$

then

$$\int_I \tilde{f} = \bigcup_{\alpha\in[0,1]} \alpha\left(\int_I \tilde{f}\right)_\alpha$$

$$= \bigcup_{\alpha\in[0,1]} \alpha\left(\int_I \tilde{f}_\alpha\right)$$

Definition 7–6 [Dubois and Prade 1982a, p. 6]

$\int_I \tilde{f}$ satisfies the *commutativity condition*

$$\text{iff } \forall\, \alpha \in [0, 1] \left(\int_I \tilde{f}\right)_\alpha = \int_I \tilde{f}_\alpha$$

Dubois and Prade [1982a, pp. 6] have proved the following properties of fuzzy integrals, which are partly a straightforward analogy of crisp analysis.

Theorem 7–1

Let $\tilde{f}$ be a fuzzy function then

$$\int_I \tilde{f} = \int_a^b \tilde{f} = -\int_b^a \tilde{f}$$

where the fuzzy integrals are fuzzy sets with the membership functions

$$\mu_{-\int_a^b f}(u) = \mu_{\int_a^b f}(-u)\ \forall\, u$$

Theorem 7–2

Let I and I' be two adjacent intervals $I = [a, b]$, $I' = [b, c]$ and a fuzzy function $\tilde{f}: [a, c] \to \tilde{P}(\mathbb{R})$. Then

$$\int_a^c \tilde{f} = \int_a^b \tilde{f} \oplus \int_b^c \tilde{f}$$

where $\oplus$ denotes the extended addition of fuzzy sets, which is defined in analogy to the subtraction of fuzzy numbers (see chapter 5).

Let $\tilde{f}$ and $\tilde{g}$ be fuzzy functions. Then $\tilde{f} \oplus \tilde{g}$ is pointwise defined by

$$(\tilde{f} \oplus \tilde{g})(u) = \tilde{f}(u) \oplus \tilde{g}(u), \qquad u \in X$$

(This is a straightforward application of the extension principle from chapter 5.1.)

Theorem 7–3

Let $\tilde{f}$ and $\tilde{g}$ be fuzzy functions whose supports are bounded. Then

$$\int_I (\tilde{f} \oplus \tilde{g}) \supseteq \int_I \tilde{f} \oplus \int_I \tilde{g} \tag{7.1}$$

$$\int_I (\tilde{f} \oplus \tilde{g}) = \int_I \tilde{f} \oplus \int_I \tilde{g} \tag{7.2}$$

iff the *commutativity condition* is satisfied for $\int \tilde{f}$ and $\int_I \tilde{g}$.

7.3.2 Integration of a (Crisp) Real-Valued Function over a Fuzzy Interval

We now consider a case for which Dubois and Prade [1982a, p. 106] proposed a quite interesting solution: A fuzzy domain $\tilde{\mathcal{D}}$ of the real line $\mathbb{R}$

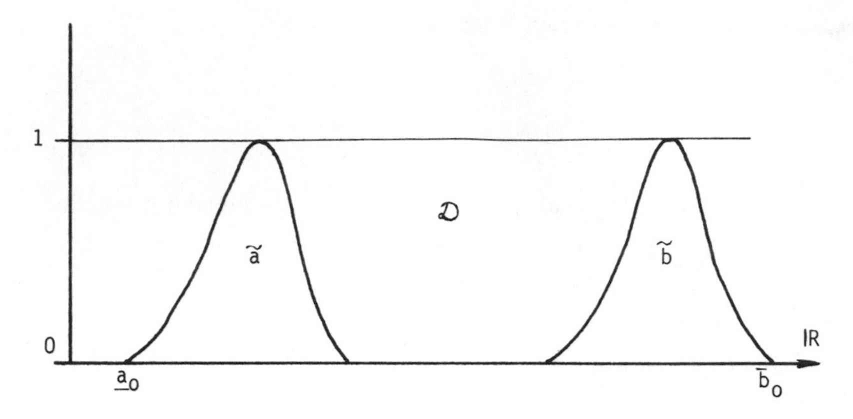

Figure 7–5. Fuzzily bounded interval.

is assumed to be bounded by two normalized convex fuzzy sets the membership functions of which are $\mu_{\tilde{a}}(x)$ and $\mu_{\tilde{b}}(x)$, respectively. (See figure 7–5). $\mu_{\tilde{a}}(x)$ and $\mu_{\tilde{b}}(x)$ can be interpreted as the degrees (of confidence) to which x can be considered a lower or upper bound of $\tilde{\mathcal{T}}$. If $\underline{a}_0$ and $\bar{b}_0$ are the lower/upper limits of the supports of $\tilde{a}$ or $\tilde{b}$ then a_0 or b_0 are related to each other by $\underline{a}_0 = \inf S(\tilde{a}) \leq \sup S(\tilde{b}) = \bar{b}_0$.

Definition 7–7

Let f be a real valued function which is integrable in the interval $J = [a_0, b_0]$ then according to the extension principle the membership function of the integral $\int_{\tilde{\mathcal{T}}} f$ is given by

$$\mu_{\int_{\tilde{\mathcal{T}}} f}(z) = \sup_{x,y \in J} \min\left(\mu_{\tilde{a}}(x), \mu_{\tilde{b}}(y)\right)$$

$$z = \int_x^y f$$

Let $F(x) = \int_c^x f(y)\,dy$, $c \in J$ (F antiderivative of f). Then using the extension principle again the membership function of $F(\tilde{a})$, $\tilde{a} \in \tilde{P}(\mathbb{R})$, is given by

$$\mu_{F(\tilde{a})}(z) = \sup_{x: z = F(x)} \mu_{\tilde{a}}(x)$$

Proposition 7–1 [Dubois and Prade 1982b, p. 106]

$$\int_{\tilde{\mathscr{D}}} f = F(\tilde{b}) \ominus F(\tilde{a})$$

where $\ominus$ denotes the extended subtraction of fuzzy sets.

Proofs of proposition 7–1 and of the following propositions can be found in Dubois and Prade [1982b, pp. 107–109].

A possible interpretation of proposition 7–1 is: If $\tilde{a}$ and $\tilde{b}$ are normalized convex fuzzy sets, then $\int_{\tilde{\mathscr{D}}} f$ is the interval between "worst" and "best" values for different levels of confidence indicated by the respective degrees of membership (see also [Dubois and Prade 1988a, pp. 34–36].)

Example 7–6

Let

$$\tilde{a} = \{(4, .8), (5, 1), (6, .4)\}$$
$$\tilde{b} = \{(6, .7), (7, 1), (8, .2)\}$$
$$f(x) = 2, \qquad x \in [a_0, b_0] = [4, 8]$$

Then

$$\int_{\tilde{\mathscr{D}}} f(x)\, dx = \int_{\tilde{a}}^{\tilde{b}} 2dx = 2x \Big|_{\tilde{a}}^{\tilde{b}}$$

The detailed computational results are:

(a, b)	$\int_a^b 2dx$	$\min(\mu_x(a), \mu_x(b))$
(4, 6)	4	.7
(4, 7)	6	.8
(4, 8)	8	.2
(5, 6)	2	.7
(5, 7)	4	1.0
(5, 8)	6	.2
(6, 6)	0	.4
(6, 7)	2	.4
(6, 8)	4	.2

Hence choosing the maximum of the membership values for each value of the integral yields $\int_{\tilde{\mathscr{D}}} f = \{(0, .4), (2, .7), (4, 1), (6, .8), (8, .2)\}$.

Some properties of the integral discussed above are listed in propositions 7–2 to 7–4 below. Their proofs as well as descriptions of other approaches to "fuzzy integration" can again be found in Dubois and Prade [1982a, pp. 107–108].

Proposition 7–2

Let f and g be two functions $f, g: I \to \mathbb{R}$, integrable on I. Then

$$\int_{\tilde{a}}^{\tilde{b}} (f+g) \subseteq \int_{\tilde{a}}^{\tilde{b}} f \oplus \int_{\tilde{a}}^{\tilde{b}} g$$

where $\oplus$ denotes the extended addition (see chapter 5).

Example 7–7

Let

$$f(x) = 2x - 3$$
$$g(x) = -2x + 5$$

$$\tilde{a} = \{(1, .8), (2, 1), (3, .4)\}$$
$$\tilde{b} = \{(3, .7), (4, 1), (5, .3)\}$$

So

$$\int_a^b f(x)\,dx = [x^2 - 3x]_a^b$$

$$\int_a^b g(x)\,dx = [-x^2 + 5x]_a^b$$

$$\int_a^b f(x) + g(x)\,dx = [2x]_a^b$$

In analogy to example 7–6 we obtain

$$\int_{\tilde{a}}^{\tilde{b}} f = \{(0, .4), (2, .7), (4, .4), (6, 1), (10, .3), (12, .3)\}$$

$$\int_{\tilde{a}}^{\tilde{b}} g = \{(-6, .3), (-4, .3), (-2, .1), (0, .8), (2, .7)\}$$

Applying the formula for the extended addition according to the extension principle (see section 5–3) yields

$$\int_{\tilde{a}}^{\tilde{b}} f + \int_{\tilde{a}}^{\tilde{b}} g = \{(-6, .3), (-4, .3), (-2, .4), (0, .7), (2, .7), (4, 1), (6, .8), (8, .7), (10, .3), (12, .3), (14, .3)\}$$

Similarly to example 7–6 we compute

$$\int_{\tilde{a}}^{\tilde{b}} (f+g) = \{(0, .4), (2, .7), (4, 1), (6, .8), (8, .3)\}$$

Now we can easily verify that

$$\int_{\tilde{a}}^{\tilde{b}} f \oplus \int_{\tilde{a}}^{\tilde{b}} g \supseteq \int_{\tilde{a}}^{\tilde{b}} (f+g)$$

Proposition 7–3

If $f, g: I \rightarrow R^{+}$ or $f, g: I \rightarrow R^{-}$

then equality holds

$$\int_{\tilde{a}}^{\tilde{b}} (f+g) = \int_{\tilde{a}}^{\tilde{b}} f \oplus \int_{\tilde{a}}^{\tilde{b}} g$$

Proposition 7–4

Let $\tilde{\mathcal{T}} = (\tilde{a}, \tilde{b})$, $\tilde{\mathcal{T}}' = (\tilde{a}, \tilde{c})$ and $\tilde{\mathcal{T}}'' = (\tilde{c}, \tilde{b})$. Then the following relationships hold:

$$\int_{\tilde{\mathcal{T}}} f \subseteq \int_{\tilde{\mathcal{T}}'} f_1 \oplus \int_{\tilde{\mathcal{T}}''} f_2 \tag{7.3}$$

$$\int_{\tilde{\mathcal{T}}} f = \int_{\tilde{\mathcal{T}}'} f_1 \oplus \int_{\tilde{\mathcal{T}}''} f_2 \qquad \text{iff } \tilde{c} \in \mathbb{R} \tag{7.4}$$

7.4 Fuzzy Differentiation

In analogy to integration differentiation can be extended to fuzzy mathematical structures.

The results will, of course, depend on the type of function considered. In terms of section 7.1 we will focus our attention on functions that are not fuzzy themselves but which only "carry" the possible fuzziness of their arguments. Differentiation of fuzzy functions is considered by Dubois and Prade [1980a, p. 116 and 1982b, p. 227].

Here we shall consider only differentiation of a differentiable function $f: \mathbb{R} \supseteq [a, b] \rightarrow \mathbb{R}$ at a "fuzzy point." A "fuzzy point" $\tilde{X}_0$ [Dubois and Prade 1982b, p. 225] is a convex fuzzy subset of the real line $\mathbb{R}$ (see definition 2–4).

In the following, fuzzy points will be considered for which the support is contained in the interval $[a, b]$, that is, $S(\tilde{x}) \subseteq [a, b]$

Such a fuzzy point can be interpreted as the possibility distribution of a point x whose precise location is only approximately known.

The uncertainty of the knowledge about the precise location of the point induces an uncertainty about the derivative $f'(x)$ of a function $f(x)$ at this point. The derivative might be the same for several x belonging to $[a, b]$. The possibility of $f'(\widetilde{X}_0)$ is therefore defined [Zadeh 1978] to be the supremum of the values of the possibilities of $f'(x) = t$, $x \in [a, b]$.

The "derivative" of a real-valued function at a fuzzy point can be interpreted as the fuzzy set $f'(\widetilde{X}_0)$, the membership function of which expresses the degree to which a specific $f'(x)$ is the first derivative of a function f at point $\widetilde{X}_0$.

Definition 7–8

The membership function of the fuzzy set "*derivative* of a real-valued function at a fuzzy point $\widetilde{X}_0$" is defined by the extension principle as

$$\mu_{f'(\tilde{X}_0)}(y) = \sup_{x \in f'^{-1}(y)} \mu_{\tilde{X}_0}(x)$$

where $\widetilde{X}_0$ is the fuzzy number that characterizes the fuzzy location.

Example 7–8

Let

$$f(x) = x^3$$
$$\widetilde{X}_0 = \{(-1, .4), (0, 1), (1, .6)\} \quad \text{be a fuzzy location.}$$

Because of $f'(x) = 3x^2$ we obtain $f'(\widetilde{X}_0) = \{(0, 1), (3, .6)\}$ as derivative of a real-valued function at the fuzzy point $\widetilde{X}_0$.

Proposition 7–5

The extended sum $\oplus$ of the derivatives of two real-valued functions f and g at the fuzzy point $\widetilde{X}_0$ is defined by

$$\mu_{(f'+g')(\tilde{x}_0)}(y) = \sup_{x:y=f'(x)+g'(x)} \mu_{\tilde{x}_0}(x)$$

Hence

$$f'(\widetilde{X}_0) \oplus g'(\widetilde{X}_0) \supseteq (f' + g')\widetilde{X}_0$$

Proposition 7–6 [Dubois and Prade 1982b, p. 227]

If f' and g' are continuous and both nondecreasing or nonincreasing

$$f'(\widetilde{X}_0) \oplus g'(\widetilde{X}_0) = (f' + g')(\widetilde{X}_0)$$

Proposition 7–7 *(Chain rule of differentiation)*

1. $(f \cdot g)'(\widetilde{X}_0) = (f'g + fg')(\widetilde{X}_0) \subseteq [f'(\widetilde{X}_0) \odot g(\widetilde{X}_0)] \oplus [f(\widetilde{X}_0) \odot g'(\widetilde{X}_0)]$
2. If f, g, f', and g' are continuous, f and g are both positive, and f' and g' are both nondecreasing (f, g is negative and f', g' is nondecreasing) then

$$(f \cdot g)'(\widetilde{X}_0) = [f'(\widetilde{X}_0) \odot g(\widetilde{X}_0)] \oplus [f(\widetilde{X}_0) \odot g'(\widetilde{X}_0)]$$

Exercises

1. Determine the maximizing set of

$$f(x) = \begin{cases} 2x^2 - 3 & -2 \leq x \leq 2 \\ 5 & \text{else} \end{cases}$$

2. Show that computing $\mu \int_a^b b_f$ according to the extension principle yields the usual integral if $\widetilde{f}$ is a crisp function.
3. Let $\widetilde{f}(x) = (f(x), s(x), t(x))_{LR}$ with

$$f(x) = nx$$
$$s(x) = \frac{1}{|x|+1}$$
$$t(x) = \frac{1}{1 + \sin^2 x}$$
$$L(x) = \frac{1}{1 + 2|x|^3}$$
$$R(x) = \frac{1}{1 + x^{1/2}}$$

Determine $\widetilde{f}(x)$ explicitly for $x = .5$, $x = 1$, and $x = 2$. Compute the integral $\widetilde{I}(a, b)$.

4. Let $f(x) = 2x^3 + (x - 1)^2$

$$\widetilde{X}_0 = \{(-1, .5), (0, .8), (1, 1), (2, .6), (3, .4)\}$$

Compute $f'(\widetilde{X}_0)$. Verify that proposition 7–6 holds.

5. Let $\widetilde{X}_0 = \{(-1, .4), (0, 1), (1, .6)\}$

$$f(x) = x^3 + 2 \qquad g(x) = 2x + 3$$

Compute $f'(\widetilde{X}_0)$. Verify that proposition 7–6 holds.

8 POSSIBILITY THEORY, PROBABILITY THEORY, AND FUZZY SET THEORY

Since L. Zadeh proposed the concept of a fuzzy set in 1965 the relationships between probability theory and fuzzy set theory have been discussed. Both theories seem to be similar in the sense that both are concerned with some type of uncertainty and both use the [0, 1] interval for their measures as the range of their respective functions. (At least as long as one considers normalized fuzzy sets only!) Other uncertainty measures, which were already mentioned in chapter 4, also focus on uncertainty and could therefore be included in such a discussion. The comparison between probability theory and fuzzy set theory is difficult primarily for two reasons:

1. The comparison could be made on very different levels, that is, mathematically, semantically, linguistically, and so on.
2. Fuzzy set theory is not or is no longer a uniquely defined mathematical structure, such as Boolean algebra or dual logic. It is rather a very general family of theories (consider, for instance, all the possible operations defined in chapter 3 or the different types of membership functions). In this respect, fuzzy set theory could rather be compared with the different existing theories of multivalued logic.

Further, there does not yet exist and probably never will exist, a unique context-independent definition of what fuzziness really means. On the other hand, neither is probability theory uniquely defined. There are different definitions and different linguistic appearances of "probability."

In recent years some specific interpretations of fuzzy set theory have been suggested. One of them, possibility theory, used to correspond, roughly speaking, to the min-max version of fuzzy set theory, that is, to fuzzy set theory in which the intersection is modeled by the min-operator and the union by the max-operator. This interpretation of possibility theory, however, is no longer correct. Rather it has been developed into a well-founded and comprehensive theory. After the basic articles by L. Zadeh [Zadeh 1978, 1983] most of the advances in possibility theory are due to Dubois and Prade. (See, for instance, their excellent book on this topic [Dubois and Prade 1988]).

We shall first describe the essentials of possibility theory and then compare it with other theories of uncertainty.

8.1 Possibility Theory

8.1.1 Fuzzy Sets and Possibility Distributions

Possibility theory focuses primarily on imprecision, which is intrinsic in natural languages and is assumed to be rather "possibilistic" than probabilistic. Therefore the term *variable* is very often used in a more linguistic sense than in a strictly mathematical one. This is one reason why the terminology and the symbolism of possibility theory differs in some respects from that of fuzzy set theory. In order to facilitate the study of possibility theory, we will therefore use the common possibilistic terminology but always show the correspondence to fuzzy set theory.

Suppose, for instance, we want to consider the proposition "X is $\tilde{F}$," where X is the name of an object, a variable, or a proposition. For instance, in "X is a small integer," X is the name of a variable. In "John is young," John is the name of an object. $\tilde{F}$(i.e., "small integer" or "young") is a fuzzy set characterized by its membership function $\mu_{\tilde{F}}$.

One of the central concepts of possibility theory is that of a possibility distribution (as opposed to a probability distribution). In order to define a possibility distribution, it is convenient first to introduce the notion of a fuzzy restriction. To visualize a fuzzy restriction the reader should imagine an elastic suitcase which acts on the possible volume of its contents as a constraint. For a hardcover suitcase, the volume is a crisp number. For a soft valise, the volume of its contents depends to a certain degree on the strength that is used to stretch it. The variable in this case would be the volume of the valise; the values this variable can assume may be $u \in U$, and the degree to which the variable (X) can assume different values of u is expressed by $\mu_{\tilde{F}}(u)$. Zadeh [1975 et al., p. 2; Zadeh 1978, p. 5] defines these relationships as follows.

Definition 8–1

Let $\tilde{F}$ be a fuzzy set of the universe U characterized by a membership function $\mu_{\tilde{F}}(u)$. $\tilde{F}$ is a *fuzzy restriction* on the variable X if $\tilde{F}$ acts as an elastic constraint on the values that may be assigned to X, in the sense that the assignment of the values u to X has the form

$$X = u: \mu_{\tilde{F}}(u)$$

$\mu_{\tilde{F}}(u)$ is the degree to which the constraint represented by $\tilde{F}$ is satisfied when u is assigned to X. Equivalently, this implies that $1 - \mu_{\tilde{F}}(u)$ is the degree to which the constraint has to be stretched in order to allow the assignment of the values u to the variable X.

Whether a fuzzy set can be considered as a fuzzy restriction or not obviously depends on its interpretation: This is only the case if it acts as a constraint on the values of a variable, which might take the form of a linguistic term or a classical variable.

Let $\tilde{R}(X)$ be a fuzzy restriction associated with X such as defined in definition 8–1. Then $\tilde{R}(X) = \tilde{F}$ is called a *relational assignment equation* which assigns the fuzzy set $\tilde{F}$ to the fuzzy restriction $\tilde{R}(X)$.

Let us now assume that $A(X)$ is an implied attribute of the variable X. For instance, $A(X) =$ "age of Jack" and $\tilde{F}$ is the fuzzy set "young." The proposition "Jack is young" (or better "the age of Jack is young") can then be expressed as

$$\tilde{R}(A(X)) = \tilde{F}$$

Example 8–1 [see Zadeh 1978, p. 5]

Let p be the proposition "John is young" in which young is a fuzzy set of the universe $U = [0, 100]$ characterized by the membership function

$$\mu_{\text{young}}(u) = S(u; 20, 30, 40)$$

where u is the numerical age and the S-function is defined by

$$S(u;\alpha,\beta,\gamma)=\begin{cases}1 & \text{for } u\leq\alpha\\ 1-2\left(\dfrac{u-\alpha}{\gamma-\alpha}\right)^2 & \text{for } \alpha\leq u\leq\beta\\ 2\left(\dfrac{u-\gamma}{\gamma-\alpha}\right)^2 & \text{for } \beta\leq u\leq\gamma\\ 0 & \text{for } u\geq\gamma\end{cases}$$

In this case, the implied attribute $A(X)$ is Age (John) and the translation of "John is young" has the form

$$\text{John is young} \rightarrow \tilde{R}(\text{Age (John)}) = \text{young}$$

Zadeh (1978) related the concept of a fuzzy restriction to that of a possibility distribution as follows.

> Consider a numerical age, say $u = 28$, whose grade of membership in the fuzzy set "young" is approximately 0.7. First we interpret 0.7 as the degree of compatibility of 28 with the concept labelled young. Then we postulate that the proposition "John is young" converts the meaning of 0.7 from the degree of compatibility of 28 with young to the degree of possibility that John is 28 given the proposition "John is young." In short, the compatibility of a value of u with young becomes converted into the possibility of that value of u given "John is young." [Zadeh 1978, p. 6]

The concept of a possibility distribution can now be defined as follows:

Definition 8–2 [Zadeh 1978, p. 6]

Let $\tilde{F}$ be a fuzzy set in a universe of discourse U which is characterized by its membership function $\mu_{\tilde{F}}(u)$, which is interpreted as the compatibility of $u \in U$ with the concept labeled $\tilde{F}$.

Let X be a variable taking values in U and $\tilde{F}$ act as a fuzzy restriction, $\tilde{R}(X)$, associated with X. Then the proposition "X is $\tilde{F}$," which translates into $\tilde{R}(X) = \tilde{F}$ associates a *possibility distribution*, π_x, with X which is postulated to be equal to $\tilde{R}(X)$.

The possibility distribution function, $\pi_x(u)$, characterizing the possibility distribution π_x is defined to be numerically equal to the membership function $\mu_{\tilde{F}}(u)$ of $\tilde{F}$, that is,

$$\pi_x \mathrel{\hat{=}} \mu_{\tilde{F}}$$

The symbol $\hat{=}$ will always stand for "denotes" or "is defined to be." In order to stay in line with the common symbol of possibility theory we will denote a possibility distribution with π_x rather than with $\tilde{\pi}_x$, even though it is a fuzzy set.

Example 8–2 [Zadeh 1978, p. 7]

Let U be the universe of positive integers and $\tilde{F}$ be the fuzzy set of small integers defined by

$$\tilde{F} = \{(1, 1), (2, 1), (3, .8), (4, .6), (5, .4), (6, .2)\}$$

Then the proposition "X is a small integer" associates with X the possibility distribution

$$\pi_x = \tilde{F}$$

in which a term such as (3, .8) signifies that the possibility that X is 3, given that X is a small integer, is .8.

Even though definition 8–2 does not assert that our intuition of what we mean by possibility agrees with the min-max fuzzy set theory, it might help to realize their common origin. It might also make more obvious the difference between possibility distribution and probability distribution.

Zadeh [1978, p. 8] illustrates this difference by a simple but impressive example.

Example 8–3

Consider the statement "Hans ate X eggs for breakfast." $X = \{1, 2, \ldots\}$. A possibility distribution as well as a probability distribution may be associated with X. The possibility distribution $\pi_x(u)$ can be interpreted as the degree of ease with which Hans can eat u eggs while the probability distribution might have been determined by observing Hans at breakfast for 100 days. The values of $\pi_x(u)$ and $P_x(u)$ might be as shown in the following table:

u	1	2	3	4	5	6	7	8
$\pi_x(u)$	1	1	1	1	.8	.6	.4	.2
$P_x(u)$	.1	.8	.1	0	0	0	0	0

We observe that a high degree of possibility does not imply a high degree of probability. If, however, an event is not possible, it is also improbable. Thus, in a way the possibility is an upper bound for the probability. A more detailed discussion of this "*possibility/probability consistency principle*" can be found in Zadeh [1978].

This principle is not intended as a crisp principle, from which exact probabilities or possibilities can be computed but rather as a heuristic principle, expressing the principle relationship between possibilities and probabilities.

8.1.2 Possibility and Necessity Measures

In chapter 4 a possibility measure was already defined (definition 4–2) for the case that A is a crisp set. If $\tilde{A}$ is a fuzzy set a more general definition of a possibility measure has to be given [Zadeh 1978, p. 9].

Definition 8–3

Let $\tilde{A}$ be a fuzzy set in the universe U are π_x a possibility distribution associated with a variable X which takes values in U. The *possibility measure*, $\pi_x(\tilde{A})$, of $\tilde{A}$ is then defined by

$$\begin{aligned} \text{poss}\,\{X \text{ is } \tilde{A}\} &\mathrel{\hat{=}} \pi(\tilde{A}) \\ &\mathrel{\hat{=}} \sup_{u \in U} \min\,\{\mu_{\tilde{A}}(u), \pi_x(u)\} \end{aligned}$$

Example 8–4 [Zadeh 1978]

Let us consider the possibility distribution induced by the proposition "X is a small integer" (see example 8–2).

$$\pi_x = \{(1, 1), (2, 1), (3, .8), (4, .6), (5, .4), (6, .2)\}$$

and the crisp set $A = \{3, 4, 5\}$.

The possibility measure $\pi(\tilde{A})$ is then

$$\pi(\tilde{A}) = \max\,(.8, .6, .4) = .8$$

If $\tilde{A}$, on the other hand, is assumed to be the fuzzy set "integers which are not small," defined as

$$\tilde{A} = \{(3, .2), (4, .4), (5, .6), (6, .8), (7, 1), \ldots\}$$

then the possibility measure of "X is not a small integer" is

$$\text{poss}(X \text{ is not a small integer}) = \max\{.2, .4, .4, .2\} = .4$$

Similar to probability theory there exist also conditional possibilities. Such a conditional possibility distribution can be defined as follows [Zadeh 1981b, p. 81].

Definition 8–4

Let X and Y be variables in the universes U and V, respectively. The *conditional possibility distribution* of X given Y is then induced by a proposition of the form "If X is $\tilde{F}$ then Y is $\tilde{G}$" and is denoted by $\pi_{(Y/X)}(v/u)$.

Proposition 8–1

Let $\pi_{(Y/X)}$ be the conditional possibility distribution functions of X and Y, respectively. The joint possibility distribution function of X and Y, $\pi_{(X,Y)}$ is then given by

$$\pi_{(X,Y)}(u,v) = \min\{\pi_X(u), \pi_{(Y/X)}(v/u)\}$$

Not quite settled yet seems to be the question of how to derive the conditional possibility distribution functions from the joint possibility distribution function. Different views on this question are presented by Zadeh [1981b, p. 82], Hisdal [1978], and Nguyen [1978].

Fuzzy measures as defined in definition 4–2 express the degree to which a certain subset of a universe, Ω, or an event is possible. Hence, we have

$$g(\emptyset) = 0 \text{ and } g(\Omega) = 1$$

As a consequence of condition 2 of definition 4–2, that is,

$$A \subseteq B \Rightarrow g(A) \leq g(B)$$

we have

$$g(A \cup B) \geq \max(g(A), g(B)) \quad \text{and} \tag{8.1}$$

$$g(A \cap B) \leq \min(g(A), g(B)) \quad \text{for } A, B \subseteq \Omega \tag{8.2}$$

Possibility measures (definition 4–2) are defined for the limiting cases:

$$\pi(A \cup B) = \max(\pi(A), \pi(B)) \tag{8.3}$$

$$\pi(A \cap B) = \min(\pi(A), \pi(B)) \tag{8.4}$$

If $\complement A$ is the complement of A in Ω, then

$$\pi(A \cup \complement A) = \max(\pi(A), \pi(\complement A)) = 1 \tag{8.5}$$

which expresses the fact that either A or $\complement A$ is completely possible.

In possibility theory another additional measure is defined, which uses the conjunctive relationship and, in a sense, is dual to the possibility measure:

$$N(A \cap B) = \min(N(A), N(B)) \tag{8.6}$$

N is called the necessity measure. $N(A) = 1$ indicates that A is necessarily true (A is sure). The dual relationship of possibility and necessity requires that

$$\pi(A) = 1 - N(\complement A); \qquad \forall A \subseteq \Omega \tag{8.7}$$

Necessity measures satisfy the condition

$$\min(N(A), N(\complement A)) = 0 \tag{8.8}$$

The relationships between possibility measures and necessity measures satisfy also the following conditions [Dubois and Prade 1988, p. 10]:

$$\pi(A) \geq N(A), \qquad \forall A \subseteq \Omega \tag{8.9}$$

$$\begin{aligned} N(A) > 0 &\Rightarrow \pi(A) = 1 \\ \pi(A) < 1 &\Rightarrow N(A) = 0 \end{aligned} \tag{8.10}$$

Here Ω is always assumed to be finite.

Example 8–5

Let us assume that we know, from past experience, the performance of 6 students in written examinations. Table 8–1 exhibits the possibility functions for the grades A through E and students 1 through 6.

First we observe that the membership function for the grades of student 4 is not a possibility function, since $g(\Omega) \neq 1$.

We can now ask different questions:

1. How reliable is the statement of student 1 that he will obtain a B in his next exam?
   In this case "A" is $\{B\}$ and "$\complement A$" is $\{A, C, D, E\}$.

Table 8–1. Possibility functions.

Student \ Grade	*A*	*B*	*C*	*D*	*E*
1	.8	1	.7	0	0
2	1	.8	.6	.1	0
3	.6	.7	.9	.1	0
4	0	.8	.9	.5	0
5	0	0	.3	1	.2
6	.3	1	.3	0	0

Hence, $\pi(A) = 1$

$$N(A) = \min\{1 - \pi_i\}$$
$$= \min\{.2, .3, 1, 1\} = .2.$$

Hence, the possibility of student 1 getting a B is $\pi = 1$, the necessity $N = .2$.

2. If we want to know the truth of the statement "Either student 1 or 2 will achieve an A or a B," our Ω has to be defined differently. It now contains the elements of the first two rows. The result would be

$$\pi(A) = \pi(\text{student 1 A or B or Student 2 A or B}) = 1$$
$$N(A) = .3$$

3. Let us finally determine the credibility of the statement "student 1 will get a C."

In this case $\pi(A) = .7$

$$N(A) = 0.$$

8.2 Probability of Fuzzy Events

By now it should have become clear that possibility is not a substitute for probability but rather another kind of uncertainty.

Let us now assume that an event is not crisply defined except by a possibility distribution (a fuzzy set) and that we are in classical situation of stochastic uncertainty, that is, that the happening of this (fuzzily described) event is not certain and that we want to express the probability of its happening. Two views on this probability can be adopted: Either this probability should be a scalar (measure) or this probability can be considered as a fuzzy set also. We shall consider both views briefly.

8.2.1 Probability of a Fuzzy Event as a Scalar

In classical probability theory an event, A, is a member of an α-field a, of subsets of a sample space Ω. A probability measure P is a normalized measure over a measurable space (Ω, a) that is, P is a real-valued function which assigns to every A in a a probability, $P(A)$ such that

1. $P(A) \geq 0 \qquad A \in a$
2. $P(\Omega) = 1$
3. If $A_i \in a$, $i \in I \subset \mathbb{N}$, pairwise disjoint then

$$P(\bigcup_{i \in I} A_i) = \sum_{i \in I} P(A_i)$$

If Ω is, for instance, a euclidean n-space and a the σ-field of Borel sets in $\mathbb{R}^n$ then the probability of A can be expressed as

$$P(A) = \int_A dP$$

If $\mu_A(x)$ denotes the characteristic function of a crisp set of A and $E_p(\mu_A)$ the expectation of $\mu_A(x)$ then

$$P(A) = \int_{R^n}^{\mu_A} (x)\, dP = E_P(\mu_A)$$

If $\mu_A(x)$ does not denote the characteristic function of a crisp set but rather the membership function of a fuzzy set the basic definition of the probability of A should not change. Zadeh [1968] therefore defined the probability of a fuzzy event $\widetilde{A}$ (i.e., a fuzzy set $\widetilde{A}$ with membership function $\mu_{\widetilde{A}}(x)$) as follows.

Definition 8–5

Let $(\mathbb{R}^n, a, P)$ be a probability space in which a is the σ-field of Borel sets in $\mathbb{R}^n$ and P is a probability measure over $\mathbb{R}^n$. Then a *fuzzy event* in $\mathbb{R}^n$ is a fuzzy set $\widetilde{A}$ in $\mathbb{R}^n$ whose membership function $\mu_{\widetilde{A}}(x)$ is Borel measurable.

The *probability of a fuzzy event* $\widetilde{A}$ is then defined by the Lebesque-Stieltjes integral

$$P(\widetilde{A}) = \int_{R^n}^{\mu_A} (x)\, dP = E(\mu_{\widetilde{A}})$$

In Zadeh [1968] the similarity of the probability of fuzzy events and the probability of crisp events is illustrated. His suggestions, though very plausible, were not yet axiomatically justified in 1968. Smets [1982] showed, however, that an axiomatic justification can be given for the case of crisp probabilities of fuzzy events within nonfuzzy environments. Other authors consider other cases, such as fuzzy probabilities, which we will not investigate in this book.

We shall rather turn to the definition of the probability of a fuzzy event as a fuzzy set, which corresponds quite well to some approaches we have discussed, for example, for fuzzy integrals.

8.2.2 Probability of a Fuzzy Event as a Fuzzy Set

In the following we shall consider sets with a finite number of elements. Let us assume that there exists a probability measure P defined on the set of all crisp subsets of (the universe) X, the Borel set. $P(x_i)$ shall denote the probability of element $x_i \in X$.

Let $\widetilde{A} = \{(x, \mu_{\widetilde{A}}(x) \mid x \in X\}$ be a fuzzy set representing a fuzzy event. The degree of membership of element $x_i \in \widetilde{A}$ is denoted by $\mu_{\widetilde{A}}(x_i)$. α-level sets or α-cuts such as already defined in definition 2–3 shall be denoted by A_α.

Yager [1979, 1984] suggests that it is quite natural to define the probability of an α-level set as $P(A_\alpha) = \Sigma_{x \in A_\alpha} P(x)$. On the basis of this the probability of a fuzzy event is defined as follows [Yager 1984].

Definition 8–6

Let A_α be the α-level set of a fuzzy set $\widetilde{A}$ representing a fuzzy event. Then the *probability of a fuzzy event* $\widetilde{A}$ can be defined as

$$P_Y(\widetilde{A}) = \{(P(A_\alpha), \alpha) \mid \alpha \in [0, 1]\}$$

with the interpretation "the probability of at least an α degree of satisfaction to the condition $\widetilde{A}$."

The subscript Y of P_Y indicates that P_Y is a definition of probability due to Yager which differs from Zadeh's definition which is denoted by P. It should be very clear that Yager considers α, which is used as the degree of membership of the probabilities $P(A_\alpha)$ in the fuzzy set $P_Y(\widetilde{A})$, as a kind of significance level for the probability of a fuzzy event.

On the basis of private communication with Klement, Yager also suggests another definition for the probability of a fuzzy event, which is derived as follows.

Definition 8–7

The *truth of the proposition* "the probability $\tilde{A}$ is at least w" is defined as the fuzzy set $P_y^*(\tilde{A})$ with the membership function

$$P_y^*(\tilde{A})(w) = \sup_{\alpha} \{\alpha \mid P(A_\alpha) \geq w\}, \qquad w \in [0, 1]$$

The reader should realize that now the "indicator" of significance of the probability measure is w and no longer α! The reader should also be aware of the fact that we have used Yager's terminology denoting the values of the membership function by $P_y^*(\tilde{A})(w)$. This will facilitate reading Yager's work [1984].

If we denote the complement of $\tilde{A}$ by $\complement\tilde{A} = \{(x, 1 - \mu_{\tilde{A}}(x)) \mid x \in X\}$ and the α-level sets of $\complement\tilde{A}$ by $(\complement\tilde{A})_\alpha$ then $P_y^*(\complement A)(w) = \sup_\alpha\{\alpha \mid P(\complement\tilde{A})_\alpha \geq w\}$, $w \in [0, 1]$ can be interpreted as the truth of the proposition "the probability of not $\tilde{A}$ is at least w."

Let us define $\bar{P}_y^*(\tilde{A}) = 1 - P_y^*(\complement\tilde{A})$. If $\bar{P}_y^*(\tilde{A})(w)$ is interpreted as the truth of the proposition "probability of $\tilde{A}$ is at most w," then we can argue as follows: The "and" combination of "the probability of $\tilde{A}$ is at least w" and "the probability of $\tilde{A}$ is at most w" might be considered as "the probability of $\tilde{A}$ is exactly w." If $P_y^*(\tilde{A})$ and $\bar{P}_y^*(\tilde{A})$ are considered as possibility distributions then their conjunction is their intersection (modeled by applying the min-operator to the respective membership functions). Hence the following definition [Yager 1984]:

Definition 8–8 [Yager 1984]

Let $P_y^*(\tilde{A})$ and $\bar{P}_y^*(\tilde{A})$ be defined as above. The possibility distribution associated with the proposition "the probability of $\tilde{A}$ is exactly w" can be defined as

$$\bar{P}_y(\tilde{A})(w) = \min\{P_y^*(\tilde{A})\ (w),\ \bar{P}_y^*(\tilde{A})\ (w)\}$$

Example 8–5

Let $\tilde{A} = \{(x_1, 1), (x_2, .7), (x_3, .6), (x_4, .2)\}$ be a fuzzy event with the probability defined for the generic elements: $P_1 = .1$, $P_2 = .4$, $P_3 = .3$, and $P_4 = .2$; $p\{x_2\}$ is .4, where the element x_2 belongs to the fuzzy event $\tilde{A}$ with a degree of .7.

First we compute $P_y^*(\tilde{A})$. We start by determining the α-level sets A_α for all $\alpha \in [0, 1]$. Then we compute the probability of the crisp events A_α and give the intervals of w for which $P(A_\alpha) \geq w$. We finally obtain $P_y^*(\widetilde{A})$ as the respective supremum of α.

The computing is summarized in the following table:

α	A_α	$P(A)$	w	$P_y^*(\widetilde{A}) = \sup \alpha$
[0, .2]	$\{x_1, x_2, x_3, x_4\}$	1	[.8, 1]	.2
[.2, .6]	$\{x_1, x_2, x_3\}$	.8	[.5, .8]	.6
[.6, .7]	$\{x_1, x_2\}$	.5	[.1, .5]	.7
[.7, 1]	$\{x_1\}$	.1	[0, .1]	1

Analogously, we obtain for $\bar{P}_y^*(\widetilde{A}) = 1 - P_y^*(\complement\widetilde{A})$,

α	$(\complement\widetilde{A})_\alpha$	$P(\complement\widetilde{A})_\alpha$	w	$\bar{P}_y^*(\complement\widetilde{A})$	$\bar{P}_y^*(\widetilde{A}) = 1 - P_y^*(\complement\widetilde{A})$
0	$\{x_1, x_2, x_3, x_4\}$	1	[.9, 1]	0	1
[0, .3]	$\{x_2, x_3, x_4\}$	.9	[.5, .9]	.3	.7
[.3, .4]	$\{x_3, x_4\}$	.5	[.2, .5]	.4	.6
[.4, .8]	$\{x_4\}$	.2	[0., 2]	.8	.2
[.8, 1]	0	0	0	1	0

The probability $\bar{P}_y(\widetilde{A})$ of the fuzzy event $\widetilde{A}$ is now determined by the intersection of the fuzzy sets $P_y^*(\widetilde{A})$ and $\bar{P}_y^*(\widetilde{A})$ modeled by the min-operator as in definition 8–8:

$$\bar{P}_y(\widetilde{A})(w) = \begin{cases} 0\ , & w = 0 \\ .2, & w \in [0, .2] \\ .6, & w \in [.2, .8] \\ .2, & w \in [.8, 1] \end{cases}$$

Figure 8–1 illustrates the fuzzy sets $P_y^*(\widetilde{A})(w)$, $\bar{P}_y^*(\widetilde{A})$, and $\bar{P}_y(\widetilde{A})(w)$.

8.3 Possibility vs. Probability

Questions concerning the relationship between fuzzy set theory and probability theory are very frequently raised, particularly by "newcomers" to the area of fuzzy sets. There are probably two major reasons for this: On one hand, there are certain formal similarities between fuzzy set theory (in particular when using normalized fuzzy sets) and probability theory; on the other hand, in the past probabilities have been the only means for

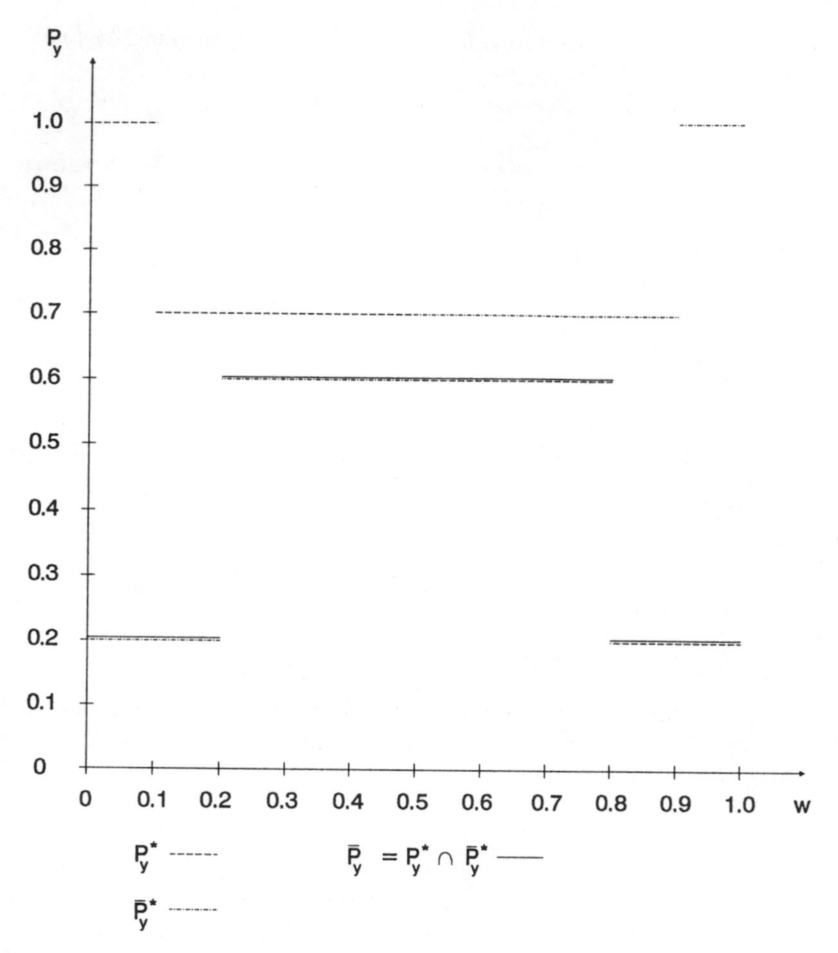

Figure 8–1. Probability of a fuzzy event.

expressing "uncertainty." It seems appropriate and helpful, therefore, to shed some more light on this question.

In the introduction to chapter 8 it was already mentioned that such comparison is difficult because of the lack of unique definitions of fuzzy sets. This lack of a unique definition is due in part to the variety of suggested possibilities for mathematically defining fuzzy sets as well as operations on them as indicated in chapters 2 and 3. It is also due to the many different kinds of fuzziness that can be modeled with fuzzy sets, as described in chapter 1.

Another problem is the selection of the aspects with respect to which these theories shall be compared (see introduction to chapter 8!).

In section 8.1 possibility theory was briefly explained. There it was mentioned that possibility theory is more than the min-max version of fuzzy set theory. It was also shown that the "uncertainty measures" used in possibility theory are the possibility measure and the necessity measure, two measures which in a certain sense are dual to each other. In comparing possibility theory with probability theory we shall first consider only possibility functions—and measures (neglecting the existence of dual measures)—of possibility theory. At the end of the chapter we shall investigate somewhat the relationship between possibility theory and probability theory.

Let us now turn to probabilities and try to characterize and classify available notions of probabilities. Three aspects shall be of main concern:

1. the linguistic expression of probability,
2. the different information context of different types of probabilities, and
3. the semantic interpretation of probabilities and its axiomatic and mathematical consequences.

Linguistically, we can distinguish explicit from implicit formulations of probability. With respect to the information content, we can distinguish between classificatory probabilities (given E, H is probable), comparative (given E, H is more probable than K), partial (given E the probability of K is in the interval $[a, \mathrm{b}]$) and quantitative (given E the probability of H is p) probabilities.

Finally the interpretation of a probability can vary considerably: Let us consider two very important and common interpretations of quantitative probabilities. Koopman [1940, pp. 269–292; and Carnap and Stegmüller 1959] interpret (subjective) probabilities essentially as degrees of truth of statements in dual logic. Axiomatically Koopman derives a concept of probability, q, which mathematically is a Boolean ring.

Kolmogoroff [1950] interprets probabilities "statistically." He considers a set Ω and an associated σ-algebra $\mathcal{F}$, the elements of which are interpreted as events. On the basis of measure theory he defines a (probability) function $P: \mathcal{F} \rightarrow [0, 1]$ with the following properties:

$$P: I \rightarrow [0, 1] \tag{8.11}$$

$$P(\Omega) = 1 \tag{8.12}$$

$$\forall (X_i^N) \in \mathcal{F} (\forall i, j \in \mathbb{N}: i \neq j \rightarrow X_i \cap X_j = \emptyset)\, P(\bigcup_{i \in N} X_i) = \sum_{i \in N} P(X_i) \tag{8.13}$$

From those the following relationships can easily be derived:

$$X, \not{C}X \in \mathcal{F} \rightarrow P(\not{C}X) = 1 - P(X) \tag{8.14}$$

$$X, Y \in \mathcal{F} \rightarrow P(X \cup Y) = P(X) + P(Y) - P(X \cap Y) \tag{8.15}$$

where $\not{C}X$ denotes the complement of X.

Table 8–2 illustrates the difference between Koopman's and Kolmogoroff's concept of probability, taking into account the different linguistic and informational possibilities mentioned above.

Now we are ready to compare "fuzzy sets" with "probabilities" or at least one certain version of fuzzy set theory with one of probability theory. Implicit probabilities are not comparable to fuzzy sets since fuzzy set models try particularly to model uncertainty explicitly. Comparative and partial probabilities are more comparable to probabilistic statements using "linguistic variables," which we will cover in chapter 9.

Hence the most frequently used versions we shall compare now are quantitative, explicit Kolmogoroff probabilities with possibilities.

Table 8–2. Koopman's vs. Kolmogoroff's probabilities.

Koopman	*Kolmogoroff*
D, D', H, H' are statements of dual logic, Q is a non-negative real number (generally $Q \in [0, 1]$)	W is a set of events, W_1 are subsets of W.
Classificatory:	
1. Implicit: D supports H	1. W_1 is nonempty subset of W
2. Explicit: H is probable on the basis of D	2. If one throws the dice W times probably no W_1 is empty.
Comparative:	
1. Implicit: D supports H more than D' supports H'	1. For W times throwing the dice W_1 is of equal size as W_j.
2. H is more probable given D than H' is, given D'.	2. Throwing a coin W times W_1 is as probable as W_j.
Quantitative:	
1. The degree of support for H on the basis of D is G.	1. The ratio of the number of events in W_1 and W is Q.
2. The probability for H given D is Q.	2. The probability that the result of throwing A dice is I when throwing the dice M times is Q_1.

Table 8–3. Relationship between Boolean algebra, probabilities, and possibilities.

	Boolean algebra	*Probabilities (quantitative explicit)*	*Possibilities*
Domain	Set of (logic) statements	σ-algebra	Any universe X
Range of values membership	{0, 1}	[0, 1]	[0, 1] fuzzy: $0 < \mu < \infty$ real
Special constraints		$\sum_{\Omega} p(u) = 1$	
Union (independent, noninteractive)	max	$\sum_{w_1}$	max
Intersection	min	Π	min
Conditional equal to joint?	yes	no	often
What can be used for inference?	conditional	conditional or joint	conditional, often joint

Table 8–3 depicts some of the main mathematical differences between three areas that are similar in many respects.

Let us now return to the "duality" aspect of possibility measures and necessity measures.

A probability measure, $P(A)$, satisfies the additivity axiom, that is, $\forall A$, $B \subseteq \Omega$ for which $A \cap B = \emptyset$

$$P(A \cup B) = P(A) + P(B) \tag{8.16}$$

This measure is monotonic in the sense of condition 2 of definition 4–2. Equation (8.12) is the probabilistic equivalent to (8.1) and (8.2).

The possibility theory conditions (8.5) and (8.8) imply

$$N(A) + N(\complement A) \leq 1 \tag{8.17}$$

$$\pi(A) + \pi(\complement A) \geq 1 \tag{8.18}$$

which is less stringent than the equivalent relation

$$P(A) + P(\complement A) = 1 \tag{8.19}$$

of probability theory.

In this sense possibility corresponds more to evidence theory [Shafer

1976] than to classical probability theory, in which the probabilities of an element (a subset) are uniquely related to the probability of the contrary element (complement). In Shafer's theory, which is probabilistic in nature, this relationship is also relaxed by introducing an "upper probability" and a "lower probability" which are as "dual" to each other as are possibility and necessity.

In fact, possibility and necessity measures can be considered as limiting cases of probability measures in the sense of Shafer, that is,

$$N(A) \leq P(A) \leq \pi(A) \qquad \forall A \subseteq \Omega \tag{8.20}$$

This in turn links intuitively again with Zadeh's "possibility/probability consistency principle" mentioned in section 8.1.1.

Concerning the theories considered in this chapter we can conclude: Fuzzy set theory, possibility theory, and probability theory are no substitutes, but they complement each other. While fuzzy set theory has quite a number of "degrees of freedom" with respect to intersection and union operators, kinds of fuzzy sets (membership functions) etc., the latter two theories are well-developed and uniquely defined with respect to operation and structure. Fuzzy set theory seems to be more adaptable to different contexts. This, of course, also implies the need to adapt the theory to a context if one wants it to be an appropriate modeling tool.

Exercises

1. Let U and $\tilde{F}$ be defined as in example 8–2. Determine the possibility distribution associated with the statement "X is not a small integer."
2. Define a probability distribution and a possibility distribution which could be associated with the proposition "cars drive X mph on American freeways."
3. Compute the possibility measures (definition 8–3) for the following possibility distributions:

$$A = \{6, 7, \ldots, 13, 14\}$$

"X is an integer close to 10"

$$\pi_{\tilde{A}} = \{(8, .6), (9, .8), (10, 1), (11, .8), (12, .6)\}$$

alternatively:

$$\pi_{\tilde{B}} = \{(6, .4), (7, .5), (8, .6), (9, .8), (10, 1), (11, .8), (12, .6), (13, .5), (14, .4)\}$$

Discuss the results.

4. Discuss the relationships between general measures, fuzzy measures, probability measures, and possibility measures.
5. Determine Yager's probability of a fuzzy event for the event "*X* is an integer close to 10" as defined in exercise 3 above.
6. List examples for each of the kinds of probabilistic statements given in Table 8–2.
7. Analyze and discuss the assertion that $\bar{P}_y^*(\widetilde{A})(w)$ can be interpreted as the truth of the proposition "the probability of $\widetilde{A}$ is at most w."

II APPLICATIONS OF FUZZY SET THEORY

Applications of fuzzy set theory can already be found in many different areas. One could probably classify those applications as follows:

1. Applications to mathematics, that is, generalizations of traditional mathematics such as topology, graph theory, algebra, logic, and so on.
2. Applications to algorithms such as clustering methods, control algorithms, mathematical programming, and so on.
3. Applications to standard models such as "the transportation model," "inventory control models," "maintenance models," and so on.
4. And finally applications to real-world problems of different kinds.

In this book, the first type of "applications" will be covered by looking at fuzzy logic and approximate reasoning.

The second type of applications will be illustrated by considering fuzzy clustering, fuzzy linear programming, and fuzzy dynamic programming. The third type will be covered by looking at fuzzy versions of standard operations research models and at multicriteria approaches. The fourth type, eventually, will be illustrated on the one hand by describing OR-models as well as empirical research in chapter 14. On the other hand, chapter 10 has entirely been devoted to fuzzy control and expert systems, the area in which fuzzy set theory has probably been applied to the largest extent and also closest to real applications. This topic is treated in still more detail in the second volume of this book [Zimmermann 1987].

9 FUZZY LOGIC AND APPROXIMATE REASONING

9.1 Linguistic Variables

"In retreating from precision in the face of overpowering complexity, it is natural to explore the use of what might be called *linguistic* variables, that is, variables whose values are not numbers but words or sentences in a natural or artificial language.

The motivation for the use of words or sentences rather than numbers is that linguistic characterizations are, in general, less specific than numerical ones" [Zadeh 1973a, p. 3].

This quotation presents in a nutshell the motivation and justification for fuzzy logic and approximate reasoning. Another quotation might be added, which is much older. The philosopher B. Russell [Russell 1923] noted:

> All traditional logic habitually assumes that precise symbols are being employed. It is therefore not applicable to this terrestrial life but only to an imagined celestial existence. [Russell 1923]

One of the basic tools for fuzzy logic and approximate reasoning is the notion of a linguistic variable which in 1973 was called a *variable of higher order* rather than a fuzzy variable and defined as follows [Zadeh 1973a, p. 75].

Definition 9–1

A *linguistic variable* is characterized by a quintuple $(x, T(x), U, G, \tilde{M})$ in which x is the name of the variable; $T(x)$ (or simply T) denotes the term set of x, that is, the set of names of *linguistic values* of x, with each value being a fuzzy variable denoted generically by x and ranging over a universe of discourse U which is associated with the base variable u; G is a syntactic rule (which usually has the form of a grammar) for generating the name, X, of values of x; and M is a *semantic rule* for associating with each X its meaning, $\tilde{M}(X)$ which is a fuzzy subset of U. A particular X, that is a name generated by G, is called a *term*. It should be noted that the base variable u can also be vector-valued.

In order to facilitate the symbolism in what follows, some symbols will have two meanings wherever clarity allows this: x will denote the name of the variable ("the label") and the generic name of its values. The same will be true for X, and $\tilde{M}(X)$.

Example 9–1 [Zadeh 1973a, p. 77]

Let X be a linguistic variable with the label "Age" (i.e., the label of this variable is "Age" and the values of it will also be called "Age") with $U = [0, 100]$. Terms of this linguistic variable, which are again fuzzy sets, could be called "old," "young," "very old," and so on. The base-variable u is the age in years of life. $\tilde{M}(X)$ is the rule that assigns a meaning, that is, a fuzzy set, to the terms.

$$\tilde{M}(\text{old}) = \{(u, \mu_{\text{old}}(u)) \mid u \in [0, 100]\}$$

where

$$\mu_{\text{old}}(u) = \begin{cases} 0 & u \in [0, 50] \\ \left(1 + \left(\frac{u-50}{5}\right)^{-2}\right)^{-1} & u \in [50, 100] \end{cases}$$

$T(x)$ will define the term set of the variable x, for instance, in this case,

$$T(\text{Age}) = \{\text{old, very old, not so old, more or less young, quite young, very young}\}$$

where $G(x)$ is a rule which generates the (labels of) terms in the term set. Figure 9–1 sketches the above-mentioned relationships.

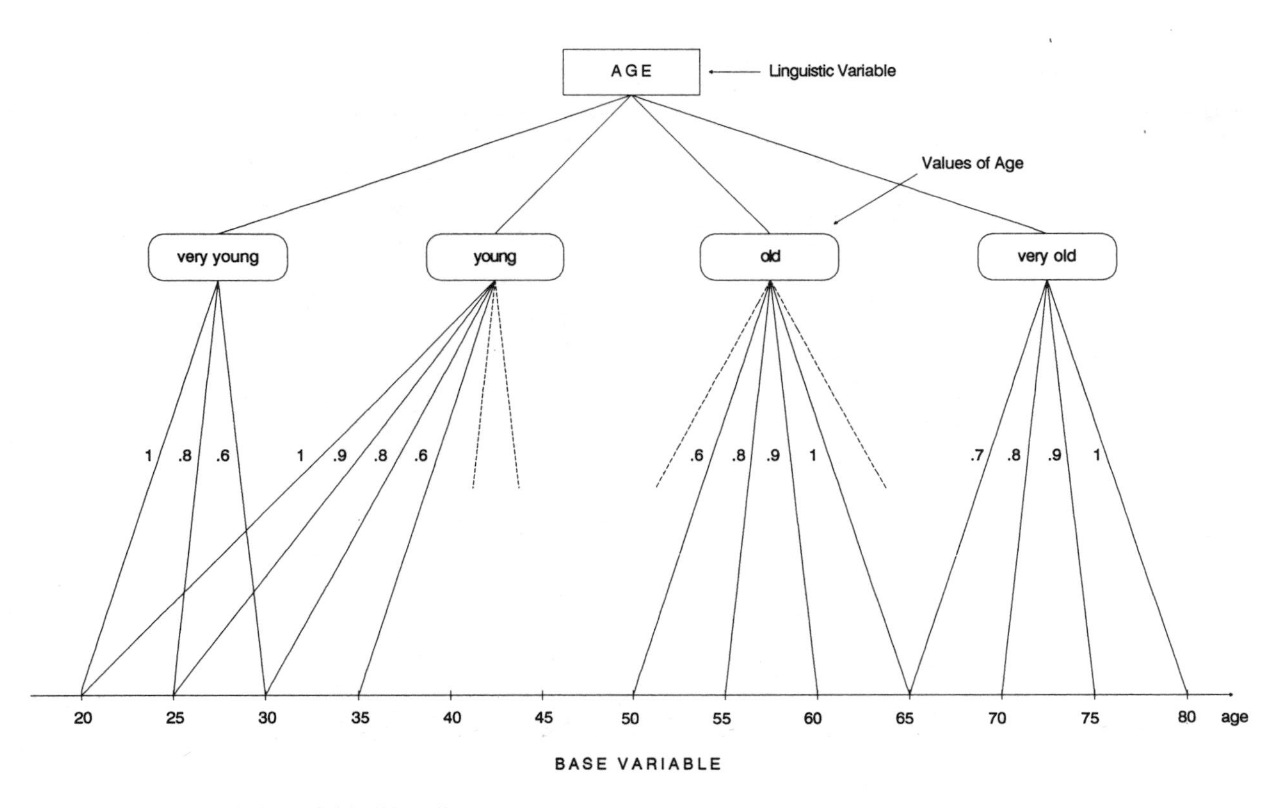

Figure 9–1. Linguistic variable ''Age.''

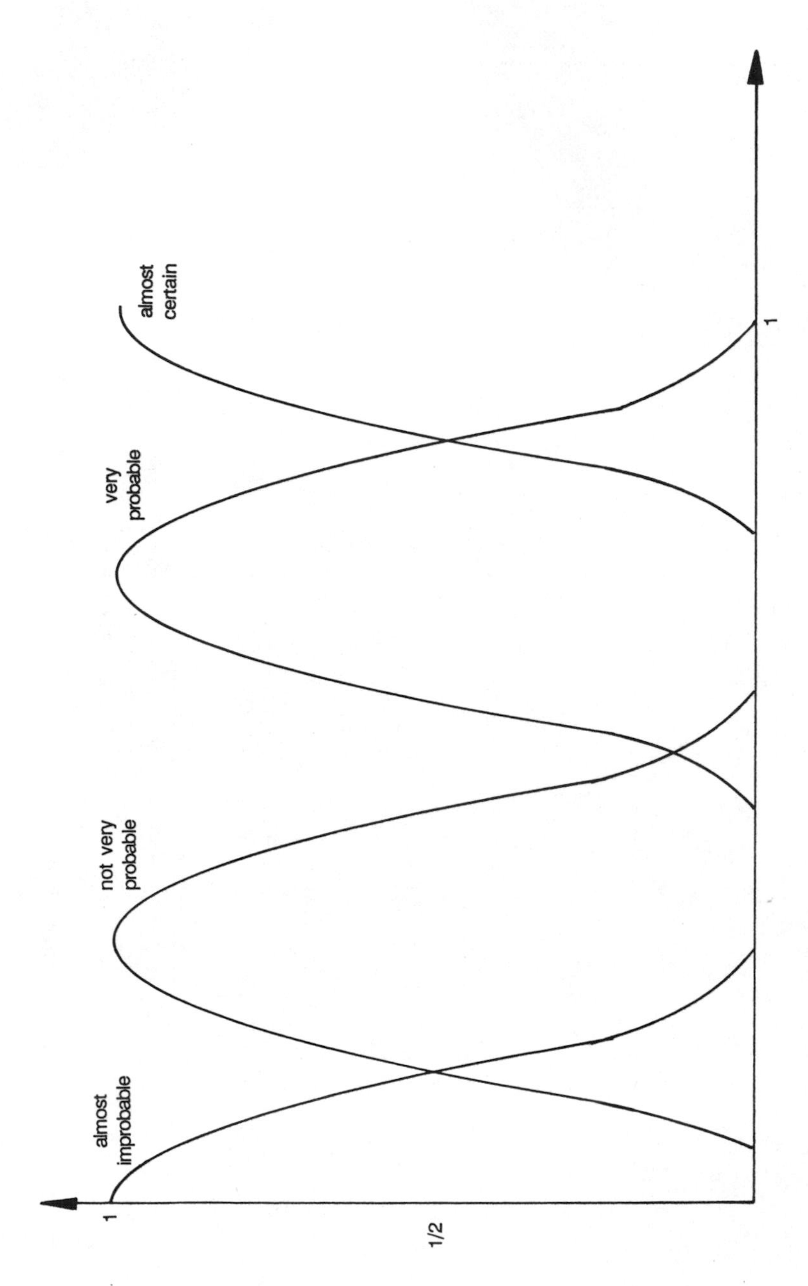

Figure 9–2. Linguistic variable "Probability."

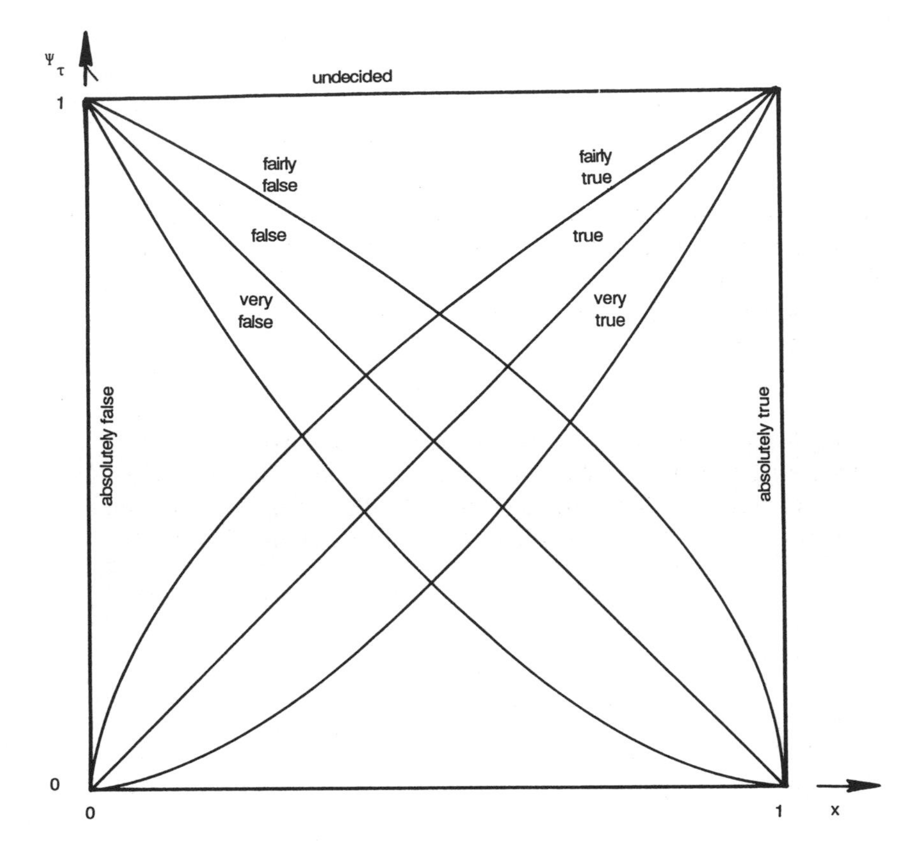

Figure 9–3. Linguistic variable "Truth."

Two linguistic variables of particular interest in fuzzy logic and in (fuzzy) probability theory are the two linguistic variables "Truth" and "Probability." The linguistic variable "Probability" is depicted exemplarily in figure 9–2.

The term set of the linguistic variable "Truth" has been defined differently by different authors. Baldwin [1979, p. 316] defines some of the terms as shown in figure 9–3. Here

$$\mu_{\text{very true}}(\nu) = (\mu_{\text{true}}(\nu))^2 \qquad \nu \in [0, 1]$$
$$\mu_{\text{fairly true}}(\nu) = (\mu_{\text{true}}(\nu))^{1/2} \qquad \nu \in [0, 1]$$

and so on. Zadeh [1973a, p. 99] suggests for the term *true* the membership function

$$\mu_{\text{true}}(v) = \begin{cases} 0 & \text{for } 0 \leq v \leq a \\ 2 \cdot \left(\dfrac{v-a}{1-a}\right)^2 & \text{for } a \leq v \leq \dfrac{a+1}{2} \\ 1 - 2 \cdot \left(\dfrac{v-1}{1-a}\right)^2 & \text{for } \dfrac{a+1}{2} \leq v \leq 1 \end{cases}$$

where $v = (1 + a)/2$ is called the *crossover* point, and $a \in [0, 1]$ is a parameter that indicates the subjective judgment about the minimum value of v in order to consider a statement as "true" at all.

The membership function of "false" is considered as the mirror image of "true," that is,

$$\mu_{\text{false}}(v) = \mu_{\text{true}}(1 - v) \qquad 0 \leq v \leq 1$$

Figure 9–4 [Zadeh 1973a, p. 99] shows the terms *true* and *false*.

Of course the membership functions of true and false, respectively, can also be chosen from the finite universe of truth values. The term set of the linguistic variable "Truth" is then defined as [Zadeh 1973a, p. 99]

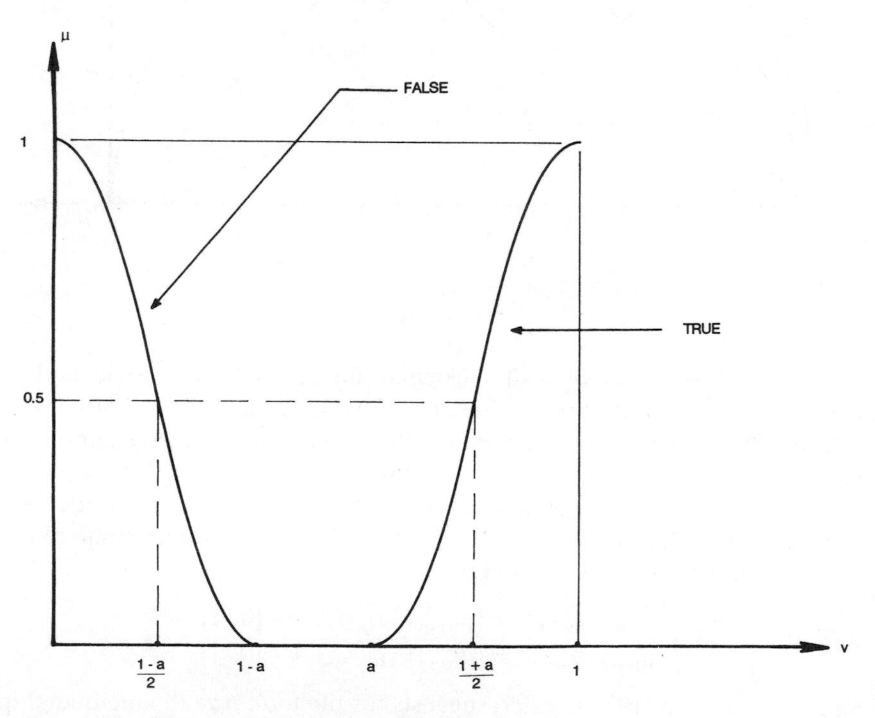

Figure 9–4. Terms "True" and "False."

$T(\text{Truth}) = \{$true, not true, very true, not very true, ..., false, not false, very false, ..., not very true and not very false, ...$\}$

The fuzzy sets (possibility distribution) of those terms can essentially be determined from the term *true* or the term *false* by applying appropriately the below mentioned modifiers (hedges).

Definition 9–2

A linguistic variable x is called *structured* if the term set $T(x)$ and the meaning $\widetilde{M}(x)$ can be characterized algorithmically. For a structured linguistic variable, $\widetilde{M}(x)$ and $T(x)$ can be regarded as algorithms which generate the terms of the term set and associate meanings with them. Before we illustrate this by an example we need to define what we mean by a "hedge" or a "modifier."

Definition 9–3

A *linguistic hedge or a modifier* is an operation that modifies the meaning of a term or, more generally, of a fuzzy set. If $\widetilde{A}$ is a fuzzy set then the modifier m generates the (composite) term $\widetilde{B} = m(\widetilde{A})$.

Mathematical models frequently used for modifiers are:

$$\text{concentration: } \mu_{\text{con}(\widetilde{A})}(u) = (\mu_{\widetilde{A}}(u))^2$$
$$\text{dilation: } \mu_{\text{dil}(\widetilde{A})}(u) = (\mu_{\widetilde{A}}(u))^{1/2}$$

contrast intensification:

$$\mu_{\text{int}(\widetilde{A})}(u) = \begin{cases} 2(\mu_{\widetilde{A}}(u))^2 & \text{for } \mu_{\widetilde{A}}(u) \in [0, .5] \\ 1 - 2(1 - \mu_{\widetilde{A}}(u))^2 & \text{otherwise} \end{cases}$$

Generally the following linguistic hedges (modifiers) are associated with above-mentioned mathematical operators.

If $\widetilde{A}$ is a term (a fuzzy set) then

$$\text{very } \widetilde{A} = \text{con } (\widetilde{A})$$
$$\text{more or less } \widetilde{A} = \text{dil } (\widetilde{A})$$
$$\text{plus } \widetilde{A} = \widetilde{A}^{1.25}$$
$$\text{slightly } \widetilde{A} = \text{int [plus } \widetilde{A} \text{ and not (very } \widetilde{A})]$$

were "and" is interpreted possibilistically.

Example 9–2 [Zadeh 1973a, p. 83]

Let us reconsider from example 9–1 the linguistic variable "Age." The term set shall be assumed to be

$$T(\text{Age}) = \{\text{old, very old, very very old, } \ldots\}$$

The term set can now be generated recursively by using the following rule (algorithm):

$$T^{i+1} = \{\text{old}\} \cup \{\text{very } T^i\}$$

that is,

$$\begin{aligned} T^0 &= \emptyset \\ T^1 &= \{\text{old}\} \\ T^2 &= \{\text{old, very old}\} \\ T^3 &= \{\text{old, very old, very very old}\} \end{aligned}$$

For the semantic rule we only need to know the meaning of "old" and the meaning of the modifier "very" in order to determine the meaning of an arbitrary term of the term set. If one defines "very" as the concentration, then the terms of the term set of the structured linguistic variable "Age" can be determined, given that the membership function of the term "old" is known.

Definition 9–4 [Zadeh 1973a, p. 87]

A *Boolean linguistic variable* is a linguistic variable whose terms, X, are Boolean expressions in variables of the form X_p, $m(X_p)$ where X_p is a primary term and m is a modifier. $m(X_p)$ is a fuzzy set resulting from acting with m on X_p.

Example 9–3

Let "Age" be a Boolean linguistic variable with the term set

$$T(\text{Age}) = \{\text{young, not young, old, not old, very young, not young and not old, young or old, } \ldots\}$$

Identifying "and" with the intersection, "or" with the union, "not" with the complementation, and "very" with the concentration we can derive the meaning of different terms of the term set as follows:

$$\begin{aligned}
\widetilde{M}(\text{not young}) &= \neg\ \text{young} \\
\widetilde{M}(\text{not very young}) &= \neg\ (\text{young})^2 \\
\widetilde{M}(\text{young or old}) &= \text{young} \cup \text{old} \\
&\text{etc.}
\end{aligned}$$

Given the two fuzzy sets (primary terms)

$$\widetilde{M}(\text{young}) = \{(u, \mu_{\text{young}}(u)) \mid u \in [0, 100]\}$$

where

$$\mu_{\text{young}}(u) = \begin{cases} 1 & \mu \in [0, 25] \\ \left(1 + \left(\frac{u-25}{5}\right)^2\right)^{-1} & u \in [25, 100] \end{cases}$$

and

$$\widetilde{M}(\text{old}) = \{(u, \mu_{\text{old}}(u)) \mid u \in [0, 100]\}$$

where

$$\mu_{\text{old}}(u) = \begin{cases} 0 & u \in [0, 50] \\ \left(1 + \left(\frac{u-50}{5}\right)^{-2}\right)^{-1} & u \in (50, 100] \end{cases}$$

then the membership function of the term "young or old" would, for instance, be

$$\mu_{\text{young or old}}(u) = \begin{cases} 1 & \text{if } u \in [0, 25] \\ \left(1 + \left(\frac{u-25}{5}\right)^2\right)^{-1} & \text{if } u \in [25, 50] \\ \max\left\{\left(1 + \left(\frac{u-25}{5}\right)^2\right)^{-1}, \left(1 + \left(\frac{u-50}{5}\right)^{-2}\right)^{-1}\right\} & \text{if } u \in [50, 100] \end{cases}$$

9.2 Fuzzy Logic

9.2.1 Classical Logics Revisited

Logics as bases for reasoning can be distinguished essentially by their three topic-neutral (context-independent) items: truth values, vocabulary (operators), and reasoning procedure (tautologies, syllogisms).

In Boolean logic, truth values can be 0 (false) or 1 (true) and by means of these truth values the vocabulary (operators) is defined via truth tables.

Let us consider two statements, A and B, either of which can be true or false, that is, have the truth value 1 or 0. We can construct the following truth tables:

A	B	$\wedge$	$\vee$	$x\vee$	$\Rightarrow$	$\Leftrightarrow$	?
1	1	1	1	0	1	1	1
1	0	0	1	1	0	0	1
0	1	0	1	1	1	0	0
0	0	0	0	0	1	1	0

There are $2^{2^2} = 16$ truth tables, each defining an operator. Assigning meanings (words) to these operators is not difficult for the first 4 or 5 columns: the first obviously characterizes the "and," the second the "inclusive or," the third the "exclusive or," and the fourth and fifth the implication and the equivalence. We will have difficulties, however, interpreting the remaining 9 columns in terms of our language. If we have three statements rather than two, this task of assigning meanings to truth tables becomes even more difficult.

So far it has been assumed that each statement, A and B, could clearly be classified as true or false. If this is no longer true then additional truth values, such as "undecided" or similar, can and have to be introduced, which leads to the many existing systems of multivalued logic. It is not difficult to see how the above-mentioned problems of two-valued logic in "calling" truth tables or operators increase as we move to multivalued logic. For only two statements and three possible truth values there are already $3^{3^2} = 729$ truth tables! The uniqueness of interpretation of truth tables, which is so convenient in Boolean logic, disappears immediately because many truth tables in three-valued logic look very much alike.

The third topic-neutral item of logical systems is the reasoning procedure itself which generally bases on tautologies such as

modus ponens: $(A \wedge (A \Rightarrow B)) \Rightarrow B$
modus tollens: $((A \Rightarrow B) \wedge \neg B) \Rightarrow \neg A$
syllogism: $((A \Rightarrow B) \wedge (B \Rightarrow C)) \Rightarrow (A \Rightarrow C)$
contraposition: $(A \Rightarrow B) \Rightarrow (\neg B \Rightarrow \neg A)$

Let us consider the modus ponens which could be intrepreted as: "If A is true and if the statement "If A is true then B is true" is also true then B is true."

The term *true* is used at different places and in two different senses: All but the last "true's" are material true's, that is, they are taken as a matter of fact, while the last "true" is a topic-neutral logical "true." In Boolean logic, however, these "true's" are all treated the same way [see Mamdani and Gaines 1981, p. xv]. A distinction between material and logical (necessary) truth is made in so called extended logics: Modal logic [Hughes and Cresswell 1968] distinguishes between necessary and possible truth, tense logic between statements that were true in the past and those that will be true in the future. Epistemic logic deals with knowledge and belief and deontic logic with what ought to be done and what is permitted to be true. Modal logic, in particular, might be a very good basis for applying different measures and theories of uncertainty as indicated in chapter 4.

Another extension of Boolean logic is predicate calculus, which is a set theoretic logic using quantifiers (all, etc.) and predicates in addition to the operators of Boolean logic.

Fuzzy logic [Zadeh 1973a, p. 101] is an extension of set theoretic multivalued logic in which the truth values are linguistic variables (or terms of the linguistic variable truth).

Since operators, like $\vee$, $\wedge$, $\neg$, $\Rightarrow$ in fuzzy logic are also defined by using truth tables, the extension principle can be applied to derive definitions of the operators. So far, possibility theory (see section 8.1) has primarily been used in order to define operators in fuzzy logic, even though other operators have also been investigated (see for instance Mizumoto and Zimmermann [1982], and could also be used. In this book we will limit considerations to possibilistic interpretations of linguistic variables and we will also stick to the original proposals of Zadeh [1973a]. To the interested reader, however, we suggest supplemental study of alternative approaches such as [Baldwin 1979], [Baldwin and Pilsworth 1980], [Giles 1979, 1980], and others.

If $v(A)$ is a point in $V=[0, 1]$, representing the truth value of the proposition "u is A" or simply A, then the truth value of not A is given by

$$v(\text{not } A) = 1 - v(A)$$

Definition 9–5

If $\tilde{v}(A)$ is a normalized fuzzy set, $\tilde{v}(A) = \{(v_i, \mu_i) \mid i = 1, \ldots, n, v_i \in [0, 1]\}$ then by applying the extension principle, the *truth value of $\tilde{v}$(not A)* is defined as

$$\tilde{v}(\text{not } A) = \{(1 - v_i, \mu_i) \mid i = 1, \ldots, n, \quad v_i \in [0, 1]\}$$

In particular "false" is interpreted as "not true," that is

$$\tilde{v}(\text{false}) = \{(1 - v_i, \mu_i) \mid i = 1, \ldots, n, \quad v_i \in [0, 1]\}$$

Example 9–4

Let us consider the terms *true* and *false*, respectively, defined as the following possibility distributions:

$$\tilde{v}(\text{true}) = \{(.5, .6), (.6, .7), (.7, .8), (.8, .9), (.9, 1), (1, 1)\}$$
$$\tilde{v}(\text{false}) = \tilde{v}(\text{not true}) = \{(.5, .6), (.4, .7), (.3, .8), (.2, .9), (.1, 1), (0, 1)\}$$

then

$$\tilde{v}(\text{very true}) = \{(.5, .36), (.6, .49), (.7, .64), (.8, .81), (.9, 1), (1, 1)\}$$
$$\tilde{v}(\text{very false}) = \{(.5, .36), (.4, .49), (.3, .64), (.2, .81), (.1, 1), (0, 1)\}$$

It has already been mentioned that fuzzy logic is essentially considered as an application of possibility theory to logic. Hence the logical operators "and," "or," and "not" are defined accordingly.

Definition 9–6

For numerical truth values $v(A)$ and $v(B)$ the logical operations *and*, *or*, *not* and *implied* are defined as

$$\begin{aligned}
v(A) \wedge v(B) &= v(A \text{ and } B) = \{(v, \min\{\mu_A(v), \mu_B(v)\})\} \\
v(A) \vee v(B) &= v(A \text{ or } B) = \{(v, \max\{\mu_A(v), \mu_B(v)\})\} \\
\neg v(A) &= \{(v, 1 - \mu_A(v))\} \\
v(A) \Rightarrow v(B) &= v(A \Rightarrow B) = \neg v(A) \vee v(B) \\
&= \{(v, \max\{(1 - \mu_A(v)), \mu_B(v)\})\}
\end{aligned}$$

If

$$\begin{aligned}
\tilde{v}(A) &= \{(v_i, \alpha_i)\}, \quad \alpha_i \in [0, 1], v_i \in [0, 1] \\
\tilde{v}(B) &= \{(w_j, \beta_j)\}, \quad \beta_j \in [0, 1], w_j \in [0, 1] \\
& \qquad i = 1, \ldots, n; j = 1, \ldots, m
\end{aligned}$$

then

$$\tilde{v}(A \text{ and } B) = \tilde{v}(A) \wedge \tilde{v}(B) = \{(u = \min\{v_i, w_j\}, \max_{u=\min\{v_i, w_j\}} \min\{\alpha_i, \beta_j\}) \mid i = 1, \ldots, n; j = 1, \ldots, m\}$$

(This is equivalent to the intersection of two type 2 fuzzy sets). The other operators are defined accordingly.

Example 9–5

Let $\tilde{v}(A) = \text{true} = \{(.5, .6), (.6, .7), (.7, .8), (.8, .9), (.9, 1), (1, 1)\}$

then

$$\neg\tilde{v}(A) = \{(0, 1), (.1, 1), (.2, 1), (.3, 1), (.4, 1), (.5, .4), (.6, .3), (.7, .2), (.8, .1)\}$$

9.2.2 Truth Tables and Linguistic Approximation

As mentioned at the beginning of this section binary connectives (operators) in classical two- and many-valued logics are normally defined by the tabulation of truth values in truth tables. In fuzzy logic the number of truth values is, in general, infinite. Hence tabulation of the truth values for operators is not possible. We can, however, tabulate truth values, that is, terms of the linguistic variable "Truth" for a finite number of terms, such as true, not true, very true, false, more or less true, and so on.

Zadeh [1973a, p. 109] suggests truth tables for the determination of truth values for operators using a four-valued logic including the truth values true, false, undecided, and unknown. "Unknown" is then interpreted as "true or false" ($T + F$) and "undecided" is denoted by $\ominus$.

Extending the normal Boolean logic with truth values true (1) and false (0) to a (fuzzy) three-valued logic (true = T, false = F, unknown = $T + F$), with a universe of truth values being two-valued (true and false) we obtain the following truth tables where the first column contains the truth values for a statement A and the first row those for a statement B [Zadeh 1973a, p. 116]:

$\wedge$	T	F	$T+F$
T	T	F	$T+F$
F	F	F	F
$T+F$	$T+F$	F	$T+F$

Truth table for "and"

$\vee$	T	F	$T+F$
T	T	T	T
F	T	F	$T+F$
$T+F$	T	$T+F$	$T+F$

Truth table for "or"

	$\neg$
T	F
F	T
$T+F$	$T+F$

Truth table for "not"

If the number of truth values (terms of the linguistic variable truth) increases one can still "tabulate" the truth table for operators by using definition 9–6 as follows: Let us assume, that the ith row of the table represents "not true" and the jth column "more or less true." The $(i, j)^{\text{th}}$ entry in the truth table for "and" would then contain the entry for "not true $\wedge$ more or less true." The resulting fuzzy set would, however, most likely not correspond to any fuzzy set assigned to the terms of the term set of "truth." In this case one could try to find the fuzzy set of the term which is most similar to the fuzzy set resulting from the computations. Such a term would then be called *linguistic approximation*. This is an analogy to statistics, where empirical distribution functions are often approximated by well-known standard distribution functions.

Example 9–6

Let $V = \{0, .1, .2, \ldots, 1\}$ be the universe,
true $= \{(.8, .9), (.9, 1), (1, 1)\}$,
more or less true $= \{(.6, .2), (.7, .4), (.8, .7), (.9, 1), (1, 1)\}$, and
almost true $= \{(.8, .9), (.9, 1), (1, .8)\}$.

Let "more or less true" be the ith row and "almost true" the jth column of the truth table for "or."

Then "more or less true $\vee$ almost true" is the $(i, j)^{\text{th}}$ entry in the table:

more or less true $\vee$ almost true
$= \{(.6, .2), (.7, .4), (.8, .7), (.9, 1), (1, 1)\} \vee \{(.8, .9), (.9, 1), (1, .8)\}$
$= \{(.6, .2), (.7, .4), (.8, .9), (.9, 1), (1, 1)\}$

Now we can approximate the right-hand side of this equation by

$$\text{true} = \{(.8, .9), (.9, 1), (1, 1)\}$$

This yields

$$\text{"more or less true} \vee \text{almost true"} \approx \text{"true."}$$

Baldwin [1979] suggests another version of fuzzy logic—fuzzy truth tables, and their determination: The truth values on which he bases his suggestions were shown graphically in figure 9–3. They were defined as

$$\begin{aligned}
\text{true} &= \{(v,\ \mu_{\text{true}}(v) = v) \mid v \in [0, 1]\} \\
\text{false} &= \{(v,\ \mu_{\text{false}}(v) = 1 - \mu_{\text{true}}(v)) \mid v \in [0, 1]\} \\
\text{very true} &= \{(v,\ (\mu_{\text{true}}(v))^2) \mid v \in [0, 1]\} \\
\text{fairly true} &= \{(v,\ (\mu_{\text{true}}(v))^{1/2}) \mid v \in [0, 1]\} \\
\text{undecided} &= \{(v,\ 1) \mid v \in [0, 1]\}
\end{aligned}$$

Very false and fairly false were defined correspondingly, and

$$\text{absolutely true} = \{(v,\ \mu_{at}(v)) \mid v \in [0, 1]\}$$

$$\text{with } \mu_{at}(v) = \begin{cases} 1 & \text{for } v = 1 \\ 0 & \text{otherwise} \end{cases}$$

$$\text{absolutely false} = \{(v,\ \mu_{af}(v)) \mid v \in [0, 1]\}$$

$$\text{with } \mu_{af}(v) = \begin{cases} 1 & \text{for } v = 0 \\ 0 & \text{otherwise} \end{cases}$$

Hence

$$\begin{aligned}
&(\text{very})^k\text{true} \rightarrow \text{absolutely true as } k \rightarrow \infty \\
&(\text{very})^k\text{false} \rightarrow \text{absolutely false as } k \rightarrow \infty \\
&(\text{fairly})^k\text{true} \rightarrow \text{undecided as } k \rightarrow \infty \\
&(\text{fairly})^k\text{false} \rightarrow \text{undecided as } k \rightarrow \infty
\end{aligned}$$

Using figure 9–3 and the interpretations of "and" and "or" as minimum and maximum, respectively, the following truth table results [Baldwin 1979, p. 318]:

$v(P)$	$v(Q)$	$v(P$ and $Q)$	$v(P$ or $Q)$
false	false	false	false
true	false	false	true
true	true	true	true
undecided	false	false	undecided
undecided	true	undecided	true
undecided	undecided	undecided	undecided
true	very true	true	very true
true	fairly true	fairly true	true

Some more considerations and assumptions are needed to derive the truth table for the implication. Baldwin considers his fuzzy logic to rest on two pillars: the denumerably infinite multivalued logic system of Lukasiewicz logic and fuzzy set theory. "Implication statements are treated by a composition of fuzzy truth value restrictions with a Lukasiewicz logic implication relation on a fuzzy truth space. Set theoretic considerations are used to obtain fuzzy truth value restrictions from conditional fuzzy linguistic statements using an inverse truth functional modification procedure. Finally true functions modification is used to obtain the final conclusion" [Baldwin 1979, p. 309].

9.3 Approximate Reasoning

We already mentioned that in traditional logic the main tools of reasoning are tautologies such as, for instance, the modus ponens, that is $(A \wedge (A \Rightarrow B)) \Rightarrow B$ or

Premise	A is true
Implication	If A then B
Conclusion	B is true

A and B are statements or propositions (crisply defined) and the B in the conditional statement is identical to the B of the conclusion.

On the basis of what has been said in sections 9.1 and 9.2, two quite obvious generalizations of the modus ponens are

1. to allow statements that are characterized by fuzzy sets.
2. to relax (slightly) the identity of the "B's" in the implication and the conclusion.

This version of the modus ponens is then called "generalized modus ponens" [Zadeh 1973a, p. 56; Mizumoto et al. 1979, Mamdani 1977a].

Example 9–7

Let $\tilde{A}, \tilde{A}', \tilde{B}, \tilde{B}'$ be fuzzy statements, then the generalized modus ponens reads

Premise:	x is $\tilde{A}'$
Implication:	If x is $\tilde{A}$ then y is $\tilde{B}$
Conclusion:	y is $\tilde{B}'$

For instance [Mizumoto and Zimmermann 1982].

Premise:	This tomato is very red.
Implication:	If a tomato is red then the tomato is ripe.
Conclusion:	This tomato is very ripe.

In 1973 Zadeh suggested the compositional rule of inference for the above-mentioned type of fuzzy conditional inference. In the meantime other authors (for instance, Baldwin [1979], Baldwin and Pilsworth [1980], Baldwin and Guild [1980], Mizumoto et al. [1979], Mizumoto and Zimmermann [1982], Tsukamoto [1979], have suggested different methods and investigated also the modus tollens, syllogism, and contraposition. In this book, however, we shall restrict considerations to Zadeh's compositional rule of inference.

Definition 9–7 [Zadeh 1973a, p. 148]

Let $\tilde{R}(x)$, $\tilde{R}(x, y)$ and $\tilde{R}(y)$, $x \in X$, $y \in Y$ be fuzzy relations in X, $X \times Y$, and Y respectively, which act as fuzzy restrictions on x, (x, y), and y, respectively. Let $\tilde{A}$ and $\tilde{B}$ denote particular fuzzy sets in X and $X \times Y$. Then the *compositional rule of inference* asserts, that the solution of the relational assignment equations (see definition 8–1) $\tilde{R}(x) = \tilde{A}$ and $\tilde{R}(x, y) = \tilde{B}$ is given by $\tilde{R}(y) = \tilde{A} \circ \tilde{B}$, where $\tilde{A} \circ \tilde{B}$ is the composition of $\tilde{A}$ and $\tilde{B}$.

Example 9–8

Let the universe be $X = \{1, 2, 3, 4\}$.
$\tilde{A}$ = little = $\{(1, 1), (2, .6), (3, .2), (4, 0)\}$.
$\tilde{R}$ = "approximately equal" be a fuzzy relation defined by

$\tilde{R}$:	1	2	3	4
1	1	.5	0	0
2	.5	1	.5	0
3	0	.5	1	.5
4	0	0	.5	1

For the formal inference denote

$$\tilde{R}(x) = \tilde{A}, \quad \tilde{R}(x, y) = \tilde{B}, \quad \text{and } \tilde{R}(y) = \tilde{A} \circ \tilde{B}$$

Applying the max-min composition for computing $\tilde{R}(y) = \tilde{A} \circ \tilde{B}$ yields

$$\begin{aligned} \tilde{R}(y) &= \max_x \min \{\mu_{\tilde{A}}(x), \mu_{\tilde{R}}(x, y)\} \\ &= \{(1, 1), (2, .6), (3, .5), (4, .2)\} \end{aligned}$$

A possible interpretation of the inference may be the following:

x is little
x and y are approximately equal

---

y is more or less little

A direct application of approximate reasoning is the fuzzy algorithm (an ordered sequence of instructions in which some of the instructions may contain labels of fuzzy sets) and the fuzzy flow chart. We shall consider both in more detail in chapter 10. Here, however, we shall briefly describe fuzzy (formal) languages.

9.4 Fuzzy Languages

Fuzzy languages are formal languages on the basis of fuzzy logic and approximate reasoning. Several of them have been developed by now, such as LPL [Adamo 1980], FLIP [Giles 1980], Fuzzy Planner [Kling 1973], and others. They are based on LP1, FORTRAN, LISP, and other programming languages and they differ in their content as well as their aims. Here we shall sketch a meaning representing language developed by Zadeh [Zadeh 1981a].

PRUF (acronym for *P*ossibilistic *R*elational *U*niversal *F*uzzy) is a meaning representation language for natural languages and is based on possibility theory. PRUF may be employed as a language for the presentation of imprecise knowledge and as a means of making precise fuzzy propositions expressed in a natural language. In essence, PRUF bears the same relationship to fuzzy logic that predicate calculus does to two-valued logic. Thus it serves to translate a set of premises expressed in natural language into expressions in PRUF to which the rules of inference of fuzzy logic or approximate reasoning may be applied. This yields other expressions in PRUF which can then be retranslated into natural language and become the conclusions inferred from the original premises.

The main constituents of PRUF are

1. a collection of translation rules, and
2. a set of rules of inference.

The latter corresponds essentially to fuzzy logic and approximate reasoning such as described in sections 9.2 and 9.3. The former will be described in more detail after the kind of representation in PRUF has been described and some more definitions introduced.

In definition 8–1 the relational assignment equation was defined. In PRUF a possibility distribution π_x is assigned via the

$$\text{possibility assignment equation (PAE):} \quad \pi_x \hat{=} \tilde{F}$$

to the fuzzy set $\tilde{F}$. The PAE corresponds to a proposition of the form "N is $\tilde{F}$" where N is the name of a variable, a fuzzy set, a proposition, or an object. For simplicity the PAE will be written as in chapter 8 as

$$\pi_x = \tilde{F}$$

Example 9–9

Let N be the proposition "Peter is old," then N (the variable) is called "Peter," $X \in [0, 100]$ is the linguistic variable "Age," "old" is, for instance, a term of the term set of "Age" and

$$\text{Peter is old} \rightarrow \pi_{\text{Age(Peter)}} = \text{old}$$

where $\rightarrow$ stands for "translates into."

There are two special types of possibility distributions which will be needed later.

Definition 9–8

The possibility distributions π_I with

$$\pi_I(u) = 1 \qquad \text{for } u \in U$$

is called the *unity possibility distribution* π_I, and with

$$\pi_{\perp}(v) = v \qquad \text{for } v \in [0, 1]$$

is defined the *unitary possibility distribution function* [Zadeh 1981a, p. 10].

In chapter 6 (definition 6–4) the projection of a binary fuzzy relation was defined. This definition holds not only for binary relations and numerical values of the related variables but also for linguistic variables.

Different fuzzy relations in a product space $U_1 \times U_2 \times \ldots \times U_n$ can have identical projections on $U_{i_1} \times \ldots \times U_{i_k}$. Given a fuzzy relation $\widetilde{R}_q$ in $U_{i_1} \times \ldots \times U_{i_k}$ there exists, however, a unique relation $\widetilde{R}_{qL}$ which contains all other relations whose projection on $U_{i_1} \times \ldots \times U_{i_k}$ is $\widetilde{R}_q$. $\widetilde{R}_{qL}$ is then called the cylindrical extension of $\widetilde{R}_q$, the latter is the base of $\widetilde{R}_{qL}$ (see definitions 6–4, 6–5).

In PRUF the operation "particularization" is also important [Zadeh 1981a, p. 13]: "By the particularization of a fuzzy relation or a possibility distribution which is associated with a variable $\widetilde{X} \triangleq (\widetilde{X}_1, \ldots, \widetilde{X}_n)$, is meant the effect of specification of the possibility distributions of one or more subvariables (terms) of $\widetilde{X}$. Particularization in PRUF may be viewed as the result of forming the conjunction of a proposition of the form "$\widetilde{X}$ is $\widetilde{F}$," where $\widetilde{X}$ is an n-ary variable with particularizing propositions of the form "$\widetilde{X}_s = \widetilde{G}$," where $\widetilde{X}_s$ is a subvariable (term) of $\widetilde{X}$ and $\widetilde{F}$ and $\widetilde{G}$, respectively, are fuzzy sets in $U_1 \times U_2 \times \ldots U_n$ and $U_{i_l} \times \ldots \times U_{i_k}$, respectively."

Definition 9–9 [Zadeh 1981a, p. 13]

Let $\tilde{\pi}_X \triangleq \tilde{\pi}_{(X_1,\ldots,X_n)} = \widetilde{F}$ and $\tilde{\pi}_{X_s} \triangleq \tilde{\pi}_{(X_{ij},\ldots,X_{ik})} = \widetilde{G}$ be possibility distributions induced by the propositions "$\widetilde{X}$ is $\widetilde{F}$" and "$\widetilde{X}_s$ is $\widetilde{G}$," respectively. The *particularization* of $\tilde{\pi}_X$ by $\widetilde{X}_s = \widetilde{G}$ is denoted by $\tilde{\pi}_X(\tilde{\pi}_{X_s} = \widetilde{G})$ and is defined as the intersection of $\widetilde{F}$ and $\widetilde{G}$, that is,

$$\tilde{\pi}_x(\tilde{\pi}_{x_s} = \widetilde{G}) = \widetilde{F} \cap \widetilde{G}'$$

where $\widetilde{G}'$ is the cylindrical extension of $\widetilde{G}$.

Example 9–10

Consider the proposition "Porsche is an attractive car," where attractiveness of a car as a function of mileage and top speed is defined in the following table.

Attractive cars	Top speed (MpH)	Mileage (MpG)	μ
	60	30	.4
	60	35	.5
	60	40	.6
	70	30	.7
	85	25	.7
	90	25	.8
	95	25	.9
	100	20	1.0
	110	15	1.0

A particularizing proposition is "Porsche is a fast car," in which "fast" is defined in the following table:

Fast cars	Top speed (MpH)	μ
	60	.4
	70	.6
	85	.7
	90	.8
	95	.9
	100	.95
	110	1.0

"Porsche is an attractive car" can equivalently be written as "Porsche is a fast car," that is, "Top speed (Porsche) is high" and "mileage (Porsche) is high."

Using definition 9–9, the particularized relation *attractive* (π_{speed} = Fast) can readily be computed as shown in the next table:

Attractive cars	Top speed	Mileage	μ
	60	30	.4
	60	35	.4
	60	40	.4
	70	30	.6
	85	25	.7
	90	25	.8
	95	25	.9
	100	20	0.95
	110	15	1

Translation Rules in PRUF. The following types of fuzzy expressions will be considered:

1. Fuzzy propositions such as "All students are young," "*X* is much larger than *Y*," and "If Hans is healthy then Biggi is happy."
2. Fuzzy descriptors such as tall men, rich people, small integers, most, several, or few.
3. Fuzzy questions.

Fuzzy questions are reformulated in such a way that additional translation rules for questions are unnecessary. Questions such as "How *A* is *B*?" will be expressed in the form "*B* is ?*A*," where *B* is the body of the question and "?*A*" indicates the form of an admissible answer, which can be: a possibility distribution (indicated as π); a truth value (indicated as τ); a probability value (indicated as λ); a possibility value (indicated as ω).

The question "How tall is Paul?" to which a possibility distribution is expected as an answer, is phrased "Paul is ?π" (rather than "How tall is Paul ?π). "Is it true that Katrin is pretty?" would than be expressed as "Katrin is pretty ?τ" and "Where is the car ?w" as "The car is ?w."

PRUF is an intentional language, that is, an expression in PRUF is supposed to convey the intended rather than the literal meaning of the corresponding expression in a natural language. Transformations of expressions are also intended to be meaning-preserving. Translation rules are applied singly or in combination to yield an expression, *E*, in PRUF which is a translation of a given expression, *e*, in a natural language.

The most important basic categories of translation rules in PRUF are:

Type I	Rules pertaining to modification
Type II	Rules pertaining to composition
Type III	Rules pertaining to quantification
Type IV	Rules pertaining to qualification

Examples of propositions to which these rules apply are the following [Zadeh 1981a, p. 29].

Type I *X* is very small.
X is much larger than *Y*.
Eleanor was very upset.
The man with the blond hair is very tall.

Type II *X* is small and *Y* is large. (conjunctive composition)
X is small or *Y* is large. (disjunctive composition)

If X is small then Y is large. (conditional composition)
If X is small then Y is large else Y is very large. (conditional and conjunctive composition)

Type III Most Swedes are tall.
Many men are much taller than most men.
Most tall men are very intelligent.

Type IV Abe is young is not very true. (truth qualification)
Abe is young is quite probable. (probability qualification)
Abe is young is almost impossible. (possibility qualification)

Rules of Type I

This type of rule concerns the modification of fuzzy-set representing propositions by means of hedges or modifiers (see definition 9–3).

If the proposition

$$P \triangleq N \quad \text{is} \quad \tilde{F}$$

translates into the possibility assignment equation

$$\pi_{(X_1, \ldots, X_n)} = \tilde{F}$$

then the translation of the modified proposition

$$P^+ \triangleq N \quad \text{is} \quad m\tilde{F} \quad \text{is}$$
$$\pi_{(X_1, \ldots, X_n)} = \tilde{F}^+$$

where $\tilde{F}^+$ is a modification of $\tilde{F}$ by the modifier m. As mentioned in chapter 9.1 the modifier "very" was defined to be the squaring operation, "more or less" the dilation and so on.

Example 9–11

Let p be the proposition "Hans is old," where "old" may be the fuzzy set defined in example 9–1. The translation of $p^+ \triangleq$ "Hans is very old," assuming "very" to be modeled by squaring, would then be

$$\pi_{\text{Age(Hans)}} = (\text{old})^2 = \{(u, \mu_{(\text{old})^2}(u)) \mid u \in [0, 100]\}$$

where

$$\mu_{(\text{old})^2}(u) = \begin{cases} 0 & u \in [0, 50] \\ \left(\left(1 + \left(\dfrac{u - 50}{5}\right)^{-2}\right)^{-1}\right)^2 & u \in (50, 100] \end{cases}$$

Rules of Type II

Rules of type II translate compound statements of the type

$$p = q * r$$

where $*$ denotes a logical connective, for example, and (conjunction) or (disjunction), if ... then (implication), and so on. Here essentially the definitions of connectives defined in section 9.1 and 9.2 are used in PRUF.

If the statements q and r are

$$q \triangleq M \quad \text{is} \quad \tilde{F} \rightarrow \pi_{(X_1, \ldots, X_n)} = \tilde{F}$$
$$r \triangleq N \quad \text{is} \quad \tilde{G} \rightarrow \pi_{(Y_1, \ldots, Y_n)} = \tilde{G}$$

then

$$(M \quad \text{is} \quad F) \quad \text{and} \quad (N \quad \text{is} \quad G) \rightarrow \pi_{(X_1, \ldots, X_n, Y_1, \ldots, Y_n)} = \tilde{F} \times \tilde{G}$$

where

$$\tilde{F} \times \tilde{G} = \{((u, v), \mu_{\tilde{F} \times \tilde{G}}(u, v)) \mid u \in U, v \in V\}$$

and

$$\mu_{\tilde{F} \times \tilde{G}}(u, v) = \min \{\mu_F(u), \mu_G(v)\}$$

"If M is $\tilde{F}$ then N is $\tilde{G}$" $\rightarrow \pi_{(X_1, \ldots, X_n, Y_1, \ldots, Y_n)} = \tilde{F}'_L \oplus \tilde{G}'_L$ where $\tilde{F}'_L$ and $\tilde{G}'_L$ are the cylindrical extensions of $\tilde{F}$ and $\tilde{G}$ and $\oplus$ is the bounded sum defined in definition 3–9. Hence

$$\mu_{\tilde{F}'_L \oplus \tilde{G}'_L}(u, v) = \min \{1, \mu_{\tilde{F}}(u) + \mu_{\tilde{G}}(v)\}$$

Example 9–12 [Zadeh 1981a, pp. 32–33]

Assume that $u = v = 1, 2, 3$ and $M \triangleq X$, $N \triangleq Y$, and

$$\tilde{F} \triangleq \text{small} \triangleq \{(1, 1), (2, .6), (3, .1)\}$$
$$\tilde{G} \triangleq \text{large} \triangleq \{(1, .1), (2, .6), (3, 1)\}$$

then X is small and Y is large $\rightarrow$

$$\pi(x, y) = \{[(1, 1), .1)], [(1, 2), .6], [(1, 3), 1], [(2, 1), .1], [(2, 2), .6], [(2, 3), .6], [(3, 1), .1], [(3, 2), .1] [(3, 3), .1]\}$$

X is small or Y is large $\rightarrow$

$$\pi(x, y) = \{[(1, 1), 1], [(1, 2), 1], [(1, 3), 1], [(2, 1), .6], [(2, 2), .6], [(2, 3), 1], [(3, 1), .1], [(3, 2), .6], [(3, 3), 1]\}$$

If X is small then Y is large $\rightarrow$

$$\pi(x, y) = \{[(1, 1), .1], [(1, 2), .6], [(1, 3), 1], [(2, 1), .5], [(2, 2), 1], [(2, 3), 1], [(3, 1), 1], [(3, 2), 1], [(3, 3), 1]\}$$

Translation rules of type II can, of course, also be applied to propositions containing linguistic variables. In some applications it is convenient to represent fuzzy relations as tables (such as shown in section 6.1). These tables can also be processed in PRUF.

Rules of Type III

These translation rules pertain to the translation of propositions of the form

$$P \triangleq QN \quad \text{are} \quad \tilde{F}$$

where N may also be a fuzzy set and Q is a so-called quantifier, for example, a term such as most, many, few, some, and so on. Examples are:

Most children are cheerful.
Few lazy boys are successful.
Some men are much richer than most men.

A quantifier, Q, is in general a fuzzy set of which the universe is either the set of integers, the unit interval, or the real line.

Some quantifiers, such as most, many, and so on, refer to propositions of sets which may either be crisp or fuzzy. In this case the definition of a quantifier makes use of the cardinality or the relative cardinality such as defined in definition 2–5.

In PRUF the notation prop $(\tilde{F}/\tilde{G})$ is used to express the proportion of $\tilde{F}$ in $\tilde{G}$ where

$$\text{prop}\,(\tilde{F}/\tilde{G}) = \frac{\text{count}\,(\tilde{F} \cap \tilde{G})}{\text{count}\,\tilde{G}} = \frac{|\tilde{F} \cap \tilde{G}|}{|\tilde{G}|}$$

where "count" corresponds to the above-mentioned cardinality. The quantifier "most" may then be a fuzzy set

$$\tilde{Q} = \{[\text{prop}\,(\tilde{F}/\tilde{G}), \mu_{\text{most}}(u, v)] \mid u \in \tilde{F}, v \in \tilde{G}\}$$

Example 9–13

The quantifier "several" could, for instance, be represented by

$$\tilde{Q} \triangleq \text{several} \triangleq \{(3, .3), (4, .6), (5, 1), (6, .8), (7, .6), (8, .3)\}$$

Rules of Type IV

In PRUF the concept of truth serves to make statements about the relative truth of a proposition p with respect to another reference proposition (and not with respect to reality!). Truth is taken to be a linguistic variable such as defined in section 9.1. Truth is then interpreted as the consistency of proposition p with proposition q. If

$$p \triangleq N \quad \text{is} \quad F \rightarrow \pi_p = \widetilde{F}$$
$$q \triangleq N \quad \text{is} \quad G \rightarrow \pi_q = \widetilde{G}$$

then the consistency of p with q is given as

$$\begin{aligned} \text{cons}\{N \quad \text{is} \quad F \mid N \quad \text{is} \quad G\} &\triangleq \text{poss}\{N \quad \text{is} \quad F \mid N \quad \text{is} \quad G\} \\ &= \sup_{u \in U} \{\min(\mu_{\widetilde{F}}(u), \mu_{\widetilde{G}}(u))\} \end{aligned}$$

Example 9–14

Let

$$p \triangleq N \quad \text{is a small integer}$$
$$q \triangleq N \quad \text{is not a small integer}$$

where

$$\text{small integer} \triangleq \{(0, 1), (1, 1), (2, .8), (3, .6), (4, .5), (5, .4), (6, .2)\}$$

Then

$$\begin{aligned} \text{cons}\{p \mid q\} &= \sup\{[0, 0, .2, .4, .5, .4, .2]\} \\ &= .5 \end{aligned}$$

More in line with fuzzy set theory is to consider the truth of a proposition as a fuzzy number. Therefore Zadeh defines in the context of PRUF truth as follows:

Definition 9–10 [Zadeh 1981a, p. 42]

Let p be a proposition of the form "N is $\widetilde{F}$" and let r be a reference proposition, $r \triangleq N$ is $\widetilde{G}$, where $\widetilde{F}$ and $\widetilde{G}$ are subsets of U. Then the truth, τ, of p relative to r is defined as the *compatibility of r with p*, that is,

$$\begin{aligned} \tau \triangleq \text{Tr}(N \quad \text{is} \quad \widetilde{F} \mid N \quad \text{is} \quad \widetilde{G}) &\triangleq \text{comp}(N \quad \text{is} \quad \widetilde{G} \mid N \quad \text{is} \quad \widetilde{F}) \\ &\triangleq \mu_{\widetilde{F}}(\widetilde{G}) \\ &\triangleq \{(\tau, \mu_{\widetilde{F}}(\widetilde{G})) \mid \tau \in [0, 1]\} \end{aligned}$$

with

$$\mu_{\tilde{F}}(\tilde{G}) = \inf_{\tau \in [0,1]} \{\mu_{\tilde{F}}(u), \mu_{\tilde{G}}(u)\}, \qquad u \in U$$

The rule for truth qualification in PRUF can now be stated as follows [Zadeh 1981a, p. 44]: Let p be a proposition of the form

$$p \triangleq N \text{ is } \tilde{F}$$

and let q be a truth-qualified version of p expressed as

$$q \triangleq N \text{ is } \tilde{F} \text{ is } \tau$$

where τ is a linguistic truth value. q is semantically equivalent to the reference proposition, that is,

$$N \text{ is } \tilde{F} \text{ is } \tau \rightarrow N \text{ is } \tilde{G}$$

where $\tilde{F}$, $\tilde{G}$, and τ are related by

$$\tau = \mu_{\tilde{F}}(\tilde{G})$$

In analogy to truth qualification, translation rules for probability qualification and possibility qualification have been developed in PRUF.

Example 9–15

Let

$$U = N_0 = \{0, 1, 2, \ldots\}, \qquad N \in N_0$$
$$p = N \text{ is small}$$
$$r = N \text{ is approximately 4}$$

where

small = {(0, 1), (1, 1), (2, .8), (3, .6), (4, .4), (5, .2)}
approximately 4 = {(1, .1), (2, .2), (3, .5), (4, 1), (5, .5), (6, .2), (7, .1)}

Then

$$\begin{aligned}\tau &= Tr(N \text{ is small} \mid N \text{ is approximately 4})\\ &= \text{comp}(N \text{ is approximately } 4 \mid N \text{ is small})\\ &= \{(\mu_{\text{small}}(u), \mu_4(u)) \mid u \in U\}\\ &= \{(0, .2), (.2, .5), (.4, 1), (.6, .5), (.8, .2), (1, .1)\}\end{aligned}$$

9.5 Support Logic Programming [Baldwin 1986]

Support logic programming generalizes logic programming to include various forms of uncertainties. SLOP (Support Logic Programming) combines the theory of fuzzy sets and a theory of evidence [Gordon and Shortliffe 1984] to form a suitable framework for reasoning under uncertainty and approximate reasoning in expert systems, based on a logic programming style. In this system a conclusion is supported by a certain degree of evidence. The negation of the conclusion is also supported to a certain degree, and the two supports do not necessarily add up to 1.

The technique of using dual measures of uncertainty used in support logic programming was also used in other systems, for instance in PFL [Whiter 1983]. In PFL the dual uncertainty measures are based upon pi-value possibility measures calculated on true space given by Baldwin and Pilsworth [1979]. The approaches of PFL and support logic programming differ, however, in the interpretation of the uncertainty measures and in the underlying uncertainty calculus.

Here we limit the considerations to the support logic programming technique of Baldwin [1986]. It is computationally efficient and easy to use, even for those, who are not familar with expert system technologies and uncertainty techniques based on fuzzy sets. We suggest, however, that the interested reader study also the alternative approach of Whiter [1983].

9.5.1 Support Horn Clause Logic Representation

In support logic programming a conclusion and its negation is supported to a certain degree by given evidence. Baldwin called this dual measure of uncertainty the "support pair," where the first number provides the "necessary support" and the second number the "possible support" of the conclusion. The meaning of the support numbers can be interpreted as follows. If a fact is known to be true it will necessarily be supported to degree 1, if it is known to be false, its negation will necessarily be supported to 1. Uncertainty arises when a statement (or a fact) and its negation are necessarily supported to degrees x and y, respectively, where x and y are nonnegative numbers less than 1. If the sum $x + y < 1$ the support values are not probabilities and this form of uncertainty is more fuzzy than probabilistic in nature.

Definition 9–11

A *support logic program* is a set of program clauses of the form

$$A{:}-\quad Q\quad :[Sn, Sp]$$

where A is an atom and Q is a conjunction, disjunction, or mixture of atomic formulas of atoms. An atom is of the form <predicate–name> (<argument–list>). The program clause should be understood as a Prolog clause with the additional support pair $[Sn, Sp]$, where Sn denotes the necessary support of A and Sp the possible support. A is known as the head of the clause, where Q forms the body of the clause.

Rules of several types were allowed in support logic programming and were formed as Prolog-like clauses:

1. conjunctions
2. unit clauses
3. disjunctions

1. Conjunctions. A conjunction rule is of the form

$$A{:-}B1,\ B2,\ \ldots,\ Bn \quad :[Sn,\ Sp]$$

where $A, B1, \ldots, Bn$ are atoms and $[Sn, Sp]$ is the support pair, which can be interpreted as follows:

If the statements $B1, \ldots, Bn$ are all true then the concluding statement is necessarily supported to degree Sn, where the negation *NOT A* is necessarily supported to degree $(1 - Sp)$. These supports are conditional with respect to A given that *B1 AND B2 AND* ... *Bn* is true. The support pair satisfies the condition

$$Sn + (1 - Sp) \leq 1$$

The difference $(Sn - Sp)$ measures the unsureness associated with the support of the rule pair $\{(A{:-}\ B1, \ldots, Bn), (NOT\ A{:-}\ B1, \ldots, Bn)\}$, that is, of the conditionals $\{(A \mid B1, \ldots, Bn), (NOT\ A \mid B1, \ldots, Bn)\}$.

2. Unit Clauses. The rule

$$A{:}\quad [Sn,\ Sp]$$

is a unit clause, that is, a rule with an empty body.

Each assignment of the variable A is necessarily supported to Sn and *NOT A* is necessarily supported to degree $(1 - Sp)$. Again the constraint $Sn + (1 - Sp) \leq 1$ holds.

3. Disjunctions. Disjunctions were expressed by statements of the form

$$A \quad {:-} \quad B1;\ B2 \quad :[Sn,\ Sp]$$

B1 and *B2* are single atoms or a conjunction of atoms and the ; signifies the disjunction. Thus the statement *A* is necessarily supported to degree *Sn*, *NOT A* to degree $(1 - Sp)$, respectively, for each assignment to *B1* and *B2* where *B1 OR B2* is true.

It should be noticed that the sum of all necessary supports for an atom and its negation generally do not add up to 1, because the necessary support of the negation of an atom is not determined from the necessary support of the atom. Instead we express the necessary support of the negation of an atom as that amount which the necessary support of the atom is different from unity. If $Sn = Sp$ the supports can be interpreted as probabilities. For $Sn = a$ and $Sp = b$, $a < b$, $a, b \in [0, 1]$, we can, in a probabilistic sense, assume that the probability of a supported statement lies in the intervall $[a, b]$. Thus the necessary support of *NOT A* is determined as (1 − necessary support of *A*). Hence we get the following interpretation of the support of an atom and its negation:

IF	P:− Q : $[Sn, Sp]$	
THEN	nec_sup$(P \mid Q) = Sn$	*pos_sup* $(P \mid Q) = Sp$
	nec_sup$(NOT\ P \mid Q) = 1 - Sp$	*pos_sup* $(NOT\ P \mid Q) = 1 - Sn$

where nec_sup denotes the necessary support and pos_sup the possible support and Q is a conjunction, disjunction, or mixture of atomic formulas.

IF	P: $[Sn, Sp]$	
THEN	nec_sup$(P) = Sn$	pos_sup$(P) = Sp$
	nec_sup$(NOT\ P) = 1 - Sp$	pos_sup$(NOT\ P) = 1 - Sn$

Example 9–16

is_young(John): [1/4, 1/2]

This is a unit clause and states that the necessary support of John being young is 1/4, where the possible support of John being young is 1/2. With the above-given rules we can deduce that the necessary support of John being not young is 1/2 and the possible support for this fact is 3/4.

The following conjunction rule expresses more generally the support of the fact that someone of definite age is a young person:

is_young(X):− is_person(X), age (X, 30) : [1/5, 2/3]

It says, that the fact that a 30-years-old person is young is necessarily supported to degree 1/5, while the necessary support of this person not being young is 1/3.

One problem in support logic programming is the determination of the support values for statements and facts. In some cases it may be appropriate to interpret the support values in terms of a group voting model, where the necessary support value *Sn* indicates the proportion of the members of a group voting for a fact or statement and $(1-Sp)$ indicates the proportion voting against [Baldwin 1986, p. 98]. Another way of obtaining support values is the use of fuzzy sets. If we describe the concept "young person" in the above example by a fuzzy set (see section 9.1) both the necessary and the possible supports are given by the membership values for the fuzzy set "young." Suppose that a person of age 30 belongs to the fuzzy set "young" with a membership value of 3/5, that is, $\mu_{\text{young}}(30) = 3/5$. Interpreting the necessary and possible supports as special versions of the Possibility and Necessity measures of fuzzy set theory (see section 8.1 and Dubois and Prade [1980a]), we obtain the possible support

$$\text{pos\_sup}\ (X \quad \text{is young} \mid x \text{ is 30 years old}) = \mu_{\text{young}}(30) = 3/5$$

If the membership function of the complement of the fuzzy set is defined as described in chapter 2, that is, $\mu_{\text{NOT young}}(x) = 1 - \mu_{\text{young}}(x)$, then the possible support of the opposit statement is given by

$$\text{pos\_sup}(X \quad \text{is NOT young} \mid X \text{ is 30 years old}) = \mu_{\text{NOT young}}(30) = 1 - 3/5$$

which is defined as (1 – necessary support for X being young). Thus we get nec_sup (X is young $|$ X is 30 years old) $= 3/5$ and the rule can be formulated as

$$\text{is\_young}(X){:}- \quad \text{is\_person}(X),\ \text{age}(X, 30) \quad : [3/5, 3/5]$$

For the case in which the age in the example above is given as the interval or fuzzy set itself we may be interested in the supports x and y, for instance, of the following rule

$$\text{is\_young}(X){:}- \quad \text{is\_person}(X),\ \text{age}\ (X, [25, 35]) \quad :[x, y]$$

where

$$x = \textit{pos\_sup}\ (X \quad \text{is young} \mid X \text{ is in age of 25 to 35})$$
$$y = \text{nec\_sup}\ (X \text{ is NOT young} \mid X \text{ is in age of 25 to 35})$$

For this case we will interpret the necessary and possible support as the usually defined Necessity and Possibility measures of fuzzy set theory.

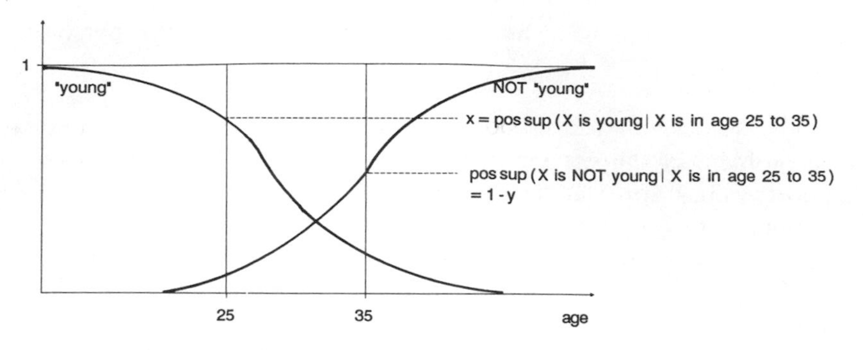

Figure 9–5. Determining support values on the basis of fuzzy sets.

Given the fuzzy set "young" the supports x and y can be computed with respect to the membership values of the value of age as shown in figure 9–5.

9.5.2 Approximate Reasoning in SLOP

Given a support logic program as a set of program clauses representing facts and rules, the programming system has to calculate the support pairs associated with solutions to queries to the system. A proof path for a final solution is determined in the normal Prolog style and a support is determined for each branch in the proof path. If more than one proof path is available then the support pairs must be combined from different proof paths to obtain the overall support for the conclusion. This is archieved in a very efficient way, using a method that ressembles the stepping-stone method commonly employed when solving transportation models. The results are then cast into rules for support logic programming.

The determination of support values for one proof path is done by determining the support pairs for disjunction and conjunction statements as well as combining those pairs when combining support pairs from different proof paths supporting the same conclusion.

If the supports of the facts A and B were

$$A: \quad [Sn1, Sp1] \qquad\qquad B: \quad [Sn2, Sp2]$$

then we are interested in the support pairs $[Sn3, Sp3]$, $[Sn4, Sp4]$ of the following compound statements:

$$(A \text{ and } B): \quad [Sn3, Sp3]$$
$$(A \text{ or } B): \quad [Sn4, Sp4]$$

Considering the necessary supports N_{ij} for the various possible conjunction statements, given the support pairs of A and B, respectively, as shown in table 9–1, we obtain several constraints for the values of N_{ij}.

Table 9–1. Support pairs.

	B $Sn2$	$NOT\ B$ $1 - Sp2$	UNSURE $-Sn2 + Sp2$
A $Sn1$	A and B N_{11}	A and $NOT\ B$ N_{12}	A N_{13}
NOT A $1 - Sp_1$	$NOT\ A$ and B N_{21}	$NOT\ A$ and $NOT\ B$ N_{22}	$NOT\ A$ N_{23}
UNSURE $Sp1 - Sn1$	B N_{31}	$NOT\ B$ N_{32}	UNSURE N_{33}

The necessary supports N_{ij} must satisfy the following equations, which are taken from the rows and colums of the table:

$$N_{11} + N_{12} + N_{13} = Sn1$$
$$N_{21} + N_{22} + N_{23} = 1 - Sp1$$
$$N_{31} + N_{32} + N_{33} = Sp1 - Sn1$$
$$N_{11} + N_{21} + N_{31} = Sn2$$
$$N_{12} + N_{22} + N_{32} = 1 - Sp2$$
$$N_{13} + N_{23} + N_{33} = Sp2 - Sn2$$

where $\sum_{i,j} N_{ij} = 1$

These equations do not yield point values for the N_{ij}. It is possible, however, to determine upper and lower bounds for them by formulating additional constraints. Suppose A and B have the necessary supports $Sn1$ and $Sn2$, respectively, then the necessary support for (A and B) cannot be greater than the minimum value of $Sn1$ and $Sn2$ since min $\{Sn1, Sn2\}$ is the maximum value that can be allocated to (A and B) with respect to the row or column constraints given in table 9–1. Furthermore, the minimum value that can be allocated to (A and B) without violating row or column constraints is $\max\{Sn1 + Sn2 - 1, 0\}$. Thus, we obtain the following new constraint for N_{11}, which is the necessary support for (A and B):

$$\max\{Sn1 + Sn2 - 1, 0\} \leq N_{11} \leq \min\{Sn1, Sn2\}$$

Conditions for the necessary supports of the other possible conjunction statements can be formulated analoguously.

In order to obtain a unique solution for the various supports on the base of the given constraints the method associated with the transportation algorithm can be used to check for uniqueness and generate alternative solutions.

Here we will discuss two approaches to finding a unique solution for this assignment problem, the MIN-Model and the Multiplication Model.

The rule of the Multiplication Model for the assignment of supports for compound statements corresponds to Dempster's rule used in the Dempster/Shafer theory of evidence [Gordon and Shortliffe 1984, p. 278]. In terms of fuzzy set theory, the "and" operation is modeled by the algebraic product operator, the "or" operation by the bounded sum, respectively.

The necessary supports Sn and possible supports Sp of (A and B) and (A or B), respectively, are computed as follows

$$Sn(A \text{ and } B) = Sn(A) \cdot Sn(B)$$
$$Sp(A \text{ and } B) = Sp(A) \cdot Sp(B)$$

and

$$Sn(A \text{ or } B) = Sn(A) + Sn(B) - Sn(A) \cdot Sn(B)$$
$$Sp(A \text{ or } B) = Sp(A) + Sp(B) - Sp(A) \cdot Sp(B)$$

In the MIN-Model the min-rule of fuzzy logic was assumed for determining the support values of the various possible compound statements. For the combinations contained in the main diagonal of table 9–1 the necessary supports were determined by the min-operator:

$$Sn(A \text{ and } B) = min\{Sn(A), Sn(B)\}$$
$$Sn(NOT\ A \text{ and } NOT\ B) = \min\{Sn(NOT\ A), Sn(NOT\ B)\}$$
$$Sn(UNSURE\ A \text{ and } UNSURE\ B) = \min\{Sn(UNSURE\ A), Sn(UNSURE\ B)\}$$

In an analoguous way we obtain the support values for disjunctive statements:

$$Sn(A \text{ or } B) = \max\{Sn(A), Sn(B)\}$$

The support values of the other combinations, for instance, in the table referring to conjunctions of statements mentioned above, are inferred with respect to the row and column constraints.

Example 9–17

Considering the following support values for A and B,

$$A: \quad [4/21, 16/21] \qquad\qquad B: \quad [7/21, 13/21]$$

the necessary supports of the following combinations can be computed by applying the determination rule of the MIN-Model

$$Sn(A \text{ and } B) = \min\{4/21, 7/21\} = 4/21$$
$$Sn(NOT\ A \text{ and } NOT\ B) = \min\{1 - Sp(A), 1 - Sp(B)\} = 5/21$$
$$Sn(UNSURE\ A \text{ and } UNSURE\ B) = \min\{Sp(A) - Sn(A), Sp(B) - Sn(B)\} = 6/21$$

For these results the row constraints for the first and second row are satisfied. Thus we obtain

$$Sn(A \text{ and } NOT\ B) = 0 \qquad Sn(A \text{ and } UNSURE\ B) = 0$$
$$Sn(NOT\ A \text{ and } B) = 0 \qquad Sn(NOT\ A \text{ and } UNSURE\ B) = 0$$

Satisfying the constraints given for the first and second column yields

$$Sn(UNSURE\ A \text{ and } B) = Sn(B) - Sn(A \text{ and } B) = 3/21.$$
$$Sn(UNSURE\ A \text{ and } NOT\ B) = Sn(NOT\ B) - Sn((NOT\ A) \text{ and } Sn(NOT\ B)) = 3/21$$

Now we should consider the rule of inference, given a supported conditional and a supported fact corresponding to the antecedent of the rule. Assuming that only one proof path is associated with the conclusion and no combination of evidence is necessary we can apply the rules for determining support values for compound statements as given in the Multiplication and the MIN-Model to infer the supports of the conclusion.

Consider the following clauses, where Q is a disjunction or conjunction or mixture of atomic formulas

$$P{:}-\ \ Q\ \ :[Sn(P \mid Q), Sp(P \mid Q)]$$
$$Q{:}\ \ [Sn(Q), Sp(Q)]$$

Applying the determination rule of the Multiplication Model the concluding supports for P were determined as follows:

The statement $((P{:}-\ Q)$ and $Q)$ is the only combination that implies P. Thus we have to compute the support pair of this conjunction.

$$Sn(P) = Sn((P{:-}\ \ Q) \text{ and } Q) = Sn(P \mid Q) \cdot Sn(Q)$$
$$Sn(NOT\ P) = Sn((NOT\ P{:-}\ \ Q) \text{ and } Q) = (1 - Sp(P \mid Q)) \cdot Sn(Q)$$

With $Sp(P) = 1 - Sn(NOT\ P)$ we obtain

$$Sn(P) = Sn(P \mid Q) \cdot Sn(Q)$$
$$Sn(NOT\ P) = 1 - [(1 - Sp(P \mid Q)) \cdot Sn(Q)]$$

Applying the MIN-Model we obtain

$$Sn(P) = Sn((P{:-}\ \ Q) \text{ and } Q) = \min\{Sn(P \mid Q), Sn(Q)\}$$

The value of $Sn(NOT\ P) = Sn((NOT\ P{:-}\ \ Q) \text{ and } Q)$ depends on the support values of various other supports and is determined with respect to the given constraints as shown in the example above.

In general, concluding statements in a support logic program may be supported by a list of support pairs, each of them resulting from a different proof path. In this case all supports must be combined to obtain the overall support values for the conclusion. Conflicts occuring while computing the final supports must, however, be resolved. For illustration purposes we consider only the method for combining evidences using the Multiplication Model. For combinations based upon the MIN-Model the reader is referred to [Baldwin 1986, p. 91].

Consider the following inference scheme for two proof paths:

$$P{:-}\ \ Q1\ :[Sn(P \mid Q1), Sp(P \mid Q1)]$$
$$P{:-}\ \ Q2\ :[Sn(P \mid Q2), Sp(P \mid Q2)]$$
$$Q1{:}\ \ [Sn(Q1), Sp(Q1)]$$
$$Q2{:}\ \ [Sn(Q2), Sp(Q2)]$$

Using the Multiplication Model we obtain the following supports for P which result for each of the different proof paths:

$$P{:}\ \ [Sn(P \mid Q1) \cdot Sn(Q1), 1 - (1 - Sp(P \mid Q1)) \cdot Sn(Q1)] = P{:}\ \ [Sn1, Sp1]$$
$$P{:}\ \ [Sn(P \mid Q2) \cdot Sn(Q2), 1 - (1 - Sp(P \mid Q2)) \cdot Sn(Q2)] = P{:}\ \ [Sn2, Sp2]$$

These resulting supports from different proof paths have to be combined to obtain unique support values $Sn(P)$ and $Sp(P)$ for the final conclusion P. The row and column constraints for the final necessary support of P are shown in table 9–2.

The second entry of the first row as well as the first cell in the second row implies (P and $NOT\ P$), which indicates a conflict. Using Shafer's renormalization method of resolving conflicts the supports of the conflicting combinations were shared among the other nonconflicting cells in

Table 9–2. Row and column constraints.

	P $Sn2$	$NOT\ P$ $1 - Sp2$	$UNSURE$ $-Sn2 + Sp2$
P $Sn1$	P and P $Sn1 \cdot Sn2$	P and $NOT\ P$ **conflict** $Sn1 \cdot (1 - Sp2)$	P $Sn1 \cdot (Sp2 - Sn2)$
$NOT\ P$ $1 - Sp1$	$NOT\ P$ and P **conflict** $(1 - Sp1) \cdot Sn2$	$NOT\ P$ and $NOT\ P$ $(1 - Sp1) \cdot (1 - Sp2)$	$NOT\ P$ $(1 - Sp1) \cdot (Sp2 - Sn2)$
$UNSURE$ $Sp1 - Sn1$	B $(Sp1 - Sn1) \cdot Sn2$	$NOT\ P$ $(Sp1 - Sn1) \cdot (1 - Sp2)$	$UNSURE$ $(Sp1 - Sn1) \cdot (Sp2 - Sn2)$

the portion of their existing supports. After reallocating conflicting supports the row and column constraints hold. Thus we obtain

$$Sn(P) = [Sn1 \cdot Sn2 + Sn1 \cdot (Sp2 - Sn2) + Sn2 \cdot (Sp1 - Sn1)]/K$$

$$Sp(P) = [(1 - Sp1) \cdot (1 - Sp2) + (1 - Sp1) \cdot (Sp2 - Sn2) + (1 - Sp2) \cdot (Sp1 - Sn1)]/K$$

where K is the normalization portion of conflicting cells

$$K = 1 - [Sn2 \cdot (1 - Sp1) - Sn1 \cdot (1 - Sp2)]$$

If more than two proof paths were available, then they would be combined two at a time.

The approach suggested for the inference in support logic programming is that of Prolog. Determining the support pairs of all goals satisfied, including subgoals, can be done in a very efficient way using the rules stated above. In support logic programming it is assumed that the same calculus is applicable, irrespective of wether the uncertainty arises from fuzzy imprecision or from stochastic uncertainty. Once the support values are given, their source is considered to be irrelevant. A description of the contents of the programming system SLOP is given in [Baldwin 1987, p. 219]. The relationship of the rules of assignment used in support logic programming to the methods of combining evidence in the general context of evidential reasoning is discussed by Baldwin [1989].

Example 9–18 [Baldwin 1986, p. 94]

A judgement problem is formulated by Baldwin [1986, p. 94]. Alternative designs for a product which differ in reliability, costs, and performance should be judged in terms of satisfying special features. For instance, a design is satisfactory with a necessary support of 0.9 if it performs well and looks modern. This problem is formulated as a support logic program.

design(X, satisfactory): – performance(X, good),
looks (X, modern) : [.9, 1]
design(X, satisfactory): – cost(X, expensive_for_market) : [0, .05]
design(X, satisfactory): – NOT reliability (X, high) : [0, .2]
performance(X, good): – engineers_report (X, satisfactory),
reliability(X, high) : [.9, 1]

looks(design_1, modern): [.8, 1]
looks(design_2, modern): [.9, 1]

reliability(design_1, high): [.7, .8]
reliability(design_2, high): [.8, .8]

engineers_report(design_1, satisfactory): [.7, 1]
engineers_report(design_2, satisfactory): [.9, 1]

cost(design_1, expensive_for_market): [.6, 1]
cost(design_2, expensive_for_market): [.3, 1]

Queries can be asked in a normal Prolog style. For example ? –design(X, satisfactory).

Using the described rules above the system infers about the supports for the different designs being satisfactory:

X = design_1: [.11, .45]

X = design_2: [.34, .76].

Exercises

1. Consider the linguistic variable "Age." Let the term *old* be defined by

$$\mu_{\text{old}}(x) = \begin{cases} 0 & \text{if } x \in [0, 40] \\ \left(1 + \left(\dfrac{x-40}{5}\right)^{-2}\right)^{-1} & \text{if } x \in (40, 100] \end{cases}$$

Determine the membership functions of the terms "very old," "not very old," "more or less old."

2. Let the term *true* of the linguistic variable "Truth" be characterized by the membership function

$$T(v; \alpha, \beta, \gamma) = \begin{cases} 0 & \text{if } v \leq \alpha \\ 2\left(\dfrac{v-\alpha}{\gamma-\alpha}\right)^2 & \text{if } \alpha \leq v \leq \beta \\ 1 - 2\left(\dfrac{v-\gamma}{\gamma-\alpha}\right)^2 & \text{if } \beta \leq v \leq \gamma \\ 1 & \text{if } v \geq \gamma \end{cases}$$

Draw the membership function of "true." Determine the membership functions of "rather true" and "very true." What is the membership function of "false" = not "true" and what of "very false"?

3. What is the essential difference between Baldwin's definition of "true" and Zadeh's definition?
4. Let the primary terms *young* and *old* be defined as in example 9–3. Determine the secondary terms *young and old*, *very young and not very old*.
5. Let "true" and "false" be defined as in example 9–4. Find the membership function of "very very true." Compare the fuzzy sets "false" and "not true."
6. Let the universe $X = \{1, 2, 3, 4, 5\}$ and "small integers" be defined as $\tilde{A} = \{(1, 1), (2, .5), (3, .4), (4, .2)\}$. Let the fuzzy relation "almost equal" be defined as follows:

$\tilde{R}$:

	1	2	3	4
1	1	.8	0	0
2	.8	1	.8	0
3	0	.8	1	.8
4	0	0	.8	1

What is the membership function of the fuzzy set B = "rather small integers" if it is interpreted as the composition $\tilde{A} \circ \tilde{R}$?

7. What is the relationship between a relational assignment equation and a possibility assignment equation?
8. Which of the definitions of "true" amounts to unity possibility distributions and which other important linguistic variables are represented by unity possibility distribution?
9. Consider example 9–10 and make propositions about cars like Mercedes, Volvo, Chevy, Rolls Royce.

10 EXPERT SYSTEMS AND FUZZY CONTROL

10.1 Fuzzy Sets and Expert Systems

10.1.1 Introduction to Expert Systems

During the last three decades, the potential of electronic data processing (EDP) has been used to an increasing degree to support human decision making in different ways. In the sixties the management informations systems (MIS) created probably exaggerated hopes of managers. Since the late 1970s and early 1980s decision support systems (DSS) found their way into management and engineering. The youngest children of these developments are the so-called knowledge-based expert systems or short expert systems which have been applied since the mid 1980s to solve management problems [Zimmermann 1987, p. 310]. It is generally assumed that expert systems will increasingly influence decision-making processes in business in the future.

If one interpretes decisions rather generally, that is, including evaluation, diagnosis, prediction, et cetera, then all three types could be classified as decision support systems that differ gradually with respect to the following properties:

1. Does the system "optimize" or just provide information?
2. Is it generally usable or just for specific purposes and areas?
3. Is it self-contained with respect to procedures and algorithms or does it "learn" and "derive" inference and decision-making rules from knowledge that is inquired from a human (expert) and analyzed within the system?

It can be expected that in the future these decision support systems will contain to an increasing degree features of all three types of the above-mentioned systems. Even though fuzzy set theory can be used in all three "prototypes" we shall concentrate on "expert systems" only because the need and problem of managing uncertainty of many kinds is most apparent there; and hence the application of fuzzy set theory is most promising and, in fact, most advanced. In operations research the modeling of problems is normally being done by the OR-specialist. The user then provides input data and the mathematical model provides the solution to the problem by means of algorithms selected by the OR-specialist.

In expert systems, the domain knowledge is typically emphasized over formal reasoning methods. "In attempting to match the performance of human experts, the key to solving the problem often lies more in specific knowledge of how to use the relevant facts than in generating a solution from some general logical principles. 'Human experts achieve outstanding performance because they are knowledgeable' " [Kastner and Hong 1984]. Conventional software engineering is based on procedural programming languages. The tasks to be programmed have to be well understood, the global flow of the procedure has to be determined, and the algorithmic details of each subtask have to be known before actual programming may proceed. Debugging often represents a huge investment in time, and there is little hope of automatically explaining how the results are derived. Later modification or improvement of a program becomes very difficult.

> Most of the human activities concerning planning, designing, analyzing, or consulting have not been considered practical for being programmed in conventional software. Such tasks require processing of symbols and meanings rather than numbers. But more importantly, it is extremely difficult to describe such tasks as a step-by-step process. When asked, an expert usually cannot procedurally describe the entire process of problem solving. However, an expert can state a general number of pieces of knowledge, without a coherent global sequence, under persistent and trained interrogation. Early AI research concentrated on how one processes relevant relations that hold true in a specific domain to solve a given problem. Important foundations have been developed that enable, in principle, any and all logical consequences to be generated from

> a given set of declared facts. Such general purpose problem solving techniques, however, usually become impractical as the toy world used for demonstration is replaced by even a simple real one. The realization that knowledge of how to solve problems in the specific domain should be a part of the basis from which inferences are drawn contributed heavily to making expert systems technology practical. [Kastner and Hong 1984]

While the typical OR-model or software package normally supports the expert, an expert system is supposed to model an expert and make his expert knowledge available to nonexperts for purposes of decision making, consulting, diagnosis, learning, or research.

The character of an expert system might become more apparent if we quote some of the system characteristics considered as being attributes of expert systems [Konopasek and Jayaraman 1984]. Attributes of expert systems include:

The expert system has separate domain-specific knowledge and problem-solving methodology and includes the concepts of the Knowledge Base and the Inference Engine.

The expert system should think the way the human expert does.

Its dynamic knowledge base should be expandable, modifiable, and facilitate "plugging in" different knowledge modules.

The interactive knowledge transfer should minimize the time needed to transfer the expert's knowledge to the knowledge base.

The expert system should interact with the language "natural" to the domain expert; it should allow the user to think in problem-oriented terms. The system should adapt to the user and not the other way around. The user should be insulated from the details of the implementation.

The principal bottleneck in the transfer of expertise—the knowledge engineer—should be eliminated.

The control strategy should be simple and user-transparent, the user should be able to understand and predict the effect of adding new items to the knowledge base. At the same time, it should be powerful enough to solve complex problems.

There should be an inexpensive framework for building and experimenting with expert systems.

It should be able to reason under conditions of uncertainty and insufficient information and should be capable of probabilistic reasoning.

An expert system should be able to explain "why" a fact is needed to complete the line of reasoning and "how" a conclusion was arrived at.

Expert systems should be capable of learning from experience.

Cutting a long story short Kastner and Hong [1984] provide this definition: "*An expert system* is a computer program that solves problems that heretofore required significant human expertise by using explicitly represented domain knowledge and computational decision procedures."

A sample of some other definitions of an expert system can be found in

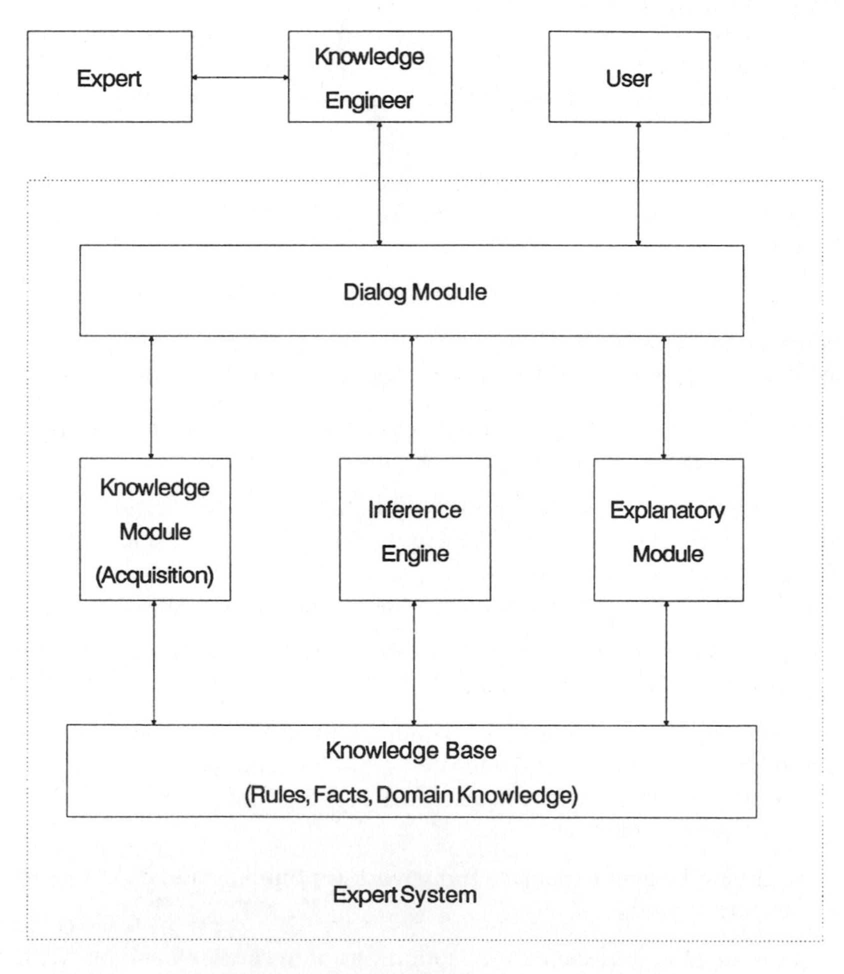

Figure 10–1. Structure of an expert system.

[Fordyce et al. 1989, p. 66]. The general structure of an expert system is shown in Figure 10–1, (see also Zimmermann [1987], p. 262). In the following, the five components of such a system are explained in more detail. The *Knowledge Acquisition Module* supports the building of an expert system's knowledge base.

> The subject of knowledge acquisition for knowledge-based systems falls conveniently into two parts depending on whether the knowledge is elicited from the experts by knowledge engineers or whether that knowledge is acquired automatically by the computer using some form of automatic learning strategy and algorithms. [Graham and Jones 1988, p. 279]

A module that aids the knowledge engineer during the process of knowledge elicitation could consist of a user-friendly rule editor, an "automatic error-checking when rules are being put in, and good online help facilities" [Ford 1987, p. 162]. (See also [Buchanan et al. [1983], p. 129]). AQUINAS is such a system; it is presented by [Boose 1989, p. 7].

Another way to acquire domain dependent knowledge is the application of machine learning techniques to automatically generate a part of the knowledge base. It is expected that rapid improvements will take place in the field of automatic knowledge acquisition in the future. The interested reader is referred to [Michalski 1986, p. 3; Morik 1989, p. 107].

The *knowledge base* contains all the knowledge about a certain domain that has been entered via the above-mentioned knowledge acquisition module. Apart from special storage requirements and system-dependent structures, the knowledge base can be exchanged in some expert systems. That means that there can be several knowledge bases, each covering a different domain, which can be "plugged in" the "shell" of the remaining expert system.

> There are basically two types of knowledge that will need to be represented in the system: *declarative knowledge* and *procedural knowledge*. The declarative part of the knowledge base describes "what" the objects (facts, terms, concepts, ...) are that are used by the expert (and the expert system). It also describes the relationships between these objects. This part of the knowledge base is sometimes referred to as the "data base" or "facts base."
>
> The procedural part of the knowledge base contains information on how these objects can be used to infer new conclusions and ultimately arrive at a solution. Since this "how-to" knowledge is usually expressed as (heuristic or other) rules, it is generally known as the rule-base. [Rijckaert et al. 1988, p. 493]

A number of techniques for representing the expert knowledge have been developed. These are described in [Barr and Feigenbaum 1981/82] in a greater detail. The four methods most frequently used in expert systems are production rules, semantic nets, frames, and predicate calculus (see [Zimmermann 1987, p. 266]). While we will investigate here the first three of these, the reader is referred to [Nilsson 1980, p. 132] for the latter.

Production Rules. Production rules are by far the most frequently used method for representing procedural knowledge in an expert system. They are usually of the form: "If a set of conditions are satisfied, then a set of consequences can be produced."

> Production rules are used to capture the expert's rule of thumb or heuristic as well as useful relations among the facts in the domain. These if–then rules provide the bulk of the domain-dependent knowledge in rule-based expert systems and a separate control strategy is used to manipulate the rules.
>
> If the car won't start and
> the car lights are dim
> then the battery may be dead.
>
> Many experts have found rules a convenient way to express their domain knowledge. Also, rule bases are easily augmented by simply adding more rules. The ability to incrementally develop an expert system's expertise is a major advantage of rule-based schemes." [Kastner and Hong 1984]

Semantic Nets. A method of encoding declarative knowledge is that of a semantic net. Concepts, categories, or phenomena are presented by a number of nodes associated with one another by links (edges). These links may represent causation, similarity, propositional assertions, and the like. On the basis of these networks, insight into structures can be gained, inferences can be made, and classifications can be obtained. In figure 10–2 a semantic net is used to represent declarative knowledge about the structure of some vehicles.

Frames. The concept of a frame for representing knowledge in an expert system is introduced by Minsky [1975]: "A frame is a structure that collects together knowledge about a particular concept and provides expectations and default knowledge about that concept." Typically, the frame is represented in the computer as a group of slots and associated values. The values may themselves be other frames.

Frame:	vehicle
classes	passenger, motorcycle, truck, bus, bicycle,...
wheels	(integer)
propelled by	motor, human feet....

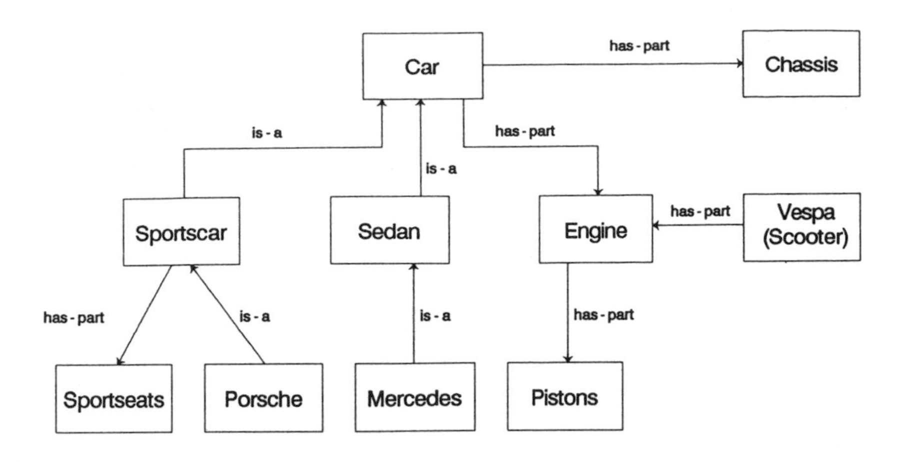

Figure 10–2. Semantic net.

.
.
.

Frame:	bicycle
is-a	vehicle
wheels	2(default)
capacity	1 person (default)

.
.
.

[Kastner and Hong 1984]

New concepts can often be represented by adding frames or by putting new information in "slots" of existing frames. Slots in frames may also be used for inference rules and empty slots might indicate missing information.

> The inclusion of procedures in frames joins together in a single representational strategy two complementary (and, historically, competing) ways to state and store facts: procedural and declarative representations. [Harmon and King 1985, p. 44]

The *inference engine* is a mechanism for manipulating the encoded knowledge from the knowledge base and to form inferences and draw conclusions. The conclusions can be deduced in a number of ways which depend on the structure of the engine and the method used to represent the knowledge. In the case of production rules for knowledge encoding, different control strategies have been used which direct input and output and select which rules to evaluate. Two very popular strategies are "forward chaining" and "backward chaining." In the former, data-driven rules are evaluated for which the conditional parts are satisfied. The latter strategy (goal-driven) selects a special rule for evaluation. The "goal" is to satisfy the conditional part of this rule. If that can not be achieved directly then subgoals are established on the basis of which a chain of rules can be established, such that eventually the conditional part of the first rule can be satisfied. Further information about inference strategies can be found in [Waterman 1986].

Table 10–1. Expert Systems.

Name	*Domain of expertise*	*Major technique*
CADIAG-2 [Adlassnig et al. 1985]	internal medicine	rules*
DENDRAL [Lindsay et al. 1980]	molecular structure elucidation	rules
EMERGE [Hudson, Cohen 1988]	chest pain analysis	rules*
ESP [Zimmermann 1989]	strategic planning	rules*
EXPERT [Weiss, Kulikowski 1981]	rheumatology, ophtalmology	rules* hierarchies
FAULT [Whalen et al. 1987]	financial accounting	rules*
MYCIN [Buchanan, Shortliffe 1984]	infectious disease diagnosis and treatment	rules
OPAL [Bensana et al. 1988]	job shop scheduling	rules*
PROSPECTOR [Benson 1986]	mineral exploration	inference network
R1/XCON [McDermott 1982]	computer configuration	rules
SPERIL [Ishizuka et al. 1982]	earthquake engineering	rules*

* includes fuzzy logic

The above-mentioned approaches can, of course, be combined. In addition to these techniques, expert systems may also contain rather sophisticated mathematical algorithms, such as cluster algorithms, optimization and search techniques like tabu search, (see [Glover and Greenberg 1989, p. 119]). This is actually already a development in the direction of decision support systems but in many cases it will make the expert system more efficient and even more user-friendly. Table 10–1 gives some indication in which area expert systems are already available and what techniques they use. This table does by no means claim to be exhaustive.

10.1.2 Uncertainty Modeling in Expert Systems

There are three main reasons for the use of fuzzy set theory in expert systems:

1. The interfaces of the expert system on the expert side as well as on the user side are with human beings. Therefore communication in a "natural" way seems to be the most appropriate; and "natural" means, generally, in the language of the expert or user. This suggests the use of linguistic variables as they were described in chapter 9.
2. The knowledge base of an expert system is a repository of human knowledge, and since much of human knowledge is imprecise in nature it is usually the case that the knowledge base of an expert system is a collection of rules and facts which, for the most part are neither totally certain, nor totally consistent [Zadeh 1983a, p. 200]. The storage of this vague and uncertain portion of the knowledge by using fuzzy sets seems much more appropriate than the use of crisp concepts and symbolism.
3. As a consequence of what has been said in point two the "management of uncertainty" plays a particularly important role. Uncertainty of information in the knowledge base induces uncertainty in the conclusions and therefore the inference engine has to be equipped with computational capabilities to analyze the transmission of uncertainty from the premises to the conclusions and associate the conclusion with some measure of uncertainty that is understandable and properly interpretable by the user. The reader should also recall from chapter 1 that imprecision in human thinking and communication is often a consequence of abundance of information, that is, the fact, that humans can often process the required amount of information efficiently only by using aggregated (generic) information. This

efficiency of human thinking, when modeled in expert systems, might also increase efficiency, that is, decrease answering time, and so on.

Most of the expert systems existing so far contain an inference engine on the basis of dual logic. The uncertainty is taken care of by Bayesian probability theory. The conclusions are normally associated with a certainty or uncertainty factor expressing stochastic uncertainty, confidence, likelihood, evidence, or belief. Only recently the designers of expert systems have become aware of the fact that all of the types of uncertainty mentioned above cannot be treated the same way and that a factor of, for example, .8, expressing the uncertainty of a conclusion does not mean very much to the user. The expert systems marked with an asterisk in table 10–1 are already using fuzzy set approaches in different ways. We shall illustrate some of them later. In addition, proposals have been published on how fuzzy set theory could be used meaningfully in expert systems.

The most relevant approaches in fuzzy set theory are fuzzy logic and approximate reasoning for the inference engine [Lesmo et al. 1982] and [Sanchez 1979]; the presentation of conditions, indicators, or symptoms by fuzzy sets, especially linguistic variables, to arrive at judgments about secondary phenomena [Esogbue and Elder 1979], Moon et al. [1977], etc.; the use of fuzzy clustering for diagnosis [Fordon and Bezdek 1979] and [Esogbue and Elder 1983]; and combinations of fuzzy set theory with other approaches, for example, Dempster's theory of evidence [Ishizuka 1982], to obtain justifiable and interpretable measures of uncertainty.

We shall describe some more recent attempts to apply fuzzy set theory to knowledge representation and inference mechanisms in expert systems.

While in a precise environment, production rules are adequate to represent procedural knowledge (as was seen in section 10.1.1), this is no longer true in a fuzzy environment. One way to deal with imprecision is to use *fuzzy production rules*, where the conditional part and/or the conclusions part contains linguistic variables (see chapter 9). An application of this knowledge-representation technique in the area of job-shop scheduling can be found in [Dubois 1989, p. 83]. Negoita gives a basic introduction into fuzzy production rules [Negoita 1985, pp. 80].

While little work is done in the field of "fuzzy semantic nets," suggestions to fuzzify frames to represent uncertain declarative knowledge, and an illustrative example, stem from [Graham and Jones 1988, p. 67]. The two main generalizations for arriving at a *fuzzy frame* are:

1. allowing slots to contain fuzzy sets as values, in addition to text, list, and numeric values,
2. allowing partial inheritance through is-a slots.

As a consequence of the representation of imprecise and uncertain knowledge it is necessary to develop adequate reasoning methods. Since 1973 when Zadeh suggested the compositional rule of inference, a lot of work has been done in the field of fuzzy inference mechanisms (Dubois and Prade [1988a], p. 67; Zimmermann [1988], p. 736). Here we shall consider – as an example – the methods used in the *fuzzy*TECH development tool, which supports the implementation of fuzzy expert systems and fuzzy control systems. With this tool, a feasibility study on how to stabilize the cruise of a car in extreme situations such as skidding or sliding was made for the German automotive industry.

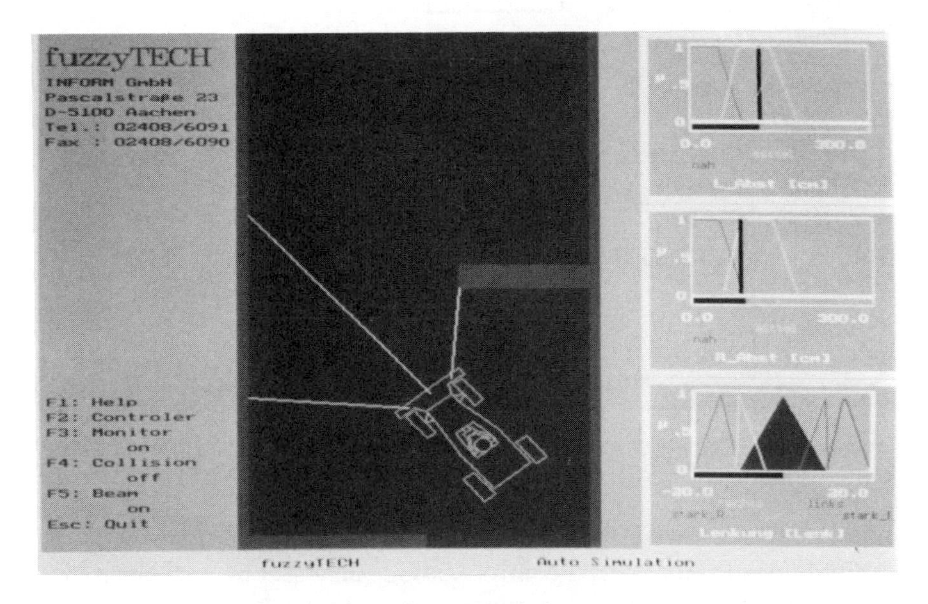

Figure 10–3. Simulation screen of the model car as presented at the FUZZ-IEEE conference in San Diego, March 1992.

For complex system structures, *fuzzy*TECH introduces the concept of normalized rule bases. With that, even very large rule bases remain lucid and easy to comprehend and can be visualized as multi-dimensional matrices. In addition to the "max-min" or "max-prod" inference methods, more advanced reasoning methods such as Fuzzy Associative Maps and compensatory operators may be used. This eases the tuning of a fuzzy controller. Figure 10–4 shows the rule base of the real model car as an example.

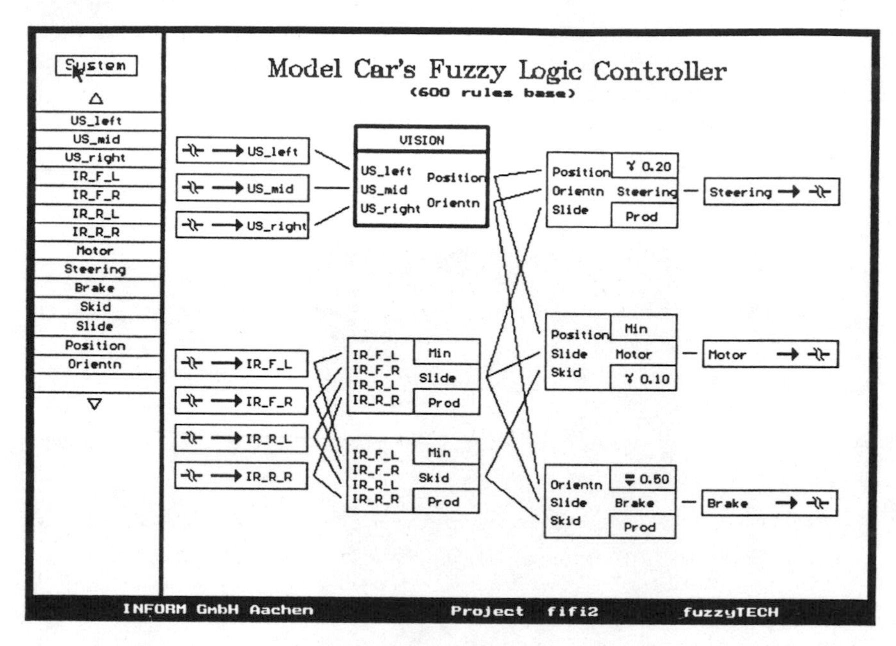

Figure 10–4. Controller structure of the real model car with *fuzzy*TECH.

For dynamic processes, a control strategy may also be developed "on-line", i.e. while the process is running in real-time. For that, the work-station / PC running the development system is connected to the target controller hardware by a serial link.

10.1.3 Applications

We shall now illustrate the use of fuzzy set theory in expert systems by sketching some example "cases" (existing expert systems and published approaches which could be used in expert systems).

Case 10–1: Linguistic Description of Human Judgments [Freksa 1982]

Freksa presents empirical results that suggest that more natural, especially linguistic representations of cognitive observations yield more informative and reliable interpretations than do traditional arithmomorphic representations. He starts from the following assumed chain of cognitive transformations.

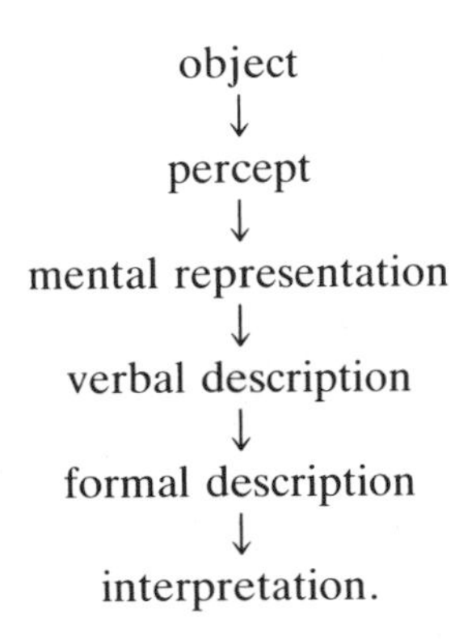

The suggested representation system for "soft observations" is supported to have the following properties [Freksa 1982, p. 302]:

1. The resolution of the representation should be flexible to account for varying precision of individual observations.
2. The boundaries of the representing objects should not necessarily be sharp and should be allowed to overlap with other representing objects.
3. Comparison between different levels of resolution of representation should be possible.
4. Comparison between subjective observations of different observers should be possible.
5. The representation should have a small "cognitive distance" to the observation.
6. It should be possible to construct representing objects empirically rather than from theoretical considerations.

The observations are expressed by simple fuzzy sets which can be described by the quadruples $\{A, B, C, D\}$, illustrated in figure 10–5, with the following interpretation: It is entirely possible that the actual feature value observed is in the range $[B, C]$; it may be possible that the actual value is in the ranges $[A, B]$ or $[C, D]$, but more easily closer to $[B, C]$ than further away; an actual value outside of $[A, D]$ is incompatible with the

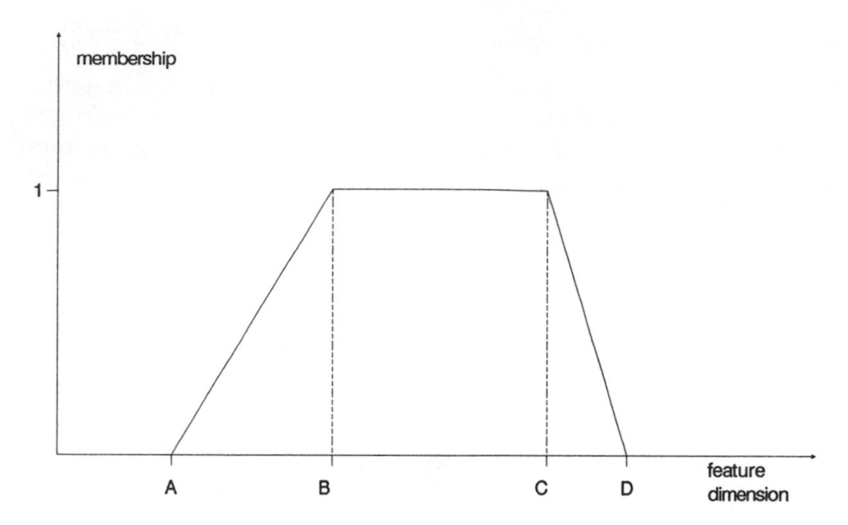

Figure 10–5. Linguistic descriptors.

observation. $[B, C]$ is called "core," $[A, B]$ and $[C, D]$ are called "penumbra" of the possibility distribution.

The construction of a repertoire of semantic representations for linguistic descriptors is done in the following way (see figure 10–6):

1. The observer selects a set of linguistic labels which allows for referencing all possible values of the feature dimension to be described.
2. The repertoire of linguistic labels is arranged linearily or hierarchically in accordance with their relative meaning in the given feature dimensions.
3. A set of examples containing a representative variety of feature values in the given feature dimension is presented to the observer. The observer marks all linguistic labels that definitively apply to the example feature value with "yes" and the labels that definitely do not apply with "no." The labels that have not been marked may be applicable, but to a lesser extent than the ones marked "yes."
4. From the data thus obtained simple membership functions are constructed by arranging the example objects according to their feature values (using the same criterion to which the linguistic labels had been arranged). These values form the domain for the assignment of membership values.

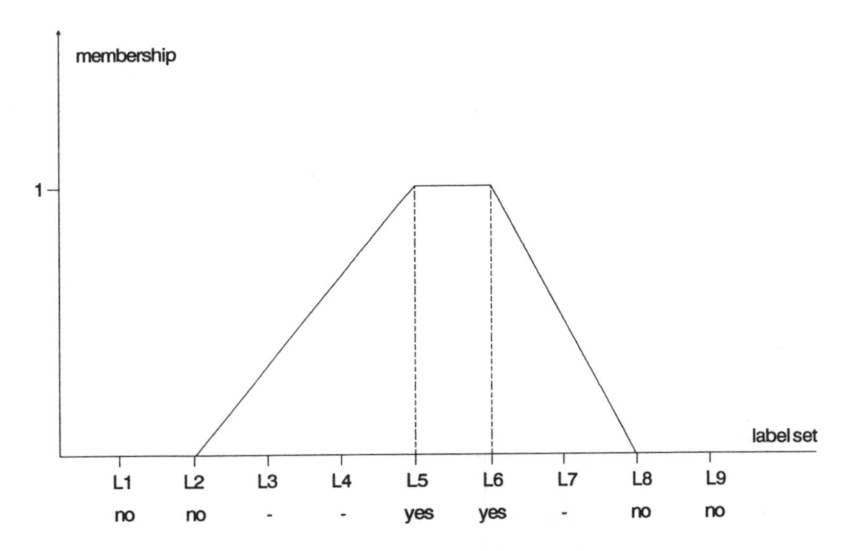

Figure 10–6. Label sets for semantic representation.

5. Finally, we assign to a given label the membership value "yes" to the range of examples in which the given label was marked "yes" for all examples and the membership value "no" to the ranges in which the given label was marked "no" for all examples. The break-off points between the regions with membership value "no" and "yes" are connected by some continuous, strongly monotonic function to indicate that the membership of label assignment increases the closer one gets to the region with membership assignment "yes" [Freska 1982, p. 303].

It is not difficult to imagine how the above technique could be used in expert systems for knowledge acquisition and for the user interface.

Case 10–2: CADIAG-2, An Expert System for Medical Diagnosis

Expert knowledge in medicine is to a large extent vague. The use of objective measurements for diagnostic purposes is only possible to a certain degree. The assignment of laboratory test results to the ranges "normal" or "pathological" is arbitrary in borderline cases, and many observations are very subjective. The intensity of pain, for instance, can only be described verbally and depends very much on the subjective estimation of the patient. Even the relationship between symptoms and diseases are generally far from crisp and unique. Adlassnig and Kolarz [1982, p. 220] mention a few typical statements from medical books which should illustrate to readers who are not medical doctors the character of available information:

Acute pancreatitis is almost always connected with sickness and vomiting.

Typically, acute pancreatitis begins with sudden aches in the abdomen.

The case history frequently reports about ulcus ventriculi and duodendi.

Bilirubinurie excludes the hemolytic icterus but bilirubin is detectable with hepatocellular or cholestatic icterus.

They designed and implemented CADIAG-2 for which they stated the following objectives [Adlassnig 1980, p. 143; Adlassnig et al. 1985]:

1. Medical knowledge should be stored as logical relationships between symptoms and diagnoses.
2. The logical relationships might be fuzzy. They are not obliged to correspond to Boolean logic.
3. Frequent as well as rare diseases are offered after analyzing the patient's symptom pattern.
4. The diagnostic process can be performed iteratively.
5. Both proposals for further investigations of the patient and reasons for all diagnostic results are put out on request.

To sketch their system let us use the following symbols:

$$\tilde{S} = \{\tilde{S}_{1,\ldots,} \tilde{S}_m\}: = \text{set of symptoms}$$

$$\tilde{D} = \{\tilde{D}_{1,\ldots,} \tilde{D}_n\}: = \text{set of diseases or diagnoses}$$

$$\tilde{P} = \{\tilde{P}_{1,\ldots,} \tilde{P}_q\}: = \text{set of patients}$$

All $\tilde{S}_i$, $\tilde{D}_j$, and $\tilde{P}_k$ are fuzzy sets characterized by their respective membership functions.

$\mu_{\tilde{S}_i}$ expresses the intensity of symptom i

$\mu_{\tilde{D}_j}$ expresses the degree of membership of a patient to $\tilde{D}_j$

$\mu_{\tilde{P}_k}$ assigns to each diagnosis a degree of membership for $\tilde{P}_k$.

Two aspects of symptom $\tilde{S}_i$ with respect to disease $\tilde{D}_j$ are of particular interest:

1. *Occurrence* of $\tilde{S}_i$ in case of $\tilde{D}_j$, and
2. *Confirmability* of $\tilde{S}_i$ for $\tilde{D}_j$

This leads to the definition of two fuzzy sets

$\tilde{O}(x)$, $x = \{0, 1, \ldots, 100\}$ for occurrence of $\tilde{S}_i$ at $\tilde{D}_j$

and

$\tilde{C}(x)$, $x = \{0, 1, \ldots, 100\}$ representing the frequency that $\tilde{S}_i$ has been confirmed for $\tilde{D}_j$

The membership functions for these two fuzzy sets are defined to be

$$\mu_{\tilde{O}}(x) = f(x; 1, 50, 99) \qquad x \in X$$
$$\mu_{\tilde{C}}(x) = f(x; 1, 50, 99) \qquad x \in Y$$

where X is the occurrence space, Y the confirmability space, and f is defined as follows (see also figure 9–4!):

$$f(x;\, \mathrm{a}, \mathrm{b}, \mathrm{c}) = \begin{cases} 0 & x \leq a \\ 2\left(\dfrac{x-a}{c-a}\right)^2 & a < x \leq b \\ 1 - 2\left(\dfrac{x-c}{c-a}\right)^2 & \text{for } b < x \leq c \\ 1 & \text{for } x > c \end{cases}$$

The $\tilde{S}_i\tilde{D}_j$ occurrence and confirmability relationships are acquired empirically from medical experts using the following linguistic variables:

i	Occurrence $\tilde{O}_i$	Confirmability $\tilde{C}_i$
1	always	always
2	almost always	almost always
3	very often	very often
4	often	often
5	unspecific	unspecific
6	seldom	seldom
7	very seldom	very seldom
8	almost never	almost never
9	never	never
	unknown	unknown

The membership functions of $\tilde{O}_i$ and $\tilde{C}_i$ are shown in figure 10–7.

They are arrived at by applying modifiers (see definition 9–3) to "never" and "always." For details of the data acquisition process see Adlassnig and Kolarz [1982, p. 226].

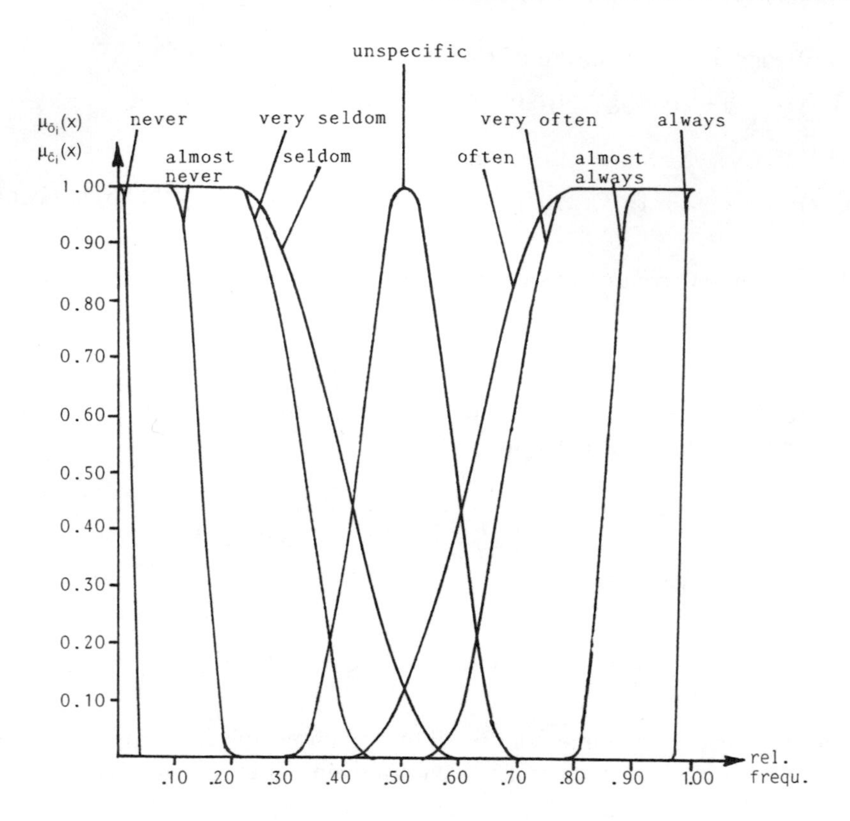

Figure 10–7. Linguistic variables for occurrence and confirmability.

Other relationships such as symptom-symptom, disease-disease, and symptom-disease are also defined as fuzzy sets (fuzzy relations). Possibilistic interpretations of relations (min-max) are used. Given a patient's symptom pattern, the symptom|disease relationships, the symptom|combination-disease relationships, and the disease|disease relationships yield fuzzy diagnostic indications that are the basis for establishing confirmed and excluded diagnosis as well as diagnostic hypotheses.

Three binary fuzzy relations are then introduced: The occurrence relation, $\widetilde{R}_{\widetilde{O}}$, the confirmability relation, $\widetilde{R}_{\tilde{c}}$, *both in* $X \otimes Y$, and the symptom relation, $\widetilde{R}_{\widetilde{S}}$, which is determined on the bases of the symptom patterns of the patients.

Finally, four different fuzzy indications are calculated by means of fuzzy relation compositions [Adlassnig and Kolarz 1982, p. 237]:

1. $\tilde{S}_i\tilde{D}_j$ *occurrence indication* $\tilde{R}_1 = \tilde{R}_{\tilde{S}} \circ \tilde{R}_{\tilde{O}}$

$$\mu_{\tilde{R}_1}(p, \tilde{D}_j) = \max_{\tilde{S}_i} \min \{\mu_{\tilde{R}\tilde{S}}(p, \tilde{S}_i), \mu_{\tilde{R}\tilde{O}}(\tilde{S}_i, \tilde{D}_j)\}$$

2. $\tilde{S}_i\tilde{D}_j$ *confirmability indication* $\tilde{R}_2 = \tilde{R}_{\tilde{S}} \circ \tilde{R}_{\tilde{C}}$

$$\mu_{\tilde{R}_2}(p, \tilde{D}_j) = \max_{\tilde{S}_i} \min \{\mu_{\tilde{R}\tilde{S}}(p, \tilde{S}_i), \mu_{\tilde{R}\tilde{C}}(\tilde{S}_i, \tilde{D}_j)\}$$

3. $\tilde{S}_i\tilde{D}_j$ *nonoccurrence indication* $\tilde{R}_3 = \tilde{R}_{\tilde{S}} \circ (1 - \tilde{R}_{\tilde{O}})$

$$\mu_{\tilde{R}_3}(p, \tilde{D}_j) = \max_{\tilde{S}_i} \min \{\mu_{\tilde{R}\tilde{S}}(p, \tilde{S}_i), 1 - \mu_{\tilde{R}\tilde{O}}(\tilde{S}_i, \tilde{D}_j)\}$$

4. $\tilde{S}_i\tilde{D}_j$ *nonsymptom indication* $\tilde{R}_4 = (1 - \tilde{R}_{\tilde{S}}) \circ \tilde{R}_{\tilde{O}}$

$$\mu_{\tilde{R}_4}(p, \tilde{D}_j) = \max_{\tilde{S}_i} \min \{1 - \mu_{\tilde{R}\tilde{S}}(p, \tilde{S}_i), \mu_{\tilde{R}\tilde{O}}(\tilde{S}_i, \tilde{D}_j)\}$$

Similar indications are determined for symptom|disease relationships, arriving at 12 fuzzy relationships $\tilde{R}_j$.

Three categories of diagnostic relationships are distinguished:

1. Confirmed diagnoses
2. Excluded diagnoses
3. Diagnostic hypotheses

Diagnoses are considered confirmed if

$$\mu_{\tilde{R}_j} = 1 \quad \text{for} \quad j = 1 \quad \text{or} \quad 6$$

or if the max-min composition of them yields 1.

For excluded diagnosis, the decision rules are more involved; and for diagnostic hypotheses, all diagnoses are used for which the maximum of the following pairs of degrees of membership are smaller .5:

$$\max \{\mu_{\tilde{R}_j}, \mu_{\tilde{R}_k}\} \leq .5 \quad \text{for}$$
$$\{j, k\} = \{1, 2\} \quad \text{or} \quad \{5, 6\} \quad \text{or} \quad \{9, 10\}$$

CADIAG-2 can be used for different purposes, for example: diagnosing diseases, obtaining hints for further examinations of patients, explanation of patient symptoms by diagnostic results.

Case 10–3: SPERIL I, an Expert System to Assess Structural Damage
[Ishizuka, Fu. Yao 1982]

Earthquake engineering has become an important discipline in areas in which the risk of earthquake is quite high.

> Frequently, the safety and reliability of a particular or a number of existing structures need to be evaluated either as part of a periodic inspection program or immediately following a given hazardous event. Because only a few experienced engineers can practice it well to date, it is planned to establish a systematic way for the damage assessment of existing structures. SPERIL is a computerized damage assessment system as designed by the authors particularly for building structures subjected to earthquake excitation. [Ishizuka et al. 1982, p. 262]

Useful information for the damage assessment comes mainly from the following two sources:

1. visual inspection at various portions of the structure
2. analysis of accelerometer records during the earthquake

The interpretation of these data is influenced to a large extent by the particular kind of structure under study. Information for damage assessment is usually being collected in a framework depicted in figure 10–8.

It is practically impossible to express the inferential knowledge of damage assessment precisely. Therefore the production rules in SPERIL I are fuzzy. A two-stage procedure is used to arrive at fuzzy sets representing the degree of damage. First the damage is assessed on a 10-point scale and then the rating is transformed into a set of terms of the linguistic variable "damage."

Let d be the damage evaluated at a 10-point scale. Then the relationship between the terms and the original ratings can be described as follows:

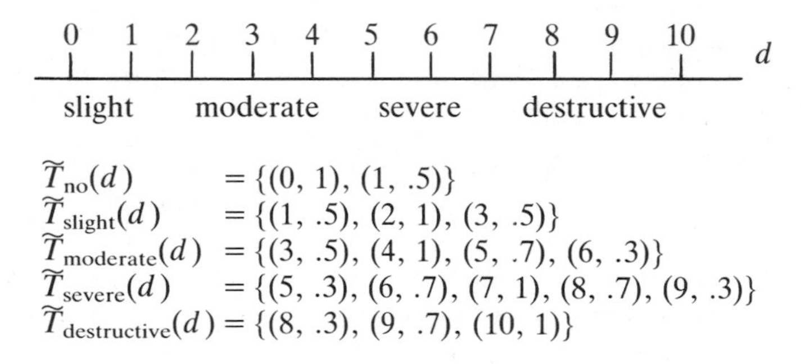

$$
\begin{aligned}
\tilde{T}_{\text{no}}(d) &= \{(0, 1), (1, .5)\} \\
\tilde{T}_{\text{slight}}(d) &= \{(1, .5), (2, 1), (3, .5)\} \\
\tilde{T}_{\text{moderate}}(d) &= \{(3, .5), (4, 1), (5, .7), (6, .3)\} \\
\tilde{T}_{\text{severe}}(d) &= \{(5, .3), (6, .7), (7, 1), (8, .7), (9, .3)\} \\
\tilde{T}_{\text{destructive}}(d) &= \{(8, .3), (9, .7), (10, 1)\}
\end{aligned}
$$

The rule associated with node 2 in figure 10–8 for instance, then would read:

IF:	MAT is reinforced concrete,
THEN IF:	STI is no,
THEN:	GLO is no with 0.6,
ELSE IF:	STI is slight,

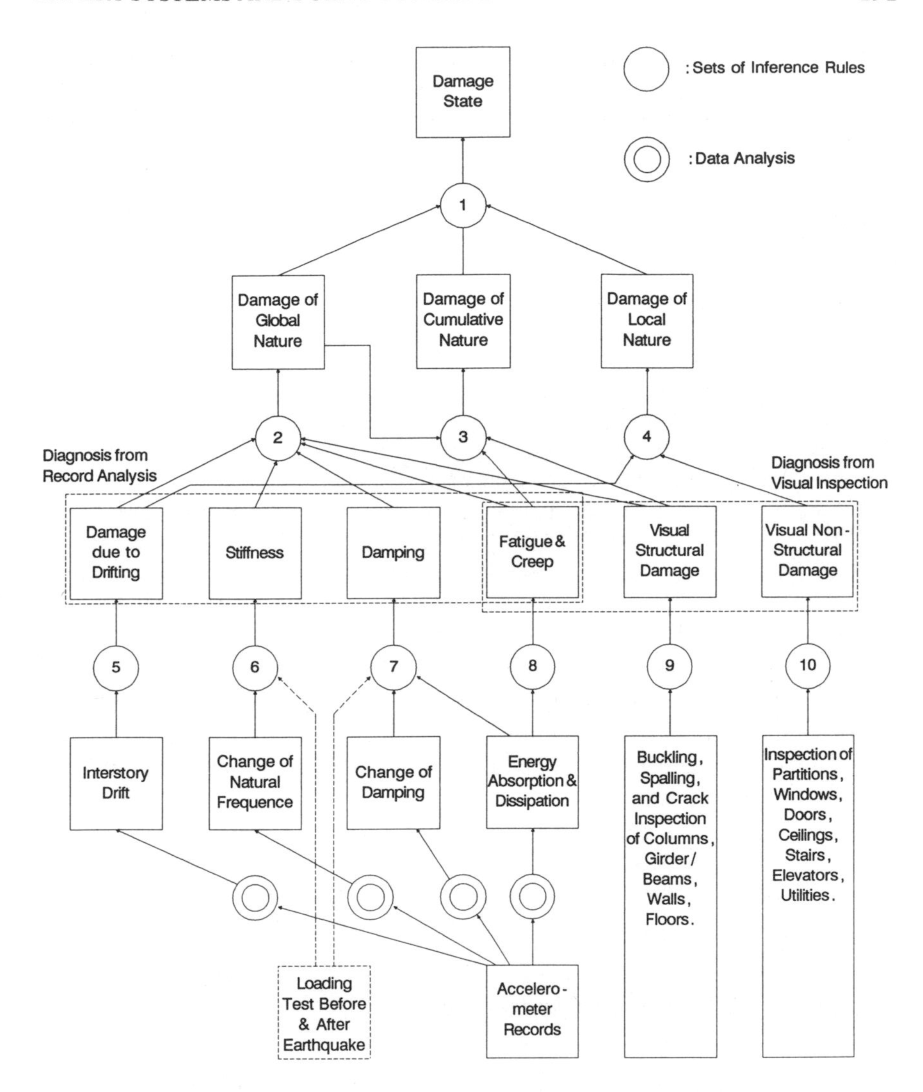

Figure 10–8. Inference network for damage assessment of existing structures. [Ishizuka et al. 1982, p. 263].

THEN: GLO is slight with 0.6,
ELSE IF: STI is moderate,
THEN: GLO is moderate with 0.6,
ELSE IF: STI is severe,

THEN: GLO is severe with 0.6,
ELSE IF: STI is destructive,
THEN: GLO is destructive with 0.6,
ELSE: GLO is unknown with 1,

where

MAT = structural material
GLO = damage of global nature,
STI = diagnosis of stiffness,
"unknown" stands for the universe set of damage grade.

To obtain a correct answer by using such knowledge, a rational inference mechanism is required to process the rules expressed with fuzzy subsets along with uncertainty in an effective manner.

To include uncertainty, first Dempster's and Shafer's probabilities were used [Dempster 1967; Shafer 1976]. Thus the conclusions were accompanied by a lower and upper probability indicating lower and upper bounds of subjective probabilities. (For details see Ishizuka et al. [1982, pp. 264–266].)

It was felt that the rules as shown for node 2 could not necessarily be expressed as crisp rules. Therefore fuzzy inference rules were introduced in order to arrive at a fuzzy damage assessment together with upper and lower probabilities. For details the reader is again referred to the above-mentioned source.

Improvements, particularly of the knowledge acquisition phase, have been suggested [Fu et al. 1982; Watada et al. 1984]. They either use fuzzy clustering or a kind of linguistic approximation.

Case 10–4: ESP, an Expert System for Strategic Planning [Zimmermann 1989]

Strategic Planning is a large heterogeneous area with changing content over time and without a closed theory such as is available in other areas of management and economics. It deals with the long-range planning of a special company and is frequently done for independent autonomous units, called Strategic Business Units (SBU) [Hax and Majluf 1984, p. 15]. One technique for analyzing the current and future business position is the business portfolio approach.

The original idea of portfolio analysis in strategic planning was to describe the structure of a corporation by the positions of SBU's in a two-dimensional portfolio matrix and to try to find strategies aimed at keeping this "portfolio" balanced. Some of the major problems encountered are:

Dimensionality: It is obvious that two dimensions are insufficient to describe adequately the strategic position of an SBU. Two dimensions are certainly preferable for didactical reasons and for presentation, but for realistic description a multidimensional vectorial positioning would be better.

Data Collection and Aggregation: Even for a two-dimensional matrix the dimensions of an SBU must be determined by a rather complex data-gathering and aggregation process. Factors such as ROI, market share, and market growth, can be obtained without too much difficult. Other factors to be considered are combinations of many aspects. It is, therefore, not surprising that intuitive aggregation and the use of scoring methods are rather common in this context, although their weaknesses are quite obvious: Aggregation procedures are kept simple for computational efficiency but they are very often not justifyable. Different factors are considered to be independent without adequate verification. A lot of subjective evaluations enter the analysis with very little control.

Strategy Assignment: In classical portfolio matrixes, broad strategic categories have been defined to which basic strategies are assigned. It is obvious that these categories are much too rough to really define operational strategies to them. One of the most important factors in determining real strategies will be the knowledge and experience of the strategic planners who transform those very general strategic recommendations into operational strategies. A knowledge that is not captured in the portfolio matrixes!

Modeling and Consideration of Uncertainty: In an area in which many ill-structured factors, weak signals, and subjective evaluations enter, and which extends so far into the future, uncertainty is obviously particularly relevant. Unfortunately, however, it is hardly considered in most of the strategic planning systems we know. The utmost that is done is to sometimes attach uncertainty factors to estimate and then to aggregate those together with the data in a rather heuristic and arbitrary way.

ESP, an Expert System for Strategic Planning, tries to improve classical approaches and to remedy some of their shortcomings. It also provides a framework in which strategic planners can analyze strategic information and develop more sophisticated strategic recommendations. Its characteristics are as follows:

Dimensionality. Multidimensional portfolio matrixes are used. For visualization, two dimensions each can be chosen, the location of SBU's are defined by vectors, however. As an example let us consider the four dimensions:

1. Technology Attractiveness
2. Technology Position
3. Market Attractiveness
4. Competitive Position.

If we combine the first two and the last two dimensions we obtain two two-dimensional portfolio matrixes which, combined, correspond to a four-dimensional matrix, see figure 10–9. If each of the two-dimensional

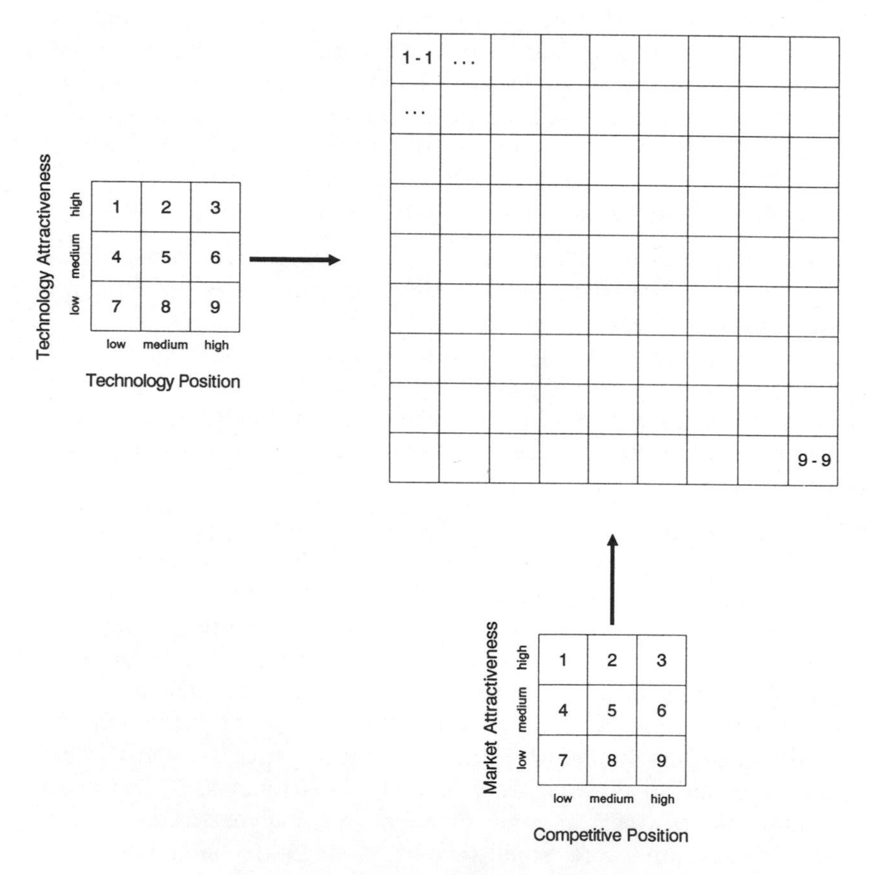

Figure 10–9. Combination of two two-dimensional portfolios.

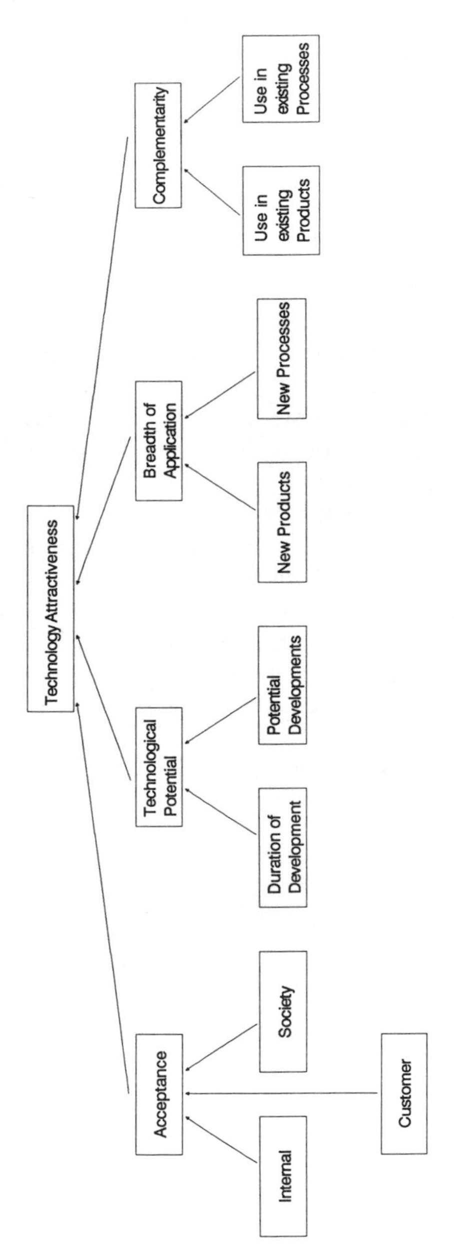

Figure 10–10. Criteria tree for technology attractiveness.

matrixes consists of nine strategic categories by having three intervals—low, medium, high—on each axis, then the combined matrix contains $9 \times 9 = 81$ strategic positions. Graphically, only the two-dimensional matrixes are shown. The positions of the combined matrix are only stored vectorially and used for more sophisticated policy assignment.

Data Collection and Aggregation. Each "dimension" is defined by a tree of subcriteria and categories. Figure 10–10 shows a part of the tree for "Technology Attractiveness."

The input given by the user consists of one linguistic variable for all criteria of the leaves (lowest subcriteria in each of the four trees). This linguistic variable denotes the respective "degree of achievement"; it can be chosen from the terms "not at all," "little," "medium," "considerably," and "full." These terms are represented by trapezoidal membership functions which are characterized by their four characteristic values on their supports, see figure 10–11.

To arrive at the root of each tree these ratings of the leaves are aggregated on every level of the tree by using the γ-operator, described in chapter 3. There the reader will find other operators (e.g., minimum, product), which can also be chosen by the user. It is suggested that this aggregation of linguistic terms, rather than of numerical values, be done by aggregating the four characteristic values of each trapezoid in order to obtain the respective characteristic value of the resulting trapezoid. For the last aggregation level of one tree—this is shown in figure 10–12. Repeating this procedure for all characteristic values of the membership

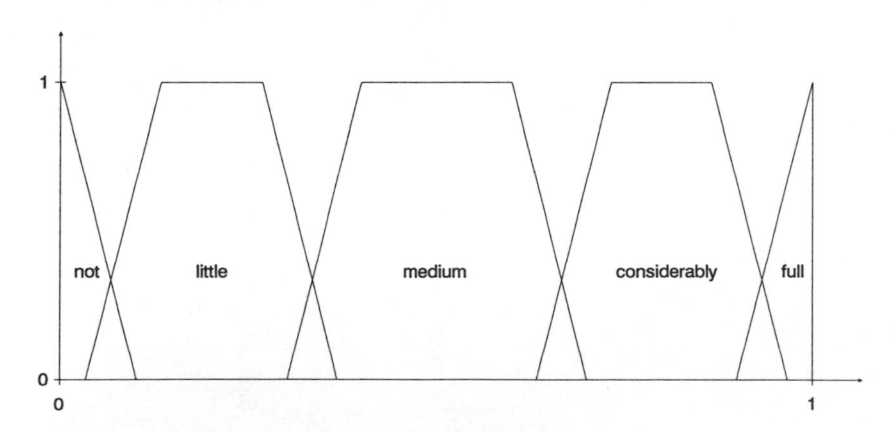

Figure 10–11. Terms of "degree of achievement."

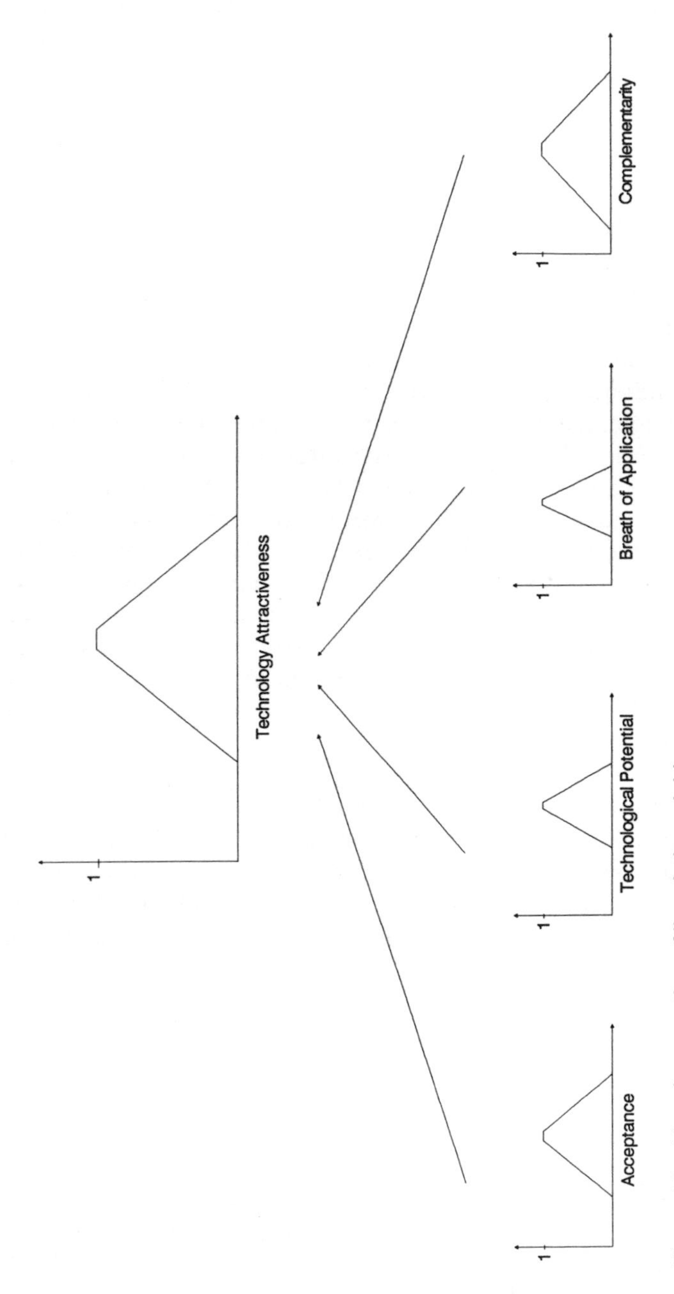

Figure 10–12. Aggregation of linguistic variables.

functions of all aggregation steps of each of the four trees leads to a trapezoidal membership function for each of the criteria.

Strategy Assignment. As already mentioned, strategy assignment is made on the basis of the vectorially described position of an SBU. Two levels can be distinguished:

A. General Policy Recommendation
   This is assigned to the position of the SBU as it is defined by the values of the roots of the trees. In our example the position would be defined by technology attractiveness, technology position, by competitive position and market attractiveness.
B. Detailed Policy Recommendation
   Policy recommendations based on the location in the portfolio matrix, which in turn is determined by the values of the roots of the evaluation trees only, can only be very rough guidelines. The same value at a root of the tree can be obtained from very different vectors of values of the nodes of the first level of the tree. The values of this vector are, therefore, used to make more specific strategic recommendations in addition to the basic policy proposal mentioned above. In the example tree shown in figure 10–10, for instance, the ratings of "Acceptance," "Technological Potential," "Breadth of Application," and "Complementarity" would be used for such a specification of the strategic recommendation.

Modeling and Consideration of Uncertainty. It is possible for the user of ESP to interact with this system by defining a special α-level which results in a rectangle in the portfolio matrixes as shown in figure 10–13. The α-level denotes the desired degree of certainty; and the corresponding area in the matrix is a visualization of the possible position of the considered SBU.

ESP: Implementation. We had intended to design ESP by using one of the available shells. It turned out, however, that none of the available shells offers all the features we needed. Therefore, a combination of a shell (in this case Leonardo 3.15) with a program (in Turbo Pascal) had to be used. The basic structure of ESP is shown in figure 10–14.

Knowledge Base I contains primarily rules which assign basic strategy recommendations to locations of SBU in multidimensional portfolio matrixes and detailed supplementary recommendations to profiles of the first levels of trees. Together with the inference engine, it provides for the

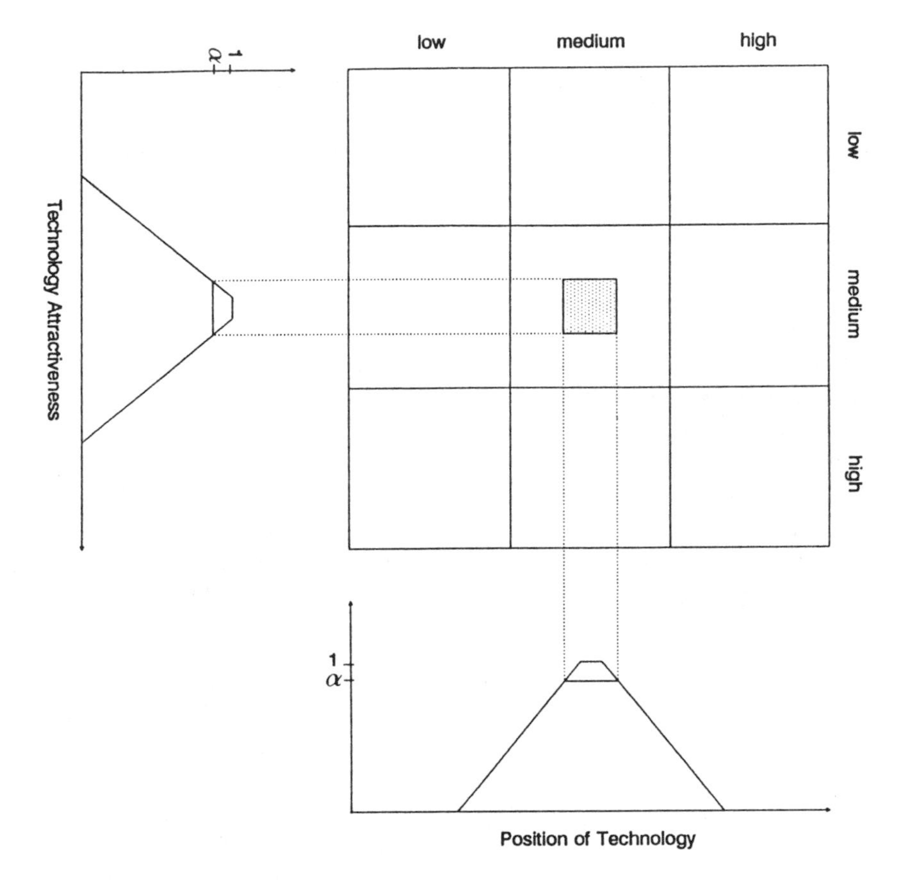

Figure 10–13. Portfolio with linguistic input.

user the "if-then"-part and the explanatory function. For this part the shell Leonardo 3.15 was used.

Knowledge Base II contains the structures of the free defineable trees that determine the location of an SBU in the different dimensions of the multidimensional matrix. The "Aggregator" computes their values and characteristic values for the linguistic values for all nodes of the trees on the basis of available structural knowledge (tree structure, α-values, and γ-values) and on the basis of data (μ-values) entered for each terminal leaf by the user. The information provided by the "Aggregator" is then used for the visual presentation of two-dimensional matrixes and profiles and also supports the explanatory module.

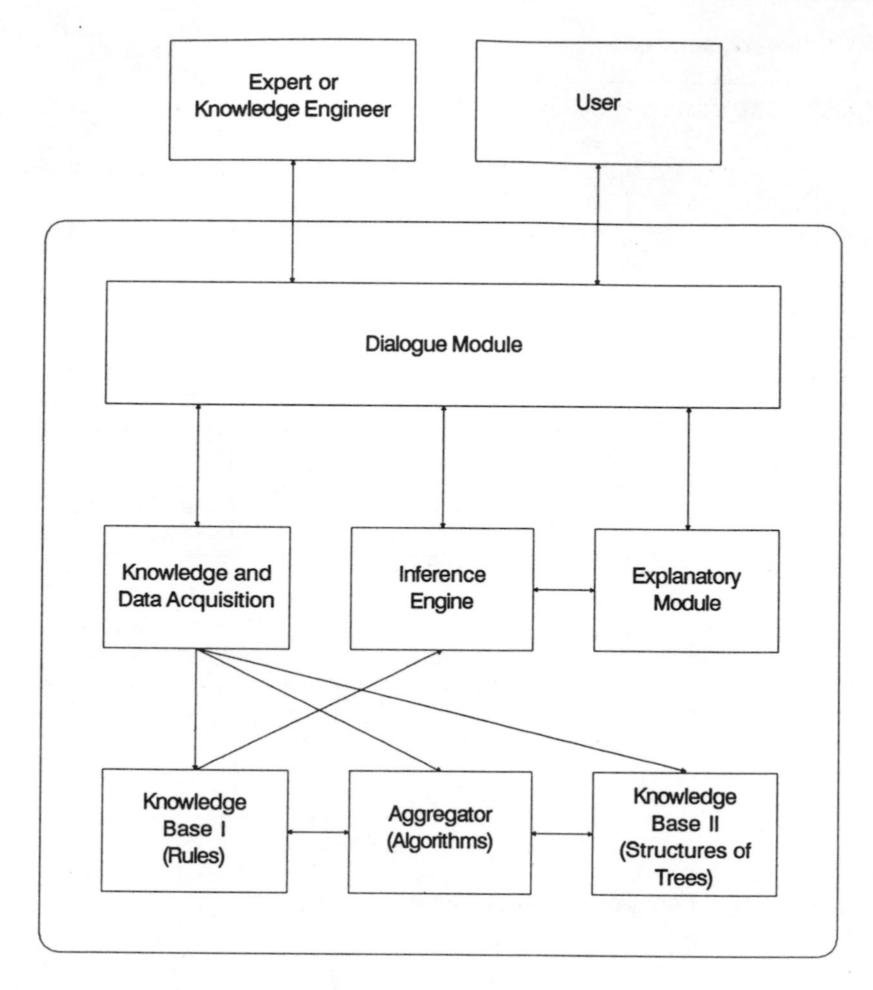

Figure 10–14. Structure of ESP.

All aggregation and visual presentation functions could not be accommodated by Leonardo 3.15. Therefore, an extra program in Turbo-Pascal and the appropriate bridge programs to Leonardo had to be written.

ESP is fully menue driven. It could be considered as a second-generation expert system that works with shallow knowledge (KB I) as well as with deep knowledge (KB II).

10.2 Fuzzy Control

10.2.1 Introduction to Fuzzy Control

Expert systems and fuzzy logic control (FLC) systems have certainly one thing in common: Both want to model human experience, human decision-making behavior. To quote some of the originators of FLC:

> The basic idea behind this approach was to incorporate the "experience" of a human process operator in the design of the controller. From a set of linguistic rules which describe the operator's control strategy a control algorithm is constructed where the words are defined as fuzzy sets. The main advantages of this approach seem to be the possibility of implementing "rule of the thumb" experience, intuition, heuristics, and the fact that it does not need a model of the process. [Kickert and Mamdani 1978]

> Certain complex industrial plants, for example, a cement kiln, can be controlled with better results by an experienced operator than by conventional automatic controllers. The control strategies employed by an operator can often be formulated as a number of rules that are simply to carry out manually but difficult to implement by using conventional algorithms. This difficulty is because human beings use qualitative rather than quantitative terms when describing various decisions to be taken as a function of different states of the process. It is this qualitative or fuzzy nature of man's way of making decisions that has encouraged control engineers to try to apply fuzzy logic to process control. [Ostergaard 1977]

> Complex industrial processes such as batch chemical reactors, blast furnaces, cement kilns, and basic oxygen steel-making are difficult to control automatically. This difficulty is due to their nonlinear, time varying behaviour, and the poor quality of available measurements. In such cases, automatic control is applied to those subsidiary variables that can be measured and controlled, for example, temperatures, pressures, and flows. The over-all process control objectives, such as the quality and quantity of product produced, has in the past been left in the hands of the human operator.
>
> In some modern plants with process control computers, plant models have been used to calculate the required controller settings automating the higher level control functions. The plant models whether they are based on physical and chemical relationships or parameter estimation methods are approximations to the real process and may require a large amount of computer time. Some successful applications have been reported, but difficulties have been experienced where processes operate over a wide range of conditions and suffer from stochastic disturbances.
>
> An alternative approach to the control of complex processes is to investigate

> the control strategies employed by the human operator. In many cases, the process operator can control a complex process more effectively than an automatic system; when he experiences difficulty this can often be attributed to the rate or manner of information display or the depth to which he may evaluate decisions. [King and Mamdani 1977]

There are, however, also clear differences between expert systems and FLC:

1. The existing FLC systems originated in control engineering rather than in artificial intelligence.
2. FLC models are all rule-based systems.
3. By contrast to expert systems FLC serves almost exclusively the control of (technological) production systems such as electrical power plants, kiln cement plants, chemical plants, and so on, that is, their domains are even narrower than those of expert systems.
4. In general, the rules of FLC systems are not extracted from the human expert through the system but formulated explicitely by the FLC designer.
5. Finally, because of their purpose, their inputs are normally observations of technological systems and their outputs control statements.

Almost all designers of FLC systems agree that the *theoretical origin* of those systems is the paper "Outline of a new approach to the analysis of complex systems and decision processes" by Zadeh [1973b]. It plays almost the same role as the Bellman-Zadeh [1970] paper on "Decision making in a fuzzy environment," for the area of decision analysis. In particular, the compositional rule of inference (see definition 9–7) is considered to be the spine of all FLC-models. The *original applied* activities centered around Queen Mary College in London. Key to that development was the work of E. Mamdani and his students in the Department of Electrical and Electronic Engineering. Richard Tong, of nearby Cambridge, was another key figure in the development of fuzzy control theory. The first application of fuzzy set theory to the control of systems was by Mamdani and Assilian [1975], who reported on the control of a laboratory model steam engine.

Nowadays, Japan has become the most active country in this respect [see Gupta and Yamakawa 1988b].

10.2.2 Process of Fuzzy Control

Before the development of FLC systems there were essentially two alternatives to process control: A process was controlled by either a human

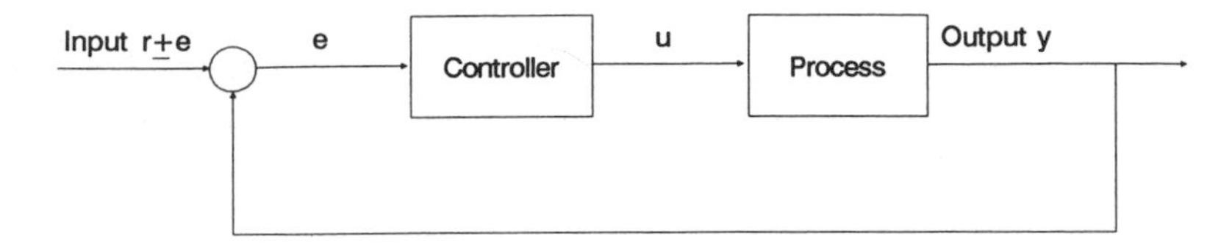

Figure 10–15. DDC-control system.

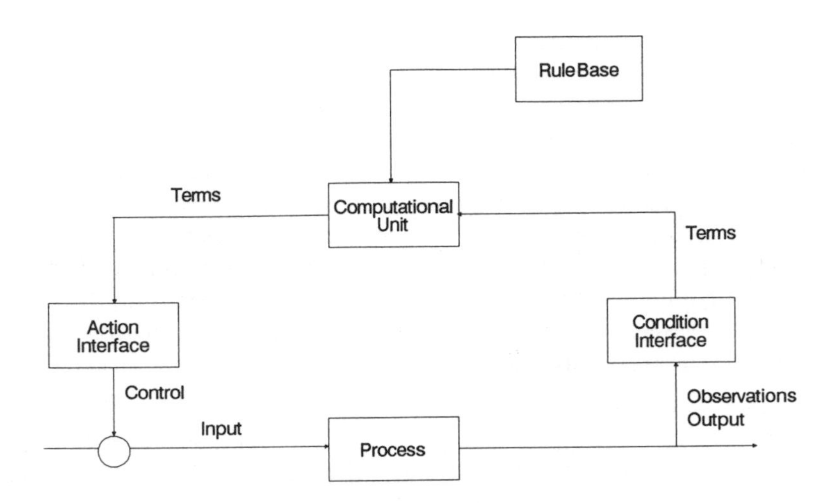

Figure 10–16. Fuzzy logic control system.

operator or a computerized DDC controller. The function of such a Direct Digital Control system can be described as follows (see figure 10–15):

> The problem consists in dimensioning a control algorithm based on the error vector $e = (e_1, e_2, \ldots, e_p)$ that generates an output vector $u = (u_1, u_2, \ldots, u_r)$ to the process so that the output vector $y = (y_1, y_2, \ldots, y_p)$ of the process is close to or eventually equal to the setpoint vector $r = (r_1, r_2, \ldots, r_p)$. In other words, we want to control the process by means of an algorithm of the following general form
>
> $$u[(k+1)T] = f(u[kT], u[(k-1)T], \ldots, u[0], e(k+1)T], e[kT], \ldots, e[0])$$
>
> where $k = 1, 2, \ldots$. And T is the sampling time." [Ostergaard 1977]

The structure of a fuzzy logical controller is depicted in figure 10–16. The process of fuzzy control can roughly be described as shown in figure 10–17 [see also Tong 1984].

Figure 10–17. Functions of FLC.

The essential design problems in FLC are therefore:

1. Define input-and control variables, that is, determine which states of the process shall be observed and which control actions are to be considered.
2. Define the condition interface, that is, fix the way in which observations of the process are expressed as fuzzy sets.
3. Design the rule base, that is, determine which rules are to be applied under which conditions.
4. Design the computational unit, that is, supply algorithms to perform fuzzy computations. Those will generally lead to fuzzy outputs.
5. Determine rules according to which fuzzy control statements can be transformed into crisp control actions.

We shall consider these five questions in turn. For the purpose of illustration we shall use as a process to be controlled the engine-boiler combination considered by Mamdani and his group [King and Mamdani 1977; Mamdani and Assilian 1975]. Most of the other FLC applications that have become known follow this pattern very closely.

1. Process Interface. Two state variables of the process are to be observed: the steam pressure in the boiler and the speed of the engine. Also two possible control actions are considered: heat input to the boiler and throttle opening at the input of the engine cylinder.

The states of the system are described by deviations from a standard, called error, and the control actions are "Heat Change" and "Throttle Change."

2. Definition of Fuzzy State Descriptions. Terms of linguistic variables (see definition 9–1) are used to describe the states of the process. In practical terms [King and Mamdani 1977, p.322]: The error value and the change of error values calculated are quantized into a number of points corresponding to the elements of a universe of discourse, and the values are then assigned as grades of membership in eight fuzzy subsets as follows:

1. PB is "positive big,"
2. PM is "positive medium,"
3. PS is "positive small,"
4. PO is "positive nil,"
5. NO is "negative nil,"
6. NS is "negative small,"
7. NM is "negative medium," and
8. NB is "negative big."

The relationship between measured error or change in error value and grade of membership are defined by look-up tables of the form given below.

The PE (Pressure Error) and SE (Speed Error) variables are quantized into 13 points, ranging from maximum negative error through zero error to maximum positive error. The zero error is further divided into negative zero error (NO—just below the set point) and positive zero error (PO—just above the set point). The subjective fuzzy sets defining these values are:

	−6	−5	−4	−3	−2	−1	−0	+0	+1	+2	+3	+4	+5	+6
PB	0	0	0	0	0	0	0	0	0	0	0.1	0.4	0.8	1.0
PM	0	0	0	0	0	0	0	0	0	0.2	0.7	1.0	0.7	0.2
PS	0	0	0	0	0	0	0	0.3	0.8	1.0	0.5	0.1	0	0
PO	0	0	0	0	0	0	0	1.0	0.6	0.1	0	0	0	0
NO	0	0	0	0	0.1	0.6	1.0	0	0	0	0	0	0	0
NS	0	0	0.1	0.5	1.0	0.8	0.3	0	0	0	0	0	0	0
NM	0.2	0.7	1.0	0.7	0.2	0	0	0	0	0	0	0	0	0
NB	1.0	0.8	0.4	0.1	0	0	0	0	0	0	0	0	0	0

The CPE (Change in Pressure Error) and CSE (Change in Speed Error) variables are similarly quantized without the further division of the zero state. The subjective definitions are:

	−6	−5	−4	−3	−2	−1	0	+1	+2	+3	+4	+5	+6
PB	0	0	0	0	0	0	0	0	0	0.1	0.4	0.8	1.0
PM	0	0	0	0	0	0	0	0	0.2	0.7	1.0	0.7	0.2
PS	0	0	0	0	0	0	0	0.9	1.0	0.7	0.2	0	0
NO	0	0	0	0	0	0.5	1.0	0.5	0	0	0	0	0
NS	0	0	0.2	0.7	1.0	0.9	0	0	0	0	0	0	0
NM	0.2	0.7	1.0	0.7	0.2	0	0	0	0	0	0	0	0
NB	1.0	0.8	0.4	0.1	0	0	0	0	0	0	0	0	0

Apart from the above definitions a further value, ANY, is allowed for all four variables, that is, PE, SE, CPE, CSE. ANY has a membership function of 1.0 at every element. The HC (Heat Change) variable is quantized into 15 points ranging from a change of −7 steps through 0 to +7 steps. The subjective definitions are:

	−7	−6	−5	−4	−3	−2	−1	0	+1	+2	+3	+4	+5	+6	+7
PB	0	0	0	0	0	0	0	0	0	0	0	0.1	0.4	0.8	1.0
PM	0	0	0	0	0	0	0	0	0	0.2	0.7	1.0	0.7	0.2	0
PS	0	0	0	0	0	0	0	0.4	1.0	0.8	0.4	0.1	0	0	0
NO	0	0	0	0	0	0	0.2	1.0	0.2	0	0	0	0	0	0
NS	0	0	0	0.1	0.4	0.8	1.0	0.4	0	0	0	0	0	0	0
NM	0	0.2	0.7	1.0	0.7	0.2	0	0	0	0	0	0	0	0	0
NB	1.0	0.8	0.4	0.1	0	0	0	0	0	0	0	0	0	0	0

Similarly, the TC (Throttle Change) variable is quantized into 5 points:

	−2	−1	0	+1	+2
PB	0	0	0	0.5	1.0
PS	0	0	0.5	1.0	0.5
NO	0	0.5	1.0	0.5	0
NS	0.5	1.0	0.5	0	0
NB	1.0	0.5	0	0	0

[Mamdani, Assilian 1981, p. 317]

3. Design of Rule-Base. The control rules are defined as fuzzy conditional statements of the type "If PE is NB then HC is PB." The min-max compositional rule of inference is then used to derive fuzzy control statements from observed observations of the states of the process. If several rules are combined by "else," this is interpreted as the union operator "max." The heater algorithm contains 15 "if-then" statements combined by "else." For instance,

If PE = NB
and CPE = not (NB or NM)
and SE = ANY
and CSE = ANY
then HC = PB
Else
if PE = NB or NM
and CPE = NS

```
        and    SE = ANY
        and   CSE = ANY
      then     HC = PM
Else
        if     PE = NS
        and   CPE = PS or NO
        and    SE = ANY
        and   CSE = ANY
      then     HC = PM
Else
        if     PE = NO
        and   CPE = PB or PM
        and    SE = ANY
        and   CSE = ANY
      then     HC = PM
```

The throttle algorithm looks accordingly such as

```
        If     PE = ANY
        and   CPE = ANY
        and    SE = NB
        and   CSE = not (NB or NM)
      then     TC = PB
Else
        if     PE = ANY
        and   CPE = ANY
        and    SE = NM
        and   CSE = PB or PM or PS
      then     TC = PS
Else
        if     PE = ANY
        and   CPE = ANY
        and    SE = NS
        and   CSE = PB or PM
      then     TC = PS
Else
        if     PE = ANY
        and   CPE = ANY
        and    SE = NO
        and   CSE = PB
      then     TC = PS
```

For didactical reasons we omit the design of the computational unit, which is only of technical interest.

5. Translation of Fuzzy Control Statements into Crisp Control Actions. If the membership functions of the fuzzy control statements are convex and unimodal, the crisp maximizing decision is used for the control action.

If the membership functions of the control statements are not unimodal, other heuristic rules are applied such as "Take the action that is midway between two peaks or at the center of a plateau."

In experimental studies it turned out that FLC performed quite well and a fast development of this area was envisaged. For reasons not well known the process slowed down, however, at the end of the 1970s. One explanation might be Tong's statement:

> This may not be apparent from the published literature, but we should realize that the controllers developed so far have had rather limited rule-bases (usually of the order of 10–20 rules). The main reason for this is that the application studies have largely restricted themselves to rather simple systems in which there are at most two inputs and two outputs. Clearly, there is only a limited amount of information that needs to be captured in order to control such systems. Controllers for more complex systems will require considerably more effort and will require something other than the ad hoc methods used so far. First, there need to be systematic and formal procedures for rule elicitation. This requires some computer-aided design software that would allow the designer to specify variables, primary linguistic terms, etc., and would then engage in a dialogue designed to extract the rules. [Tong 1984]

Tong also points to the relationships between expert systems and FLC and suggests that new advances in the development of FLC might result from allying ideas which are used in expert systems nowadays to the design of the rule base of FLC, that is, to convert the rule base into a more sophisticated inference engine as in expert systems.

> In keeping with the source of this insight I want to propose a new name for the devices we are building, namely Expert Fuzzy Controllers (henceforth EFC). They differ from FLCs in at least two important ways: they contain more complex knowledge about process control, and they use this knowledge in more complex ways. Fuzzy sets are still used to model uncertainty but EFCs are more general in scope and are capable of handling a wider variety of control problems. [Tong 1984, p. 9]

Since the mid 1980s fuzzy control has become more popular among control engineers working in industry, and more and more systems are implemented [Sugeno 1985b, p. 59]. Especially in Japan, fuzzy control seems to be the most frequent application of fuzzy set theory. A state-of-the-art report of theoretical developments in the field of fuzzy control and possible applications is given by [Pedrycz 1989].

10.2.3 Applications of Fuzzy Control

Here we will give an overview of current applications (see table 10–2) of fuzzy control to real-world problems by quoting some references and by analyzing two well-known systems in more detail.

1. Fuzzy Control of a Cement Kiln [King, Karonis 1988, pp. 323] In this case we will consider a physical process as the object of control. Let us first describe briefly the process itself.

> Cement is manufactured by heating a slurry consisting of clay, limestone, sand, and iron ore to a temperature that will permit the formation of the complex compounds of cement, dicalcium silicate (C_2S), tricalcium silicate (C_3S), tricalcium aluminate (C_3Al), and tetracalcium aluminoferrite (C_4AlF). In the first stage of the kilning process the slurry is dried and excess water is driven off. In the second stage calcining takes place with the calcium carbonate decomposing to calcium oxide and carbon dioxide. In the final stage burning takes place at 1250–1450°C and free lime (CaO) combines with the other ingredients to form the cement compounds. The end product of the burning process is referred to as clinker.
>
> The kiln consists of a long steel shell about 130 m in length and 5 m in diameter. The shell is mounted at a slight inclination to the horizontal, and is

Table 10–2. Fuzzy Control Systems.

Example	*Control problem*	*Reference*
Coke oven Gas Cooling Plant	Temperature control, maintenance guide	Tobi et al. 1989
Water Purification	Chemicals addition	Yagishita 1985
Temperature Control	Self-regulation of parameters	Peng et al. 1988
Car Control by Oral Instructions	Fuzzy algorithm	Sugeno et al. 1989
Combustion Control	Control of combustion air flow	Ono et al. 1989

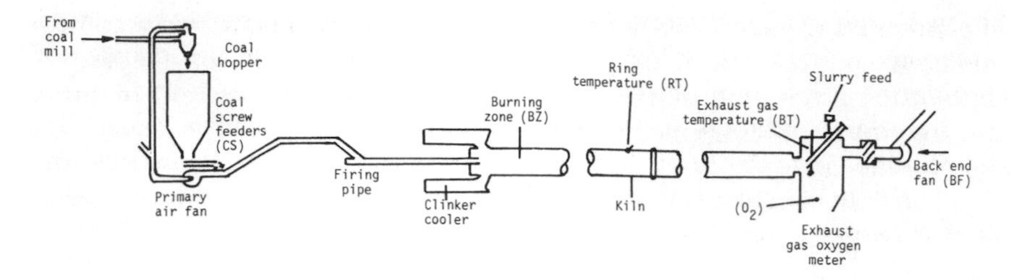

Figure 10–18. Schematic diagram of rotary cement kiln. [Umbers and King 1981, p. 371]

> lined with fire bricks. The shell rotates slowly, at approximately 1 rev/min, and the slurry is fed in at the upper or back end of the kiln. The inclination of the shell and its rotation transports the material through the kiln in about 3 hours 15 mins with a further 45 mins spent in the clinker cooler.
>
> The heat in the kiln is provided by pulverised coal mixed with air, referred to as primary air. The hot combustion gases are sucked through the kiln by an induction fan at the back end of the kiln. [Umbers and King 1981, p. 370]

Figure 10–18 illustrates the production process. The main problem in mathematically modeling a control strategy is that the relationships between input variables (measured characteristics of the process) and control variables are complex, nonlinear, and contain time lags and interrelationships, and that the kiln's response to control inputs depends on the prevailing kiln conditions. This was certainly one reason why a fuzzy control system was designed and used, which eventually even led to a commercially available fuzzy controller.

From the many possible input and control variables the following were chosen as particularly relevant. Input variables:

1. exhaust gas temperature—back-end temperature (BT);
2. intermediate gas temperature—ring temperature (RT);
3. burning-zone temperature (BZ);
4. oxygen percentage in exhaust gases (O_2); and
5. liter weight (LW)—indicates clinker quality.

The process is controlled by varying the following control variables:

1. kiln process (KS);
2. coal feed (CS)—fuel; and
3. induced draught-fan speed (BF).

The calculation of the control action was composed of the following four stages:

1. calculate the present error and its rate of change;
2. convert the error values to fuzzy variables;
3. evaluate the decision rules using the compositional rule of inference;
4. calculate the deterministic input required to regulate the process.

Concerning the control strategies used let us quote Larsen [1981, p. 337].

> The aim of the computerized kiln control system is to automate the routine control strategy of an experienced kiln operator. The applied strategies are based on detailed studies of the process operator experiences which include a qualitative model of influence of the control variables on the measured variables.

1. If the coal-feed rate is increased, the kiln drive load and the temperature in the smoke chamber will increase, while the oxygen percentage and the free lime content will decrease.
2. If the air flow is increased, the temperature in the smoke chamber and the free lime content will increase, while the kiln drive load and the oxygen percentage will decrease.

On the basis of thorough discussions with the operators, Jensen [1976] defined 75 operating conditions as fuzzy conditional statements of the type:

IF	drive load gradient is	(DL,SL,OK,SH,DH)
AND	drive load is	(DL,SL,OK,SH,DH)
AND	smoke chamber temperature is	(L,OK,H)
THEN	change oxygen percentage is	(VN,N,SN,ZN,OK,ZP,SP,P,VP)
PLUS	change air flow is	(VN,N,SN,ZN,OK,ZP,SP,P,VP)

The following fuzzy *primary terms* are used for the measured variables:

1. DL = drastically low
2. L = low
3. SL = slightly low
4. OK = ok
5. SH = slightly high
6. H = high
7. DH = drastically high

The following fuzzy *primary terms* are used for the control variables:

1. VN = very negative
2. N = negative
3. SN = small negative
4. ZN = zero negative

5.	OK = ok	8.	P = positive
6.	ZP = zero positive	9.	VP = very positive
7.	SP = small positive		

The *linguistic terms* are represented by membership functions with four discrete values in the interval [0, 1] associated with 15 discrete values of the scaled variables in the interval [−1, +1].

In order to simplify the implementation of the fuzzy logic controller, Ostergaard [1977] defined 13 *operating conditions* as fuzzy conditional statements of the type:

IF	drive load gradient is	(SN,ZE,SP)
AND	drive load is	(LN,LP)
AND	free lime content	(LO,OK,HI)
THEN	change burning zone temperature	(LN,MN,SN,ZE,SP,MP,LP)

The following fuzzy *primary terms* are used:

1.	LP = large positive	7.	SN = small negative
2.	MP = medium positive	8.	MN = medium negative
3.	SP = small positive	9.	LN = large negative
4.	ZP = zero positive	10.	HI = high
5.	ZE = zero	11.	OK = ok
6.	ZN = zero negative	12.	LO = low

The 13 operating conditions are defined by taking only some of the combinations into account, and by including also previous values of the drive load gradient, the latter being calculated from the changes in the drive load. In order to decide whether the oxygen percentage set point or the air flow should be changed, three additional fuzzy rules for each operating condition are formulated based on the actual values of the oxygen percentage and the smoke chamber temperature resulting in 39 control rules.

Details of membership functions used can be found in Holmblad and Ostergaard [1982] and results of testing the system in Umbers and King [1981] and in Larsen [1982]. We shall not describe these details here, primarily because they are not of high general interest.

Before we turn to a quite different type of control it should be mentioned, however, that the reader can find descriptions and references to more than 10 further projects of the type described here in Mamdani [1981], in Pun [1977], and in Sugeno [1985a].

2. Fuzzy Control of an Automatic Train Operation System (ATO). Here we will describe the application of fuzzy control to a public transportation system in Sendai, Japan. Since traffic conditions within its metropolitan area have become increasingly congested, Sendai's city planners commenced designing a subway system that would provide its citizens with the highest levels of comfort, safety, and efficiency. For these purposes, fuzzy control of the train operations seems appropriate to enable smooth acceleration, deceleration, and braking.

The operation strategies are divided into the two functions constant speed control (CSC) and train automatic stop control (TASC). CSC starts the train and keeps its speed below a specified limit. TASC decelerates train speed in order to stop the train at a target position.

Conventional fuzzy control turned out to have disadvantages in this special application because it does not evaluate the results of selected control commands. To overcome this, it was decided to use *predictive fuzzy control*, where the control rules are as follows:

$$R_i\text{: ``If } (u \text{ is } C_i \Rightarrow x \text{ is } A_i \text{ and } y \text{ is } B_i), \text{ then } u \text{ is } C_i\text{''}$$

In linguistic terms, this rule is: "If the state variable x is A_i and state variable y is B_i, when control action is selected for C_i at this time, then the control action C_i is issued as the output of the controller." x and y are state variables of the system and u is the control action.

The *state variables* are evaluations of the train system according to specific criteria and defined by fuzzy sets as follows [Miyamoto et al. 1987, p. 64]:

A. Safety Performance Indexes (S)
   1. Danger (SD)
   2. Safe (SS)

B. Comfort Performance Indexes (C)
   1. Good comfort (CG)
   2. Poor comfort (CP)

C. Energy Saving Performance Indexes (E)
   1. Energy-saved running (ES)
   2. Not energy-saved running (EN)

D. Traceability Performance Indexes (T)
   1. Good trace (TG)
   2. Accurate trace (TA)
   3. Low speed (TL)

E. Running Time Performance Indexes (R)
 1. In TASC zone (RT)
 2. Not in TASC zone (RF)

F. Stopgap Performance Indexes (G)
 1. Good stop (GG)
 2. Accurate stop (GA)

The exact definitions of the respective membership functions are omitted here (see for instance [Yasunobo and Miyamoto 1985, p. 10]). The possible *control actions* are

1. Pn = powering notch
2. Bn = braking notch
3. DN = difference of notches
4. N = control command notch (power notch/brake notch)

Based upon these state variables and control actions, predictive fuzzy *control rules* are formulated. Examples of such rules are given in the following:

1. CSC-Rules
 If (N is 0 $\Rightarrow$ S is SS, C is CG and E is ES), then N is 0.
 If (DN is 0 $\Rightarrow$ S is SS and T is TG), then DN is 0.
2. TASC-Rules
 If (DN is 0 $\Rightarrow$ R is RT and G is GG), then DN is 0.
 If (N is 1 $\Rightarrow$ R is RT and C is CG), then N is 1.

In the ATO system, a control action is selected by evaluating all rules every 100 ms. In simulation studies the predictive fuzzy control ATO is compared with a conventional ATO according to the criteria "Riding Comfort," "Stopgap Accuracy," "Energy Consumption," and "Running Time" [Yasunobu, Miyamoto 1985, p. 15].

(a) Riding comfort and stop gap accuracy: The results of the simulation showed that the number of notch changes in the fuzzy ATO is about a half, and the stopgap accuracy measured as the deviation of a given target is about a third compared with those of the conventional ATO. This implies that the fuzzy ATO is able to control a train with good riding comfort and stopgap accuracy.

(b) Energy consumption and running time: The investigations concerning these criteria showed a potential energy saving of over 10% and/or a shortened running time compared with the conventional ATO.

Exercises

1. What are the differences between a decision support system, an expert system, and an FLC system?
2. Construct examples of domain knowledge represented in the form of rules, frames, and networks. Discuss advantages and disadvantages of these three approaches.
3. List, describe, and define at least four different types of uncertainty mentioned in this book. Associate appropriate theoretical approaches with them.
4. An expert in strategic planning has evaluated linguistically the degree of achievement of the lowest subcriteria of the criterion "Technology Attractiveness." He denotes the corresponding trapezoidal membership functions by the vectors of the characteristic values. After the first aggregation step the evaluation of the first-level criteria results. The respective trapezoidal membership functions are given by the following vectors of the characteristic values:

   Acceptance: (.2, .3, .5, .7)

   Technological Potential: (.6, .7, .9, 1)

   Breadth of Application: (.4, .5, .6, .7)

   Complementarity: (.1, .3, .4, .6)

   Compute the four characteristic values of the criterion "Technology Attractiveness" by using the γ operator with $\gamma = .5$ and equal weights for all first-level criteria for the four respective characteristic values given above. Draw the resulting stripe in the portfolio matrix for $\alpha = .8$.
5. Consider the system described in section 10.2.2 which is used to control an engine-boiler combination. There are described the "if-then"-statements of the throttle algorithm. Assume that we know the following input values:

   (a) $PE = -3$, $SE = +1$, $CPE = +4$, $CSE = +2$

   (b) $PE = +2$, $SE = -1$, $CPE = +1$, $CSE = -4$

   Compute the membership values of the fuzzy control statements by applying the compositional rule of inference (see chapter 9). Translate these statements into crisp control actions as suggested in this chapter.

11 PATTERN RECOGNITION

11.1 Models for Pattern Recognition

Pattern recognition is one of the oldest and most obvious application areas of fuzzy set theory. The term *pattern recognition* embraces a very large and diversified literature. It includes research in the area of artificial intelligence, interactive graphic computers, computer aided design, psychological and biological pattern recognition, linguistic and structural pattern recognition, and a variety of other research topics. One could possibly distinguish between mathematical pattern recognition (primarily cluster analysis) and nonmathematical pattern recognition. One of the major differences between these two areas is that the former is far more context dependent than the latter: a heuristic computer program that is able to select features of chromosomal abnormalities according to a physician's experience will have little use for the selection of wheat fields from a photo-interpretation viewpoint. By contrast to this example, a well-designed cluster algorithm will be applicable to a large variety of problems from many different areas. The problems will again be different in structural pattern recognition, when, for instance, handwritten H's shall be distinguished from handwritten A's, and so on.

Verhagen [1975] presents a survey of definitions of pattern recognition which also cites the difficulties of an attempt to define this area properly. Bezdek [1981, p. 1] defines pattern recognition simply as "A search for structure in data."

The most effective search procedure—in those instances in which it is applicable—is still the "eyeball" technique applied by human "searchers." Their limitations, however, are strong in some directions: Whenever the dimensionality of the volume of data exceeds a limit, whenever the human senses, especially the vision, are not able to recognize data or features, the "eyeball" technique cannot be applied.

One of the advantages of human search techniques is the ability to recognize and classify patterns in a nondichotomous way. One way to initiate this strength is the development of statistical methods in mathematical pattern recognition, which in connection with high-speed computers have shown very impressive results. There are data structures, however, that are not probabilistic in nature or not even approximately stochastic. Given the power of existing EDP it seems very appropriate and promising to find nonprobabilistic nondichotomous models and structures that enable us to recognize and transmit in a usable form patterns of this type, which humans cannot find without the help of more powerful methods than "eyeball-search."

Here, obviously, fuzzy set theory offers some promise. Fuzzy set theory has already been successfully applied in different areas of pattern search and at different stages of the search process. In the references, we cite cases of linguistic pattern search [Sanchez et al. 1982], of character recognition [Chatterji 1982], of visual scene description [Jain, Nagel 1977] and of texture classification [Hajnal, Koczy 1982]. We also give references for the application of fuzzy pattern recognition to medical diagnosis [Fordon, Bezdek 1979; Sanchez et al. 1982], to earthquake engineering [Fu et al. 1982] and to pattern search in demand [Carlucci, Donati 1977].

In this book we shall restrict our attention to mathematical pattern recognition, that is, to fuzzy cluster analysis. Even there we will have to

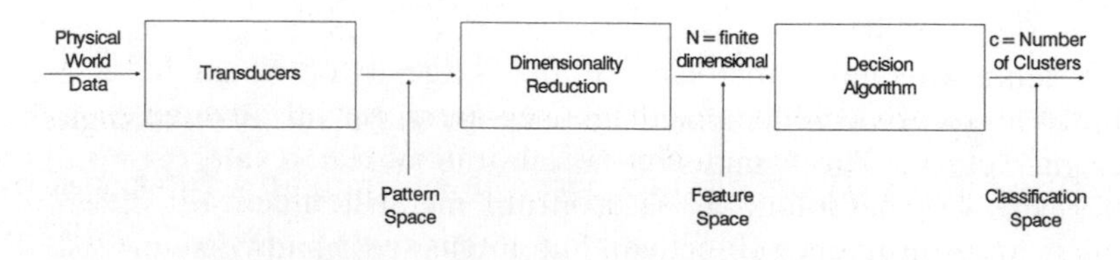

Figure 11–1. Pattern recognition.

limit detailed descriptions to only a few techniques. The interested reader will find enough references, however, to penetrate this area further if he wishes [Bezdek 1987c, p. 81].

Before we turn to cluster algorithms we shall establish a more general framework into which other pattern recognition problems and techniques will also fit. Figure 11–1 sketches the whole pattern recognition process. Let us comment on some of its components.

11.1.1 The Data

The data may be drawn from virtually any physical process or phenomenon. They may be qualitative or quantitative, numerical, pictorial, textural, linguistic, or any combination of those modes. The dimensionality may range from one-dimensional data to data in high-dimensional spaces. The "data" set will in this book generally be denoted by X.

11.1.2 Structure or Pattern Space

Observed data (it is to be hoped) carry information either about the process generating them or the phenomenon they represent. "By structure we mean the manner in which this information can be organized so that relationships between the variables in the process can be identified" [Bezdek 1981, p. 2].

The dimensionality of the pattern space which contains the structural properties, is generally lower than that of the data space.

11.1.3 Feature Space and Feature Selection

The feature space is an intermediate space between the data space and the classification process. It is generally of a much lower dimension than the data space. This is essential for applying efficient pattern-search techniques.

Feature selection searches for *internal* structure in data items, that is, for features or properties of the data which allow us to recognize and display their structure. The crucial question is: Are the features selected sufficiently representative for the physical process that generated the data to enable us to construct realistic clusters or classifiers?

11.1.4 Classification and Classification Space

This space contains the decision implemented by the classification algorithm. A classifier itself is a device, means, or algorithm by which the data space is partitioned into c "decision regions." Classification attempts to discover associations between subclasses of a population. It is obvious that the classification space has generally a very low dimensionality!

Bezdek [1981, p. 4] summarized the main features, which will also be discussed in the second part of this chapter, as follows:

Feature selection: The search for structure in data items or observations

$x_k \in X$

Cluster analysis: The search for structure in data sets, or samples, $X \subset S$.

Classification: The search for structure in data spaces or populations S.

It is worth noting that these problems exhibit a beautiful (but confounding!) sort of interactive symmetry. For example: if we could chose "optimal" features, clustering and classification would be trivial; on the other hand, we often attempt to discover the optimal features by clustering the feature variables! Conversely, if we could design an "optimal" classifier, then the features selected would be immaterial. This kind of interplay is easy to forget—but important to remember!—when one becomes engrossed in the details of a specific problem, for a successful pattern recognizer is one who has matched his or her particular data to an algorithm technique.

11.2 Fuzzy Clustering

11.2.1 Clustering Methods

Let us assume that the important problem of feature extraction, that is, the determination of the characteristics of the physical process, the image of other phenomena that are significant indicators of structural organization and how to obtain them, has been solved. Our task is then to divide n objects $x \in X$ characterized by p indicators into c, $2 \leq c < n$ categorically homogenous subsets called "clusters." The objects belonging to any one of the clusters should be similar and the objects of different clusters as dissimilar as possible. The number of clusters, c, is normally not known in advance.

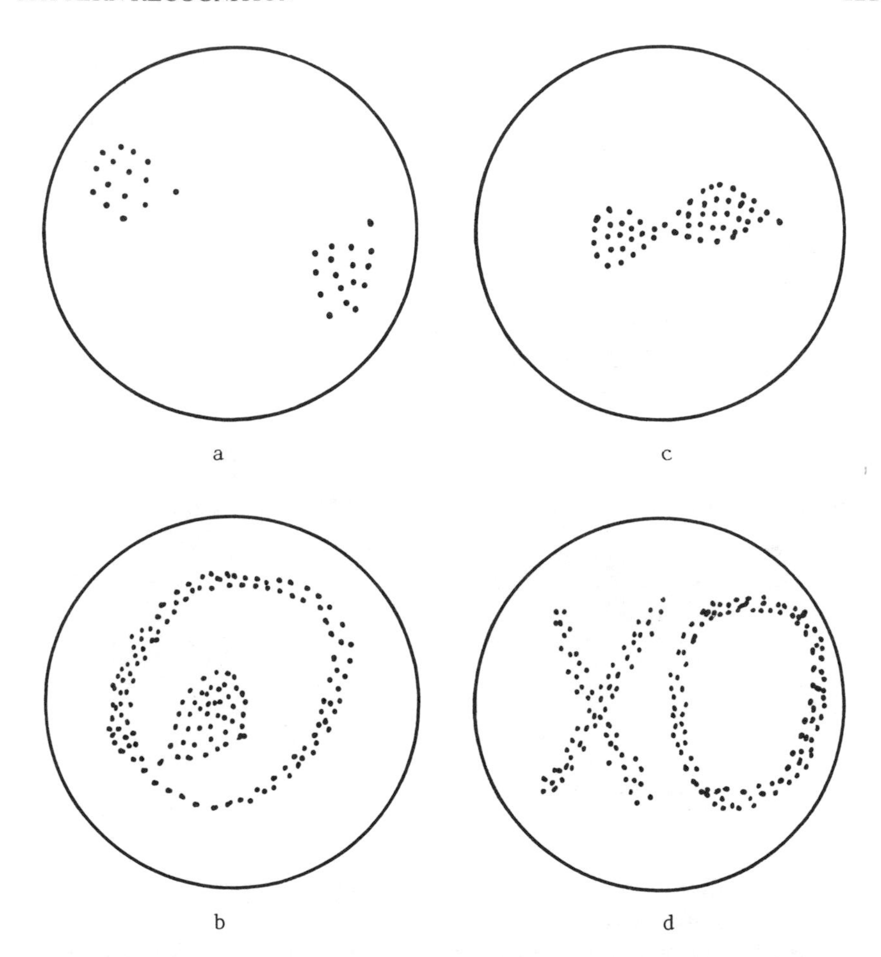

Figure 11–2. Possible data structures in the plane.

The most important to be answered before applying any clustering procedure is, which mathematical properties of the data set, for example, distance, connectivity, intensity, and so on should be used and in what way should they be used in order to identify clusters. This question will have to be answered for each specific data set since there is no universally optimal cluster criteria. Figure 11–2 shows a few possible shapes of clusters; and it should be immediately obvious that a cluster criterium which works in case 11–2a will show a very bad performance in case 11–2b or c. More examples can for instance be found in Bezdek [1981] or Roubens [1978]

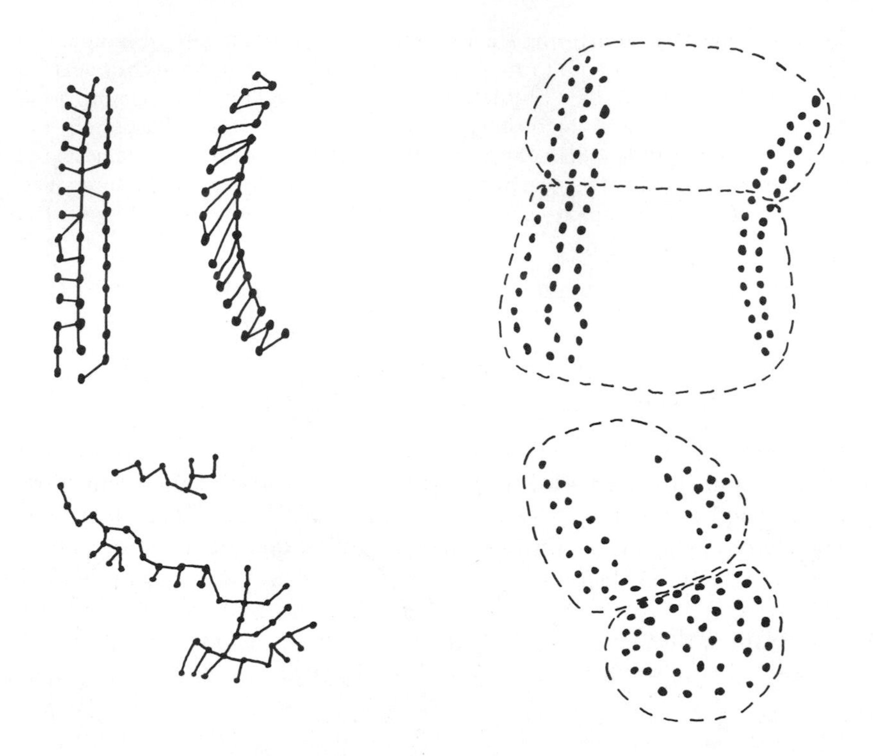

Figure 11–3. Performance of cluster criteria.

and in many other publications on cluster analysis and pattern recognition [Ismail 1988, p. 446; Gu and Dubuisson 1990, p. 213].

For further illustration of this point let us look at an example from Bezdek [1981, p. 45]. Figure 11–3 shows two data sets, which have been clustered by a distance-based objective function algorithm (the within-group sum-of-spared-error criterion) and by applying a distance-based graph-theoretic method (single-linkage algorithm). Obviously the criterion that leads to good results in one case performs very badly in the other case and vice versa. (Crisp) clustering methods are commonly categorized according to the type of clustering criterion used in hierarchical, graph-theoretic, and objective-functional methods.

Hierarchical clustering methods generate a hierarchy of partitions by means of a successive merging (agglomerative) or splitting (diverse) of clusters [Dimitrescu 1988, p. 145]. Such a hierarchy can easily be represented by a dendogram, which might be used to estimate an

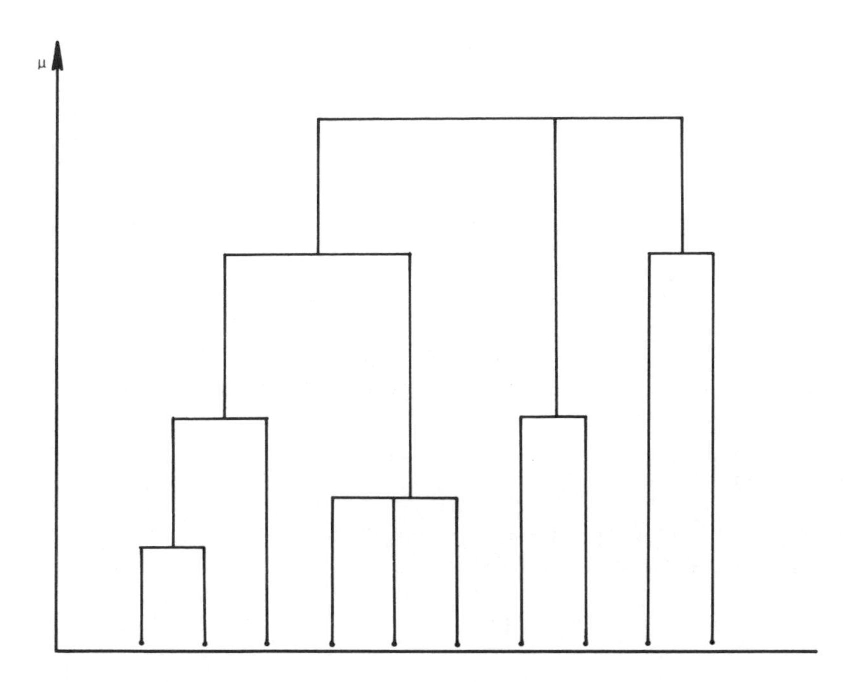

Figure 11–4. Dendogram for hierarchical clusters.

appropriate number of clusters, c, for other clustering methods. On each level of agglomeration or splitting a locally optimal strategy can be used, without taking into consideration policies used on preceding levels. These methods are not iterative; they cannot change the assignment of objects to clusters made on preceeding levels. Figure 11–4 shows a dendogram which could be the result of a hierarchical clustering algorithm. The main advantage of these methods is their conceptional and computational simplicity.

In fuzzy set theory this type of clustering method would correspond to the determination of "similarity trees" such as shown in example 6–14.

Graph-theoretic clustering methods are normally based on some kind of connectivity of the nodes of a graph representing the data set. The clustering strategy is often breaking edges in a minimal spanning tree to form subgraphs. If the graph representing the data structure is a fuzzy graph such as discussed in chapter 6 then different notions of connectivity lead to different types of clusters, which in turn can be represented as dendograms. Yeh and Bang [1975] for instance, define 4 different kinds of clusters. For the purpose of illustrating this approach we shall consider one of the types of clusters suggested there.

Definition 11–1 [Yeh and Bang 1975]

Let $\tilde{G} = [V, \tilde{R}]$ be a symmetric fuzzy graph. Then the *degree of a vertex* v is defined as $d(v) = \Sigma_{u \neq v} \mu_{\tilde{R}}(u)$. The minimum degree of $\tilde{G}$ is $\delta(\tilde{G}) = \min_{v \in V} \{d(v)\}$.

Let $\tilde{G} = [V, \tilde{R}]$ be a symmetric fuzzy graph. $\tilde{G}$ is said to be *connected* if, for each pair of vertices u and v in V, $\mu_{\tilde{R}}(u, v) > 0$. $\tilde{G}$ is called τ-*degree connected* for some $\tau \geq 0$, if $\delta(\tilde{G}) \geq \tau$ and $\tilde{G}$ is connected.

Definition 11–2

Let $\tilde{G} = [V, \tilde{R}]$ be a symmetric fuzzy graph. *Clusters* are then defined as maximal τ-degree connected subgraphs of $\tilde{G}$.

Example 11–1 [Yeh and Bang 1975, p. 145]

Let G be the symmetric fuzzy graph shown in figure 11–5. The dendogram in figure 11–6 shows all clusters for different levels of τ. For further details see Yeh [Yeh and Bang 1975].

Objective-function methods allow the most precise formulation of the clustering criterion. The "desirability" of clustering candidates is measured

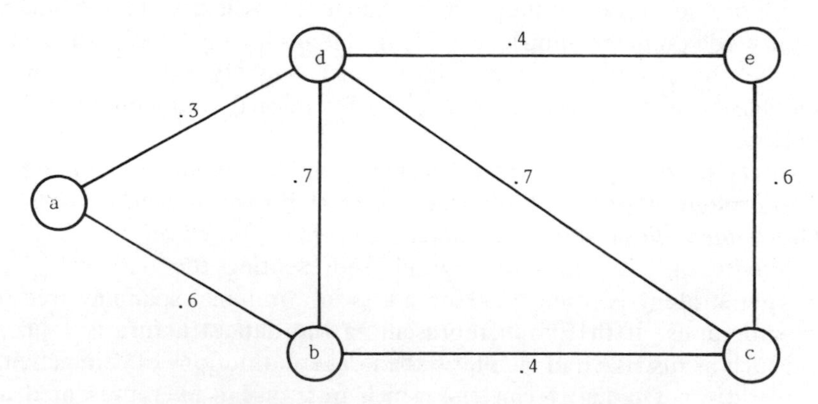

Figure 11–5. Fuzzy graph.

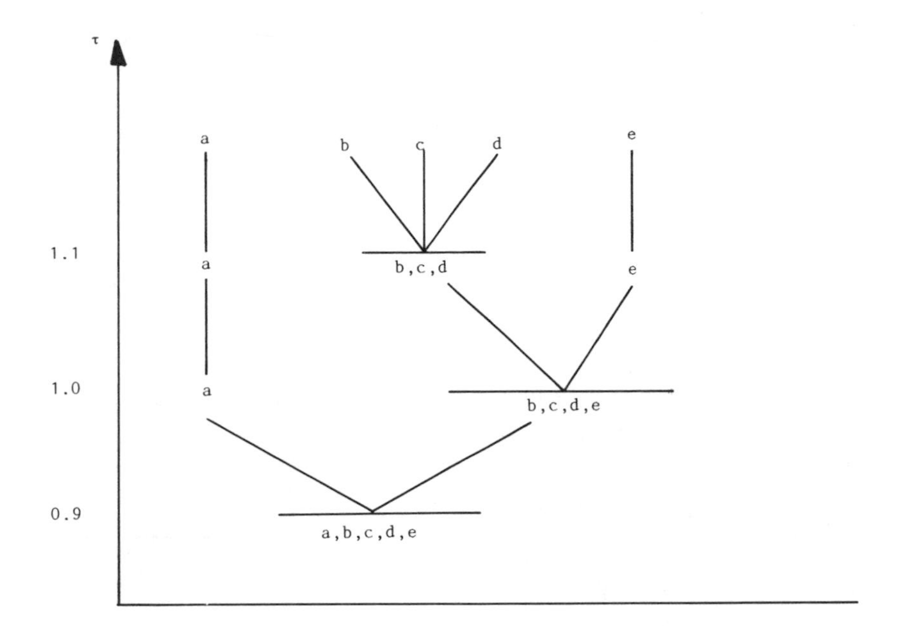

Figure 11–6. Dendogram for graph-theoretic clusters.

of each c, the number of clusters, by an objective function. Typically, local extrema of the objective function are defined as optimal clusterings. Many different objective functions have been suggested for clustering (crisp clustering as well as fuzzy clustering). The interested reader is referred in particular to the excellent book by Bezdek [1981] for more details and many references. We shall limit our considerations to one frequently used type of (fuzzy) clustering method, the so-called c-means algorithm.

Classical (crisp) clustering algorithms generate partitions such that each object is assigned to exactly one cluster. Often, however, objects cannot adequately be assigned to strictly one cluster (because they are located "between" clusters). In these cases fuzzy clustering methods provide a much more adequate tool for representing real-data structures.

To illustrate the difference between the results of crisp and fuzzy clustering methods let us look at one example used in the clustering literature very extensively: the butterfly.

Example 11–2

The data set X consists of 15 points in the plane such as depicted in figure 11–7. Clustering these points by a crisp objective-function algorithm might

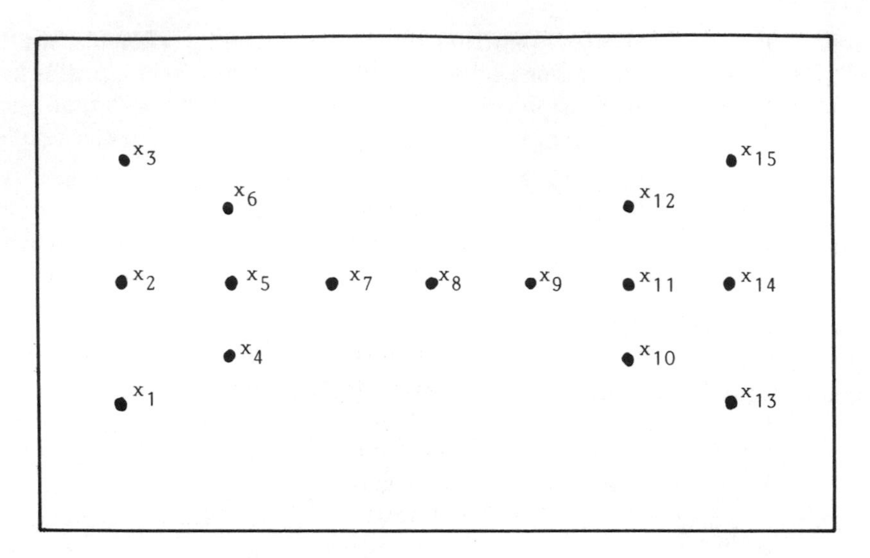

Figure 11–7. The butterfly.

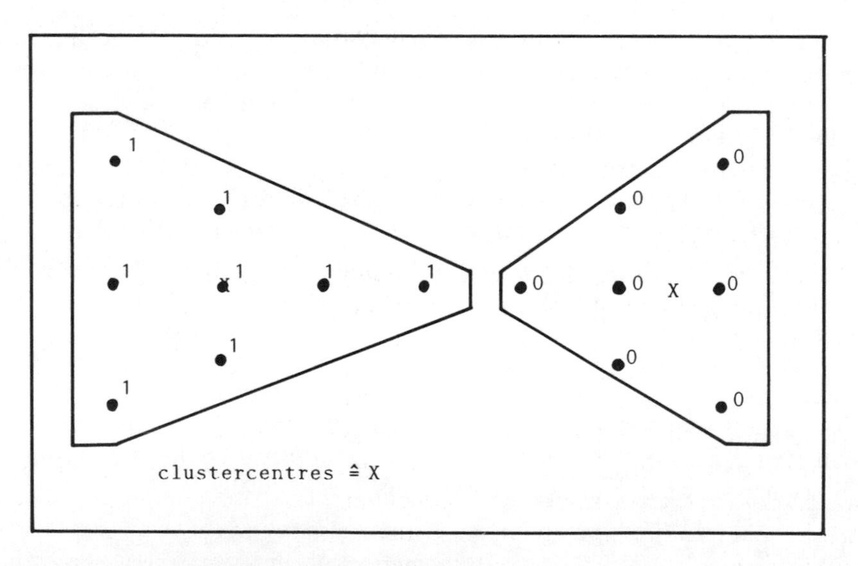

Figure 11–8. Crisp clusters of the butterfly.

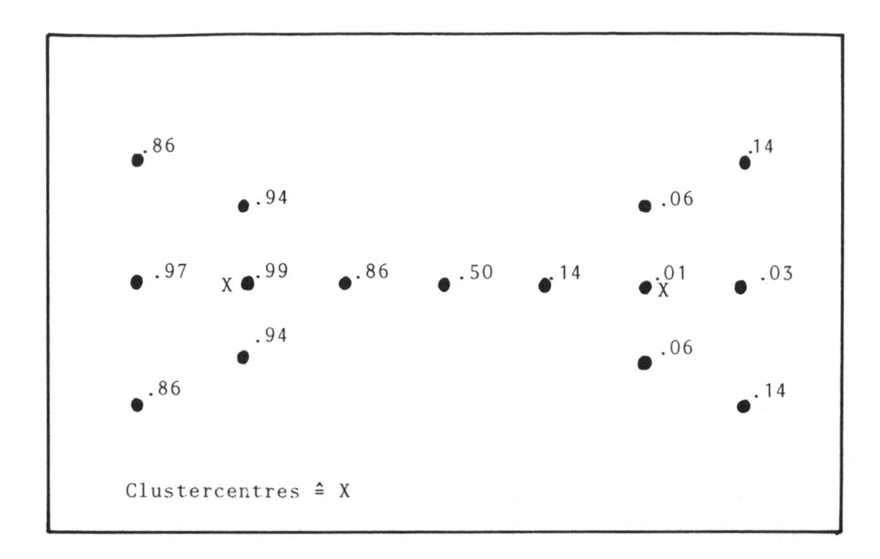

Figure 11–9. Cluster 1 of the butterfly.

yield the picture shown in figure 11–8, in which "1" indicates membership of the point to the left-hand cluster and "0" membership to the right-hand cluster. The x's indicate the centers of the clusters. Figures 11–9 and 11–10, respectively, show the degrees of membership the points might have to the two clusters when using a fuzzy clustering algorithm.

We observe that, even though the butterfly is symmetric, the clusters in figure 11–8 are not symmetric because point x, the point "between" the clusters has to be (fully) assigned to either cluster 1 or cluster 2. In figures 11–9 and 11–10 this point has the degree of membership .5 to both clusters, which seems to be more appropriate. Details of the methods used to arrive at figures 11–8 to 11–10 can be found in Bezdek [1981, p. 52] or Ruspini [1973].

Let us now consider the clustering methods themselves.

Let the data set $X = \{x_1, \ldots, x_n\} < \mathbb{R}^p$ be a subset of the real p-dimensional vector space $\mathbb{R}^p$. Each $x_k = (x_{k_1}, \ldots, x_{k_p}) \in \mathbb{R}^p$ is called a feature vector. x_{k_j} is the jth feature of observation x_k.

Since the elements of a cluster shall be as similar to each other and clusters as dissimilar as possible, the clustering process is controlled by use of similarity measures. One normally defines the "dissimilarity" or

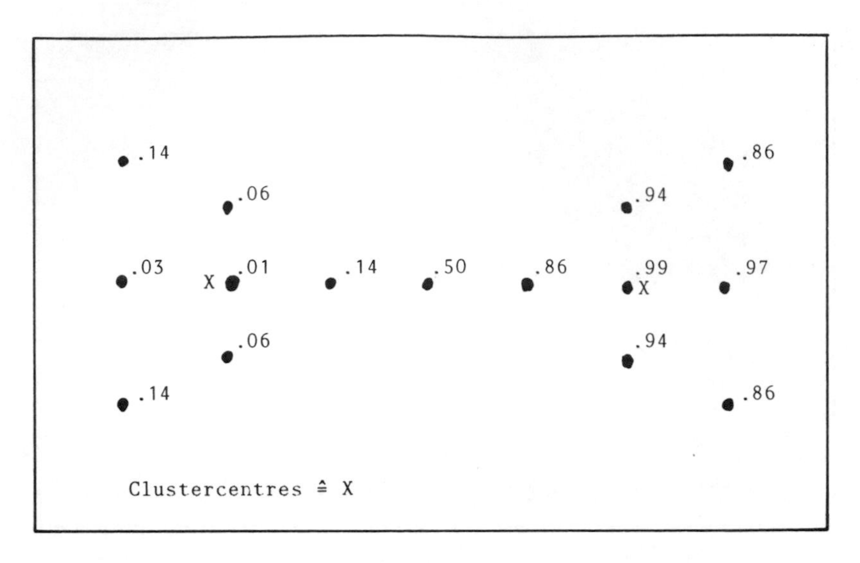

Figure 11–10. Cluster 2 of the butterfly.

"distance" of two objects x_k and x_l as a real-valued function d: $X \times X \rightarrow R^+$ which satisfies

$$d(x_k, x_l) = d_{kl} \geq 0$$
$$d_{kl} = 0 \Leftrightarrow x_k = x_l$$
$$d_{kl} = d_{lk}$$

If additionally d satisfies the triangle equality, that is,

$$d_{kl} \leq d_{kj} + d_{jl}$$

then d is a metric, a property that is not always required. If each feature vector is considered as a point in the p-dimensional space then the dissimilarity d_{kl} of two points x_k and x_l can be interpreted as the distance between these points

Each partition of the set $X = \{x_1, \ldots, x_n\}$ into crisp or fuzzy subsets $\tilde{S}_i$ $(i = 1, \ldots, c)$ can fully be described by an indicator function u_{S_i} or a membership function $\mu_{\tilde{S}_i}$ respectively. In order to stay in line with the terminology of the preceding chapters we shall use, for crisp clustering methods,

$$u_{S_i}: X \rightarrow \{0, 1\}$$

and, for fuzzy cases,

$$\mu_{\tilde{S}_i}: X \rightarrow [0, 1]$$

where u_{ik} and μ_{ik} denote the degree of membership of object x_k to the subset $\tilde{S}_i$, that is,

$$u_{ik} := u_{S_i}(x_k)$$
$$\mu_{ik} := \mu_{\tilde{S}_i}(x_k)$$

Definition 11–3

Let $X = \{x_1, \ldots, x_n\}$ be any finite set. V_{cn} is the set of all real $c \times n$ matrixes, and $2 \leq c < n$ is an integer. The matrix $U = [u_{ik}] \in V_{cn}$ is called a *crisp c-partition* if it satisfies the following conditions:

1. $u_{ik} \in \{0, 1\} \qquad 1 \leq i \leq c,\ 1 \leq k \leq n$
2. $\sum_{i=1}^{c} u_{ik} = 1 \qquad 1 \leq k \leq n$
3. $0 < \sum_{k=1}^{n} u_{ik} < n \qquad 1 \leq i \leq c$

The set of all matrixes that satisfy these conditions is called M_c.

Example 11–3

Let $X = \{x_1, x_2, x_3\}$. Then there are the following three *crisp 2-partitions*:

$$U_1 = \begin{bmatrix} 1 & 1 & 0 \\ 0 & 0 & 1 \end{bmatrix} \quad (\text{columns } x_1, x_2, x_3)$$

$$U_2 = \begin{bmatrix} 1 & 0 & 0 \\ 0 & 1 & 1 \end{bmatrix} \quad (\text{columns } x_1, x_2, x_3)$$

$$U_3 = \begin{bmatrix} 1 & 0 & 1 \\ 0 & 1 & 0 \end{bmatrix} \quad (\text{columns } x_1, x_2, x_3)$$

Obviously conditions (2) and (3) of the definition rule out the following partitions:

$$\begin{bmatrix} 1 & 1 & 0 \\ 0 & 1 & 1 \end{bmatrix} \quad (\text{columns } x_1, x_2, x_3)$$

$$\begin{bmatrix} 1 & 1 & 1 \\ 0 & 0 & 0 \end{bmatrix} \quad (\text{columns } x_1, x_2, x_3)$$

Definition 11–4

X, V_{cn}, and c are as in definition 11–3. $\tilde{U} = [\mu_{ik}] \in V_{cn}$ is called a *fuzzy-c partition* if it satisfies the following conditions [Bezdek 1981, p 26]:

1. $\mu_{ik} \in [0, 1] \qquad 1 \le i \le c,\ 1 \le k \le n$

2. $\sum_{i=1}^{c} \mu_{ik} = 1, \qquad 1 \le k \le n$

3. $0 < \sum_{k=1}^{n} \mu_{ik} < n \qquad 1 \le i \le c$

M_{fc} will denote the set of all matrixes satisfying the above conditions. By contrast to the crisp c-partition, elements can now belong to several clusters and to different degrees. Conditions (2) and (3) just require that the "total membership" of an element is normalized to 1 and that it cannot belong to more clusters than there exist.

Example 11–4

Let $X = \{x_1, x_2, x_3\}$. Then there exist infinitely many possible fuzzy 2-partitions, such as

$$\tilde{u}_1 = \begin{array}{c} \begin{matrix} x_1 & x_2 & x_3 \end{matrix} \\ \begin{bmatrix} 1 & .5 & 0 \\ 0 & .5 & 1 \end{bmatrix} \end{array}$$

$$\tilde{u}_2 = \begin{array}{c} \begin{matrix} x_1 & x_2 & x_3 \end{matrix} \\ \begin{bmatrix} .8 & .5 & .2 \\ .2 & .5 & .8 \end{bmatrix} \end{array}$$

$$\tilde{u}_3 = \begin{array}{c} \begin{matrix} x_1 & x_2 & x_3 \end{matrix} \\ \begin{bmatrix} .8 & 1 & .9 \\ .2 & 0 & .1 \end{bmatrix} \end{array}$$

and so on.

Our butterfly example (figure 11–7), for instance, could have the following partition:

$$\tilde{U} = \left\{ \begin{matrix} x_1 & x_2 & x_3 & x_4 & x_5 & x_6 & x_7 & x_8 & x_9 & x_{10} & x_{11} & x_{12} & x_{13} & x_{14} & x_{15} \\ .86 & .97 & .86 & .94 & .99 & .94 & .86 & .5 & .14 & .06 & .01 & .06 & .14 & .03 & .14 \\ .14 & .03 & .14 & .06 & .01 & .06 & .14 & .5 & .86 & .94 & .99 & .94 & .86 & .97 & .86 \end{matrix} \right\}$$

The location of a cluster is represented by its "cluster center" $v_i = (v_{i1}, \ldots, v_{ip}) \in \mathbb{R}^p$, $i = 1, \ldots, c$, around which its objects are concentrated.

Let $v = (v_1, \ldots, v_c) \in \mathbb{R}^{cp}$ be the vector of all cluster centers, where the v_i in general do not correspond to elements of X.

One of the frequently used criterion to improve an initial partition is the so-called *variance criterion*. This criterion measures the dissimilarity between the points in a cluster and its cluster center by the Euclidean distance. This distance, d_{ik}, is then [Bezdek 1981, p. 54]

$$\begin{aligned} d_{ik} &= d(x_k, v_i) \\ &= \|x_k - v_i\| \\ &= \left[\sum_{j=1}^{p} (x_{kj} - v_{ij})^2\right]^{1/2} \end{aligned}$$

The variance criterion for crisp partitions corresponds to minimizing the sum of the variances of all variables j in each cluster i, with $|S_i| = n$, yields:

$$\min \sum_{i=1}^{c} \sum_{j=1}^{p} \frac{1}{|S_i|} \sum_{x_k \in S_i} (x_{kj} - v_{ij})^2 \Leftrightarrow$$

$$\min \frac{1}{n} \sum_{i=1}^{c} \sum_{x_k \in S_i} \sum_{j=1}^{p} (x_{kj} - v_{ij})^2$$

As indicated by the above transformation, the variance criterion corresponds—except for the factor $1/n$—to minimizing the sum of the squared Euclidean distances. The criterion itself amounts to solving the following problem:

$$\min z(S_1, \ldots, S_c; v) = \sum_{i=1}^{c} \sum_{x_k \in S_i} \|x_k - v_i\|^2$$

such that

$$v_i = \frac{1}{|S_i|} \sum_{x_k \in S_i} x_k$$

Using definition 11–3, the variance criterion for crisp c-partitions can be written as

$$\min z(\widetilde{U}, v) = \sum_{i=1}^{c} \sum_{k=1}^{n} u_{ik} \|x_k - v_i\|^2$$

such that

$$v_i = \frac{1}{\sum_{k=1}^{n} u_{ik}} \sum_{k=1}^{n} (u_{ik}) x_k$$

For fuzzy c-partitions according to definition 11–4 the variance criterion amounts to solving the following problem:

$$\min z(U, v) = \sum_{i=1}^{c} \sum_{k=1}^{n} (\mu_{ik})^m \, \| x_k - v_i \|^2$$

such that

$$v_i = \frac{1}{\sum_{k=1}^{n} u_{ik}} \sum_{k=1}^{n} (u_{ik})^m x_k, \qquad m > 1$$

Here v_i is the mean of the x_k m-weighted by their degrees of membership. That means that the x_k with high degrees of membership have a higher influence on v_i than those with low degrees of membership. This tendency is strengthened by m, the importance of which we will discuss in more detail at a later time. It was shown (see, for instance [Bock 1979a, p. 144]) that, given a partition $\widetilde{U}$, v_i is best represented by the clusters $\widetilde{S}_i$ as described above.

Generalizing the criterion concerning the used norm the crisp clustering problem can be stated as follows: Let G be a $(p \times p)$ matrix, which is symmetric and positive-definite. Then we can define a general norm

$$\| x_k - v_i \|_G^2 = (x_k - v_i)^T G (x_k - v_i)$$

The possible influence of the chosen norm, determined by the choice of G, will be discussed later. This yields as formulation of the problem:

$$\min z(U, v) = \sum_{k=1}^{n} \sum_{i=1}^{c} u_{ik} \, \| x_k - v_i \|_G^2$$

such that

$$U \in M_c$$
$$v \in R^{cp}$$

This is a combinatorial optimization problem which is hard to solve, even for rather small values of c and n. In fact, the number of distinct ways to partition x into non empty subsets is

$$|M_c| = (1/c!) \left[\sum_{j=1}^{c} \binom{c}{j} (-1)^{c-j} j^n \right]$$

which for $c = 10$ and $n = 25$ are already roughly 10^{18} distinct 10-partitions of the 25 points [Bezdek 1981, p. 29].

The basic definition of the fuzzy partitioning problem for $m > 1$ is

$$\min z_m(\widetilde{U}; v) = \sum_{k=1}^{n} \sum_{i=1}^{c} (\mu_{ik})^m \; \| x_k - v_i \|_G^2$$

(P_m)

such that

$$\widetilde{U} \in M_{fc}$$
$$v \in R^{cp}$$

(P_m) is an analytical problem, which has the advantage that using differential calculus one can determine necessary conditions for local optima. Differentiating the objective function with respect to v_i (for fixed $\widetilde{U}$) and to μ_{ik} (for fixed v) and applying the condition $\Sigma_{i=1}\mu_{ik} = 1$ one obtains [see Bezdek 1981, p. 67]:

$$v_i = \frac{1}{\sum_{k=1}^{n} (\mu_{ik})^m} \sum_{k=1}^{n} (\mu_{ik})^m x_k \qquad i = 1, \ldots, c \tag{11.1}$$

$$\mu_{ik} = \frac{\left(\dfrac{1}{\| x_k - v_i \|_G^2} \right)^{1/(m-1)}}{\sum_{j=1}^{c} \left(\dfrac{1}{\| x_k - v_j \|_G^2} \right)^{1/(m-1)}}, \qquad i = 1, \ldots, c; \; k = 1, \ldots, n \tag{11.2}$$

Let us now comment on the role and importance of m: It is called the exponential weight and it reduces the influence of "noise" when computing the cluster centers in equation (11.1) (see [Windham 1982, p. 358]) and the value of the objective function $z_m(\widetilde{U}, v)$. m reduces the influence of small μ_{ik} (points further away from v_i) compared to that of large μ_{ik} (points close to v_i). The larger $m > 1$ the stronger is this influence.

The systems described by equations (11.1) and (11.2) cannot be solved analytically. There exist, however, iterative algorithms (nonhierarchical) which approximate the minimum of the objective function, starting from a given position. One of the best known algorithms for the crisp clustering problem is the (hard) c-means algorithm or (basic) ISODATA-algorithm. Similarly the fuzzy clustering problem can be solved by using the fuzzy c-means algorithm which shall be described in more detail in the following.

The *fuzzy c-means algorithm* [Bezdek 1981, p. 69]. For each $m \in (0, \infty)$ a fuzzy c-means algorithm can be designed which iteratively solves the necessary conditions (11.1) and (11.2) above and converges to a local optimum (for Proofs of convergence see Bezdek [1981] and Bock [1979]).

The algorithm comprises the following steps:

Step 1. Choose $c(2 \leq c \leq n)$, $m(1 < m < \infty)$ and the (p, p)-matrix G with G symmetric and positive-definite. Initialize $\tilde{U}^{(0)} \in M_{fc}$, set $l = 0$.

Step 2. Calculate the c fuzzy cluster centers $\{v_i^{(l)}\}$ by using $\tilde{U}^{(l)}$ from condition (11.1).

Step 3. Calculate the new membership matrix $\tilde{U}^{(l+1)}$ by using $\{v_i^{(l)}\}$ from condition (11.2) if $x_k \neq v_i^{(l)}$. Else set

$$\mu_{jk} = \begin{cases} 1 & \text{for } j = i \\ 0 & \text{for } j \neq i \end{cases}$$

Step 4. Choose a suitable matrix norm and calculate $\Delta = \| \tilde{U}^{(l+1)} - \tilde{U}^{(l)} \|_G$. If $\Delta > \varepsilon$ set $l = l + 1$ and go to step 2. If $\Delta \leq \varepsilon \rightarrow$ stop.

For the fuzzy c-means algorithm a number of parameters have to be chosen:

the number of clusters c, $2 \leq c \leq n$;

the exponential weight m, $1 < m < \infty$;
the (p, p) matrix G (G symmetric and positive-definite) which induces a norm;

the method to initialize the membership matrix $\tilde{U}^{(0)}$;

the termination criteria $\Delta = \| \tilde{U}^{(l+1)} - \tilde{U}^{(l)} \|_G \leq \varepsilon$.

Example 11–5 [Bezdek 1981, p. 74]

The data of the butterfly shown in figure 11–7 were processed with a fuzzy 2-means algorithm, using as a starting partition

$$\tilde{U}^{(0)} = \begin{bmatrix} .854 & .146 & .854 & \ldots & .854 \\ .146 & .854 & .146 & \ldots & .146 \end{bmatrix}_{2\times 15}$$

ε was chosen to be .01; the Euclidean norm was used for G; and m was set to 1.25. Termination in six iterations resulted in the memberships and cluster centers shown in figure 11–11. For $m = 2$ the resulting clusters are shown in figure 11–12.

As for other iterative algorithms for improving starting partitions the number c has to be chosen suitably. If there does not exist any information about a good c, the computations are carried out for several values of c. In a second step, the best of these partitions is selected.

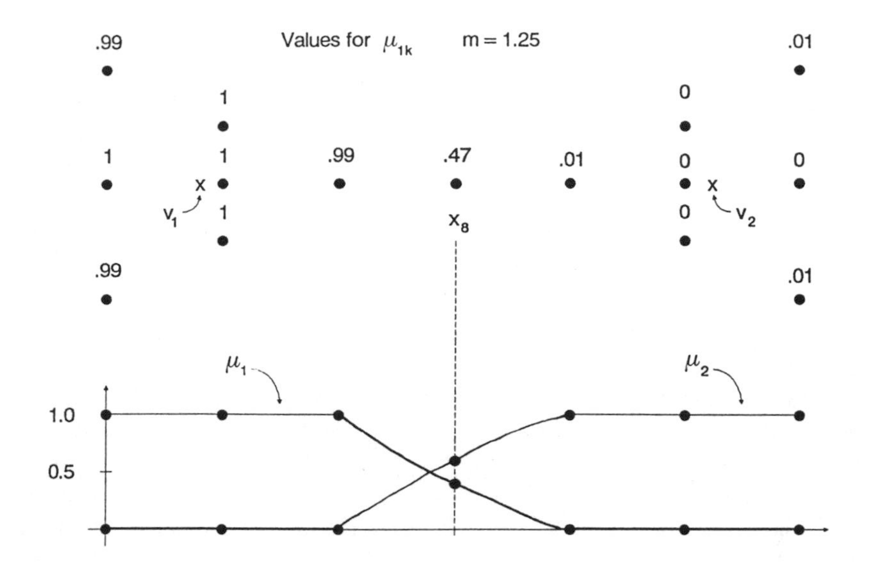

Figure 11–11. Clusters for $m = 1.25$.

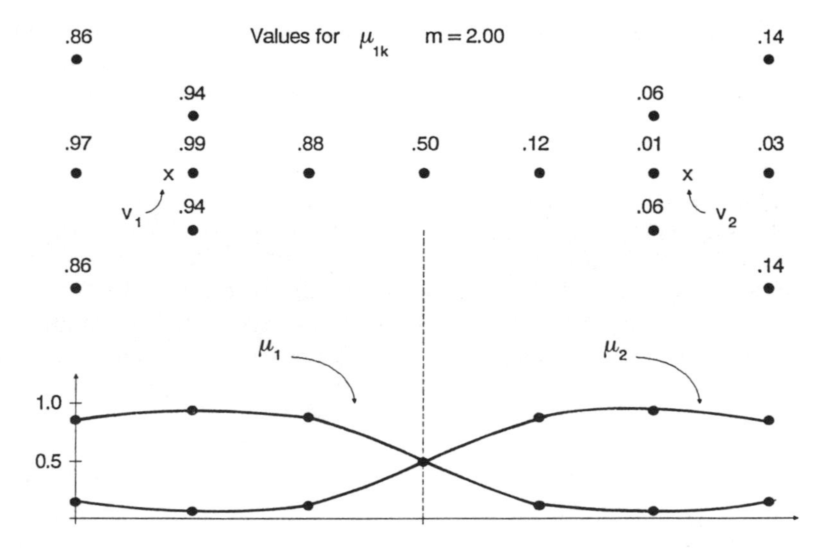

Figure 11–12. Clusters for $m = 2$.

The exponential weight m influences the membership matrix. The larger m, the fuzzier becomes the membership matrix of the final partition. For $m \to \infty$, $\tilde{U}$ approaches $\tilde{U} = [\frac{1}{c}]$. This is, of course, a very undesirable solution, because each x_k is assigned to each cluster with the same degree of membership.

Basically less fuzzy membership matrixes are preferable because higher degrees of membership indicate a higher concentration of the points around the respective cluster centers. No theoretically justified rule for choosing m exists. Usually $m = 2$ is chosen.

G determines the shape of the cluster, which can be identified by the fuzzy c-means algorithm. If one chooses the Euclidean norm N_E then G is the identity matrix I, and the shape of the clusters is assumed to be an equally sized hypersphere. Other frequently used norms are the diagonal norm or the Mahalanobisnorm for which $G_D = [\mathrm{diag}(\sigma_j^2)]^{-1}$ and $G_M = [\mathrm{cov}(x)]^{-1}$, respectively, where σ_j^2 denotes the variance of feature j.

The final partition depends on the initially chosen starting position. When choosing an appropriate c and if there exists a good clustering structure, the final partitions generated by a fuzzy c-means algorithm are rather stable.

A number of variations of the above algorithm are described in Bezdek [1981]. The interested reader is referred to this reference for further details. Numerical results for a number of algorithms are also presented in Roubens [1978].

11.2.2 Cluster Validity

"Complex algorithms stand squarely between the data for which substructure is hypothesized and the solutions they generate; hence it is all but impossible to transfer a theoretical null hypothesis about X to $\tilde{U} \in M_{fc}$, which can be used to statistically substantiate or repudiate the validity of algorithmically suggested clusters. As a result a number of scalar measures of partition fuzziness (which are interesting in their own right) have been used as heuristic validity indicants" [Bezdek 1981, p. 95].

Actually the so-called cluster validity problem concerns the quality or the degree to which the final partition of a cluster algorithm approximates the real or hypothesized structure of a set of data. Mostly this question is reduced, however, to the search for a "correct" c. Cluster validity is also relevant when deciding which of a number of starting partitions should be selected for improvement.

For measuring cluster validity in fuzzy clustering some criteria from crisp cluster analysis have been adapted to fuzzy clustering. In particular, the so-called validity functionals used express the quality of a solution by measuring its degree of fuzziness. While criteria for cluster validity are closely related to the mathematical formulation of the problem, criteria to judge the real "appropriateness" of a final partition consider primarily real rather than mathematical features.

Let us first consider some criteria taken from traditional crisp clustering.

One of the most straightforward criterion is the value of the objective function. Since it decreases monotonically with increasing number of clusters, c, that is, it reaches its minimum for $c = n$, one chooses the c^* for which a large decrease is obtained when going from c^* to $c^* + 1$. Another criterion is the rate of convergence. This is justified because experience has shown that, for a good clustering structure and for an appropriate c, a high rate of convergence can generally be obtained.

Because the "optimal" final portion depends on the initialization of the starting partition $\widetilde{U}^0$, the "stability" of the final partition with respect to different starting partitions can also be used as an indication of a "correct" number of clusters c.

All three criteria serve to determine the "correct" number of clusters. They are heuristic in nature and therefore might lead to final partitions that do not correctly identify existing clusters. Bezdek shows, for instance, that the global minimum of the objective function is not necessarily reached for the correct partition [Bezdek 1981, pp. 96 ff]. Therefore other measures of cluster validity are needed in order to judge the quality of a partition.

The following criteria calculate cluster validity functionals that assign to each fuzzy final partition a scalar which is supposed to indicate the quality of the clustering solution. When designing such criteria one assumes that the clustering structure is better identified when more points concentrate around the cluster centers, that is, the crisper (unfuzzier) is the membership matrix of the final partition generated by the fuzzy c-means algorithm.

The best-known measures for judging the fuzziness of a clustering solution are

the partition coefficient, $F(\widetilde{U}; c)$,

the partition entropy, $H(\widetilde{U}, c)$, and

the proportion exponent, $P(\widetilde{U}; c)$.

Definition 11–5 [Bezdek 1981, p. 100]

Let $\widetilde{U} \in M_{fc}$ be a fuzzy c-partition of n data points. The *partition coefficient* of $\widetilde{U}$ is the scalar

$$F(\widetilde{U}; c) = \sum_{k=1}^{n} \sum_{i=1}^{c} \frac{(\mu_{ik})^2}{n}$$

Definition 11–6 [Bezdek 1981, p. 111]

The *partition entropy* of any fuzzy c-partition $\widetilde{U} \in M_{fc}$ of X, where $|X| = n$, is for $1 \leq c \leq n$

$$H(\widetilde{U}, c) = -\frac{1}{n} \sum_{k=1}^{n} \sum_{i=1}^{c} \mu_{ik} \log_e(\mu_{ik})$$

(see definition 4–3a, b, where the entropy was already used as a measure of fuzziness.)

Definition 11–7 [Windham 1981, p. 178; Bezdek 1981, p. 119]

Let $\widetilde{U} \in (M_{fc} \setminus M_{c0})$ be a fuzzy c-partition of $X; |x| = n$; $2 \leq c < n$. For column k of $\widetilde{U}$, $1 \leq k \leq n$, let

$$\mu_k = \max_{1 \leq i \leq c} \{\mu_{ik}\}$$

$$[\mu_k^{-1}] = \text{greatest integer} \leq \left(\frac{1}{\mu_k}\right)$$

The *proportion exponent* of U is the scalar

$$P(\widetilde{U}, c) = -\log_e \left\{ \prod_{k=1}^{n} \left[\sum_{j=1}^{\mu_k^{-1}} (-1)^{j+1} \binom{c}{j} (1 - j\mu_k)^{(c-1)} \right] \right\}$$

The above-mentioned measures have the following properties:

$$\frac{1}{c} \leq F(\widetilde{U}, c) \leq 1$$

$$0 \leq H(\widetilde{U}, c) \leq \log_e(c)$$

$$0 \leq P(\widetilde{U}, c) < \infty$$

The partition coefficient and the partition entropy are similar in so far as they attain their extrema for crisp partitions $U \in M_c$:

$$F(\widetilde{U}, c) = 1 \Leftrightarrow H(\widetilde{U}, c) = 0 \Leftrightarrow \widetilde{U} \in M_c$$

$$F(\widetilde{U}, c) = \frac{1}{c} \Leftrightarrow H(\widetilde{U}, c) = \log_e(c) \Leftrightarrow \widetilde{U} = \left[\frac{1}{c}\right]$$

The (heuristic) rules for selecting the "correct" or best partitions are:

$$\max_{c} \{\max_{\widetilde{U} \in \Omega_c} \{F(\widetilde{U}, c)\}\} \qquad c = 2, \ldots, n-1$$

$$\min_{c} \{\min_{\widetilde{U} \in \Omega_c} \{H(\widetilde{U}, c)\}\} \qquad c = 2, \ldots, n-1$$

where Ω_c is the set of all "optimal" solutions for given c.

The limitations of $F(\widetilde{U}, c)$ and $H(\widetilde{U}, c)$ are mainly their monotonicity and the lack of any suitable benchmark which would allow a judgment as to the acceptability of a final partition. The monotonicity will usually tend to indicate that the "correct" partition is the 2-partition. This problem can be solved, for instance, by choosing the i^* partition for which the value of $H(\widetilde{U}, c)$ lies below the trend when going from $c* - 1$ to $c*$.

$H(\widetilde{U}, c)$ is normally more sensitive with respect to a change of the partition than is $F(\widetilde{U}, c)$. This is particularly so if m is varied.

While $F(\widetilde{U}, c)$ and $H(\widetilde{U}, c)$ depend on all $c \cdot n$ elements the proportion exponent $P(\widetilde{U}, c)$ depends on the maximum degree of membership of the n elements. $P(\widetilde{U}, c)$ converges towards ∞ with increasing μ_k and it is not defined for $\mu_k = 1$.

The heuristic for choosing a good partition is

$$\max_{c} \{\max_{\widetilde{U} \in \Omega_c} \{P_i(\widetilde{U}, c)\}\} \qquad c = 2, \ldots, n-1$$

By contrast to $F(\widetilde{U}, c)$ and $H(\widetilde{U}, c)$, $P(\widetilde{U}, c)$ has the advantage that it is not monotone in c. There exist, however, no benchmarks, such that one can judge the quality of a portion $c*$ from the value of $P(\widetilde{U}*, c*)$.

The heuristic for $P(\widetilde{U}, c)$ possibly leads to an "optimal" final partition other than the heuristics of $F(\widetilde{U}, c)$ and/or of $H(\widetilde{U}, c)$. This might necessitate the use of other decision aids derived from the data themselves or from other considerations. Bezdek [1981] describes quite a number of other approaches in his book.

Exercises

1. Describe three example problems from the areas of engineering and management each of which can be considered as a problem of pattern recognition.
2. How is the dimensionality of the data space reduced in pattern recognition?

3. What is the center of a cluster and how can it be defined?
4. Which basic types of objective-function algorithms exist in cluster analysis?
5. Consider the following fuzzy graph:

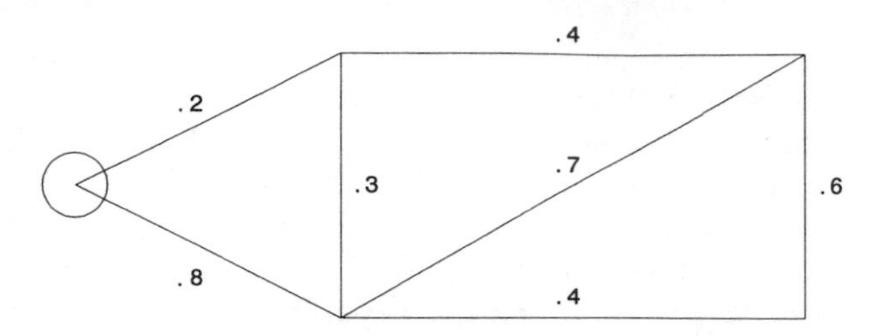

Determine the clusters of the graph in dependence of the τ-degree (*cf.* figure 11–6).
6. Let $X = \{x_1, x_2, x_3, x_4\}$ and let each x_i be a point in three-dimensional space. Determine all three-partitions that are possible and display them as shown in example 11–1.
7. Give 3 possible fuzzy three-partitions for the problem of exercise 6.
8. Let $X = \{(1, 1), (1, 3), (10, 1), (10, 3), (5, 2)\}$ be a set of points in the plane. Determine a crisp 3-partition that groups together (1, 3) and (10, 3) and which minimizes the Euclidean norm metric. Do the same for the variance criterion.
9. Determine the cluster validity of the clusters shown in figures 11–11 and 11–12 by computing the partition coefficient and the partition entropy.

12 DECISION MAKING IN FUZZY ENVIRONMENTS

12.1 Fuzzy Decisions

The term *decision* can have very many different meanings, depending on whether it is used by a lawyer, a businessman, a general, a psychologist, or a statistician. In one case it might be a legal construct, in another a mathematical model; it might also be a behavioral action or a specific kind of information processing. While some notions of a "decision" have a formal character, others try to describe decision making in reality.

In classical (normative, statistical) decision theory a decision can be characterized by a set of decision alternatives (the decision space); a set of states of nature (the state space); a relation assigning to each pair of a decision and state a result; and finally, the utility function which orders the results according to their desirability. When deciding under certainty the decision maker knows which state to expect and he chooses the decision alternative with the highest utility, given the prevailing state of nature. When deciding under risk he does not know exactly which state will occur, he only knows a probability function of the states. Then decision making becomes more difficult. We shall restrict our attention to decision making under certainty. In this instance the model of decision making is nonsymmetric in the following sense: The decision space can be described either by enumeration or by a number of constraints. The utility function orders the decision space via the one-to-one relationship of results to decision alternatives. Hence we can only have *one* utility function, supplying the order, but we may have several constraints defining the decision space.

Example 12–1

Let us assume that the board of directors wants to determine the optimal dividend. Their *objective function* (utility function) is to maximize the dividend. The *constraint* defining the decision space is that the dividend be between zero and 6%. Hence the optimal dividend is "Between 0 and 6%" *and* "maximal." (The constraint does *not* impose an order on the decision space!) The optimal dividend will obviously be 6%. Assigning a linear utility function, figure 12–1 illustrates these relationships.

In 1970 Bellman and Zadeh considered this classical model of a decision and suggested a model for decision making in a fuzzy environment that has served as a point of departure for most of the authors in "fuzzy" decision theory. They consider a situation of decision making under certainty, in which the objective function as well as the constraint(s) are fuzzy, and argue as follows: The fuzzy objective function is characterized by its membership function and so are the constraints. Since we want to satisfy

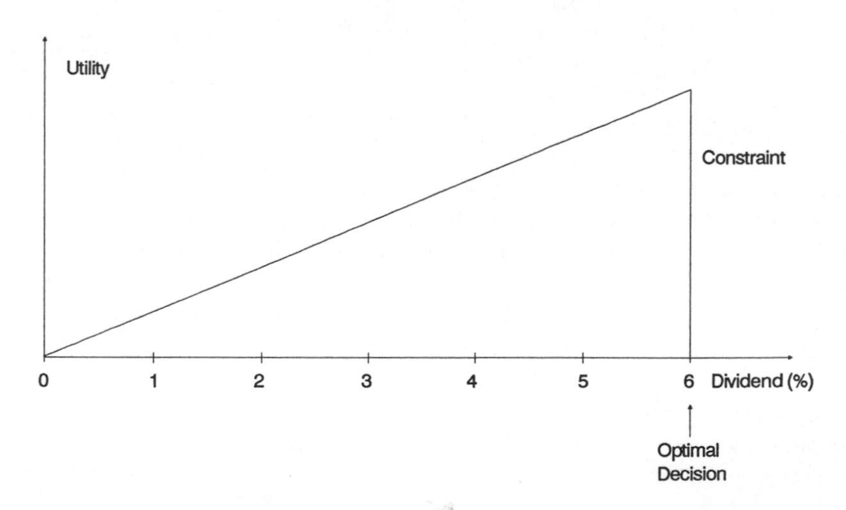

Figure 12–1. A classical decision under certainty.

(optimize) the objective function as well as the constraints, a decision in a fuzzy environment is defined by analogy to nonfuzzy environments as the selection of activities that simultaneously satisfy objective function(s) *and* constraints. According to the above definition and assuming that the constraints are "noninteractive" the logical "and" corresponds to the intersection. The "decision" in a fuzzy environment can therefore be viewed as the intersection of fuzzy constraints and fuzzy objective function(s). The relationship between constraints and objective functions in a fuzzy environment is therefore fully symmetric, that is, there is no longer a difference between the former and the latter.

This concept is illustrated by the following example [Bellman and Zadeh 1970, B-148]:

Example 12–2

Objective function "x should be substantially larger than 10," characterized by the membership function

$$\mu_{\tilde{O}}(x) = \begin{cases} 0 & x \leq 10 \\ (1 + (x-10)^{-2})^{-1} & x > 10 \end{cases}$$

Constraint "x should be in the vicinity of 11," characterized by the membership function

$$\mu_{\tilde{C}}(x) = (1 + (x-11)^4)^{-1}$$

The membership function $\mu_{\tilde{D}}(x)$ of the decision is then

$$\mu_{\tilde{D}}(x) = \mu_{\tilde{O}}(x) \wedge \mu_{\tilde{C}}(x)$$

$$\mu_{\tilde{D}}(x) = \begin{cases} \min\{(1 + (x-10)^{-2})^{-1}, (1 + (x-11)^4)^{-1}\} & \text{for } x > 10 \\ 0 & \text{for } x \leq 10 \end{cases}$$

$$= \begin{cases} (1 + (x-11)^4)^{-1} & \text{for } x > 11.75 \\ 0 & \text{for } 10 < x \leq 11.75 \\ & \text{for } x \leq 10 \end{cases}$$

This relation is depicted in figure 12–2. Let us now modify example 12–1 accordingly.

Example 12–3

The board of directors it trying to find the "optimal" dividend to be paid to the shareholders. For financial reasons it ought to be attractive and for reasons of wage negotiations it should be modest.

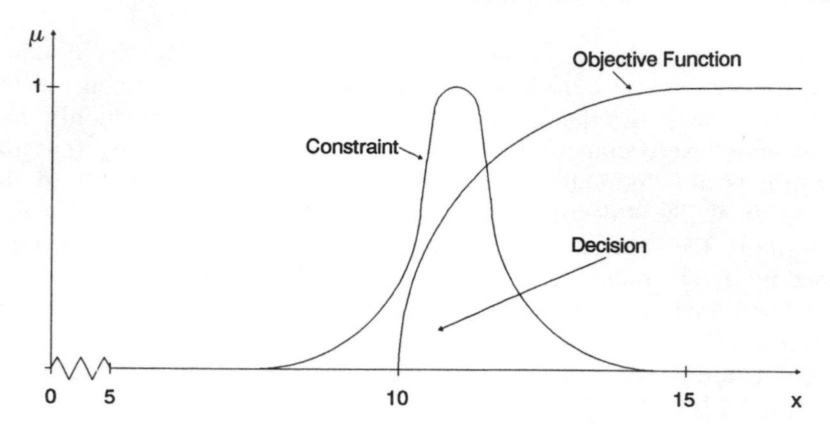

Figure 12–2. A fuzzy decision.

The fuzzy set of the objective function "attractive dividend" could for instance be defined by:

$$\mu_{\tilde{O}}(x) = \begin{cases} 1 & x \geq 5.8 \\ \frac{1}{2100}[-29x^3 - 366x^2 - 877x + 540] & 1 < x < 5.8 \\ 0 & x \leq 1 \end{cases}$$

The fuzzy set (constraint) "modest dividend" could be represented by

$$\mu_{\tilde{C}}(x) = \begin{cases} 1 & x \leq 1.2 \\ \frac{1}{2100}[-29x^3 - 243x^2 + 16x + 2388] & 1.2 < x < 6 \\ 0 & x \geq 6 \end{cases}$$

The fuzzy set "decision" is then characterized by its membership function

$$\mu_{\tilde{D}}(x) = \min \{\mu_{\tilde{O}}(x), \mu_{\tilde{C}}(x)\}$$

If the decision maker wants to have a "crisp" decision proposal, it seems appropriate to suggest the dividend with the highest degree of membership in the fuzzy set "decision." Let us call this the "maximizing decision," defined by

$$x_{\max} = \arg (\max_x \min \{\mu_{\tilde{O}}(x), \mu_{\tilde{C}}(x)\})$$

Figure 12–3 sketches this situation.

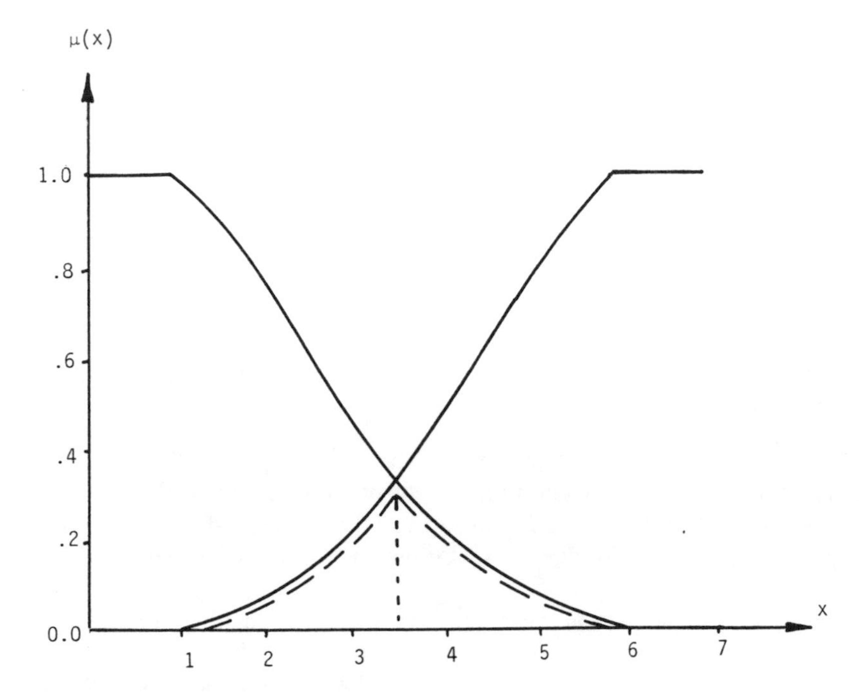

Figure 12–3. Optimal dividend as maximizing decision.

After these introductory remarks and examples we shall formally define a decision in a fuzzy environment in the sense of Bellman and Zadeh.

Definition 12–1 [Bellman and Zadeh 1970, B-148]

Assume that we are given a fuzzy goal $\tilde{G}$ and a fuzzy constraint $\tilde{C}$ in a space of alternatives X. Then $\tilde{G}$ and $\tilde{C}$ combine to form a *decision*, $\tilde{D}$, which is a fuzzy set resulting from intersection of $\tilde{G}$ and $\tilde{C}$. In symbols, $\tilde{D} = \tilde{G} \cap \tilde{C}$ and correspondingly

$$\mu_{\tilde{D}} = \min \{\mu_{\tilde{G}}, \mu_{\tilde{C}}\}$$

More generally suppose that we have n goals $\tilde{G}_1, \ldots, \tilde{G}_n$ and m constraints $\tilde{C}_1, \ldots, \tilde{G}_m$. Then, the resultant decision is the intersection of the given goals $\tilde{G}_1, \ldots, \tilde{G}_n$ and the given constraints $\tilde{C}_1, \ldots, \tilde{G}_m$. That is,

$$\tilde{D} = \tilde{G}_1 \cap \tilde{G}_2 \cap \cdots \cap \tilde{G}_n \cap \tilde{C}_1 \cap \tilde{C}_2 \cap \cdots \cap \tilde{C}_m$$

and correspondingly

$$\mu_{\tilde{D}} = \min\{\mu_{\tilde{G}_1}, \mu_{\tilde{G}_2}, \ldots, \mu_{\tilde{G}_n}, \mu_{\tilde{C}_1}, \mu_{\tilde{C}_2}, \ldots, \mu_{\tilde{C}_m}\}$$
$$= \min\{\mu_{\tilde{G}_i}, \mu_{\tilde{C}_j}\} = \min\{\mu_i\}$$

Definition 12–1 implies essentially three assumptions:

1. The "and" connecting goals and constraints in the model corresponds to the "logical and."
2. The logical "and" corresponds to the set theoretic intersection.
3. The intersection of fuzzy sets is defined in the possibilistic sense by the min-operator.

Bellman and Zadeh indicated in their 1970 paper that the min-interpretation of the intersection might have to be modified depending on the context. "In short, a broad definition of the concept of decision may be stated as: Decision = Confluence of Goals and Constraints" [Bellman and Zadeh 1970, B-149].

The question arises whether even the intersection interpretation is a generally acceptable assumption or whether "confluence" has to be interpreted in an even more general way. Let us consider the following example.

Example 12–4

An instructor at a university must decide how to grade written test papers. Let us assume that the problem to be solved in the test was a linear programming problem and that the student was free to solve it either graphically or using the simplex method. The student has done both. The student's performance is expressed—for graphical solution as well as for the algebraic solution—as the achieved degree of membership in the fuzzy sets "good graphical solution" ($\tilde{G}$) and "good simplex solution" ($\tilde{S}$), respectively. Let us assume that he reaches

$$\mu_{\tilde{G}} = 0.9 \quad \text{and} \quad \mu_{\tilde{S}} = 0.7$$

If the grade to be awarded by the instructor corresponds to the degree of membership of the fuzzy set "good solutions of linear programming problems" it would be quite conceivable that his grade $\mu_{\widetilde{LP}}$ could be determined by

$$\mu_{\widetilde{LP}} = \max\{\mu_{\tilde{G}}, \mu_{\tilde{S}}\} = \max\{0.9, 0.7\} = 0.9$$

The two definitions of decisions—as the intersection or the union of fuzzy sets—imply essentially the following: The interpretation of a decision as the intersection of fuzzy sets implies no positive compensation (trade-off) between the degrees of membership of the fuzzy sets in question, if either the minimum or the product is used as an operator. Each of them yields a degree of membership of the resulting fuzzy set (decision) which is on or below the lowest degree of membership of all intersecting fuzzy sets (see example 12–3).

The interpretation of a decision as the union of fuzzy sets, using the max-operator, leads to the maximum degree of membership achieved by any of the fuzzy sets representing objectives or constraints. This amounts to a full compensation of lower degrees of membership by the maximum degree of membership (see example 12–4).

Observing managerial decisions one finds that there are hardly any decisions with no compensation between either different degrees of goal achievement or the degrees to which restrictions are limiting the scope of decisions. The compensation, however, rarely ever seems to be "complete" such as would be assumed using the max-operator. It may be argued that compensatory tendencies in human aggregation are responsible for the failure of some classical operators (min, product, max) in empirical investigations.

Two conclusions can probably be drawn: Neither the noncompensatory "and" represented by operators that map between zero and the minimum degree of membership (min-operator, product-operator, Hamacher's conjunction operator [definition 3–15], Yager's conjunction operator [definition 3–16]) nor the fully compensatory "or" represented by the operators that map between the maximum degree of membership and 1 (maximum, algebraic sum, Hamacher's disjunction operator, Yager's disjunction operator) are appropriate to model the aggregation of fuzzy sets representing managerial decisions.

"Confluence of Goals and Constraints" should therefore be interpreted as in definition 12–2.

Definition 12–2

Let $\mu_{\tilde{C}_i}(x)$, $i = 1, \ldots, m$, $x \in X$, be membership functions of constraints, defining the decision space and $\mu_{\tilde{G}_j}(x)$, $j = 1, \ldots, n$, $x \in X$ the membership functions of objective (utility) functions or goals.

A *decision* is then defined by its membership function

$$\mu_{\tilde{D}}(x) = \circledast_i \mu_{\tilde{C}_i}(x) * \circledast_j \mu_{\tilde{G}_j}(x), \; i = 1, \ldots, m, j = 1, \ldots, n$$

where $*$, $\circledast_i$, $\circledast_j$ denote appropriate, possibly context-dependent "aggregators" (connectives).

We shall discuss the question of appropriate connectives in more detail in chapter 14. Before turning to fuzzy mathematical programming it should be mentioned that the symmetry that is a property of all definitions based on Bellman-Zadeh's concept (irrespective of the operators used) is not considered adequate by all authors (for example, see [Asai et al. 1975]).

12.2 Fuzzy Linear Programming

Linear programming models shall be considered as a special kind of decision model: the decision space is defined by the constraints; the "goal" (utility function) is defined by the objective function; and the type of decision is decision making under certainty. The classical model of linear programming can be stated as

$$\begin{aligned} &\text{maximize} \quad f(x) = c^T x \\ &\text{such that} \quad Ax \leq b \\ &\qquad\qquad\quad x \geq 0 \\ &\text{with } c, x \in \mathbb{R}^n, \; b \in \mathbb{R}^m, \; A \in \mathbb{R}^{m \times n} \end{aligned} \tag{12.1}$$

Let us now depart from the classical assumptions that all coefficients of A, b, and c are crisp numbers, that $\leq$ is meant in a crisp sense, and that "maximize" is a strict imperative!

If we assume that the LP-decision has to be made in fuzzy environments, quite a number of possible modifications of (12.1) exist. First of all the decision maker might really not want to actually maximize or minimize the objective function. Rather he might want to reach some aspiration levels which might not even be definable crisply. Thus he might want to "improve the present cost situation considerably," and so on.

Secondly, the constraints might be vague in one of the following ways: The $\leq$ sign might not be meant in the strictly mathematical sense but smaller violations might well be acceptable. This can happen if the constraints represent aspiration levels as mentioned above or if, for instance, the constraints represent sensory requirements (taste, color, smell, etc.) which cannot adequately be approximated by a crisp constraint. Of course, the coefficients of the vectors b or c or of the matrix A itself can have a fuzzy character either because they are fuzzy in nature or because perception of them is fuzzy.

Finally the role of the constraints can be different from that in classical linear programming where the violation of any single constraint by any amount renders the solution infeasable. The decision maker might accept small violations of constraints but might also attach different (crisp or fuzzy) degrees of importance to violations of different constraints. Fuzzy linear programming offers a number of ways to allow for all those types of vagueness and we shall discuss some of them below.

First of all we can either accept Bellman-Zadeh's concept of a symmetrical decision model (see definition 12–1) or we can develop specific models on the basis of a nonsymmetrical basic model of a "fuzzy" decision [Orlovsky 1980; Asai et al. 1975]. Here we shall adopt the former, more common, approach. Secondly we will have to decide how a fuzzy "maximize" is to be interpreted, or if one wants to stick to a crisp "maximize." In the latter case complications arise on how to connect a crisp objective function with a fuzzy solution space. We will discuss one approach for a fuzzy goal and one approach for a crisp objective function.

Finally one will have to decide where and how fuzziness enters the constraints. Some authors [Tanaka and Asai 1984] consider the coefficients of A, b, c as fuzzy numbers and the constraints as fuzzy functions. We shall here adapt another approach which seems to be more efficient computationally and resembles more Bellman-Zadeh's model in definition 12–1: We shall represent the goal and the constraints by fuzzy sets and then aggregate them in order to derive a maximizing decision.

In both approaches one also has to decide on the type of membership function characterizing either the fuzzy numbers or the fuzzy sets representing goal and constraints.

In classical LP the "violation" of any constraint in (12.1) renders the solution infeasible. Hence all constraints are considered to be of equal weight or importance. When departing from classical LP this is no longer true and one also has to worry about the relative weights attached to the constraints.

Before we develop a specific model of linear programming in a fuzzy environment it should have become clear, that by contrast to classical linear programming "fuzzy linear programming" is *not* a uniquely defined type of model but that many variations are possible, depending on the assumptions or features of the real situation to be modelled.

12.2.1 Symmetric Fuzzy LP

Let us now turn to a first basic model for "fuzzy linear programming." In model (12.1) we shall assume, that the decision maker can establish an aspiration level, z, for the value of the objective function he wants to achieve and that each of the constraints is modeled as a fuzzy set. Our fuzzy LP then becomes:

Find x such that

$$\begin{aligned} c^T x &\gtrsim z \\ Ax &\lesssim b \\ x &\geq 0 \end{aligned} \tag{12.2}$$

Here $\lesssim$ denotes the fuzzified version of $\leq$ and has the linguistic interpretation "essentially smaller than or equal." $\gtrsim$ denotes the fuzzified version of $\geq$ and has the linguistic interpretation "essentially greater than or equal." The objective function in (12.1) might have to be written as a minimizing goal in order to consider z as an upper bound.

We see that (12.2) is fully symmetric with respect to objective function and constraints and we want to make that even more obvious by substituting $\binom{-c}{A} = B$ and $\binom{-z}{b} = d$. Then (12.2) becomes:

Find x such that

$$\begin{aligned} Bx &\lesssim d \\ x &\geq 0 \end{aligned} \tag{12.3}$$

Each of the $(m+1)$ rows of (12.3) shall now be represented by a fuzzy set, the membership functions of which are $\mu_i(x)$. Following definition 12–1, the membership function of the fuzzy set "decision" of model (12.3) is

$$\mu_{\tilde{D}}(x) = \min_i \{\mu_i(x)\} \tag{12.4}$$

$\mu_i(x)$ can be interpreted as the degree to which x fulfills (satisfies) the fuzzy unequality $B_i x \leq d_i$ (where B_i denotes the ith row of B).

Assuming that the decision maker is interested not in a fuzzy set but in a crisp "optimal" solution we could suggest the "maximizing solution" to (12.4), which is the solution to the possibly nonlinear programming problem

$$\max_{x \geq 0} \min_i \{\mu_i(x)\} = \max_{x \geq 0} \mu_{\tilde{D}}(x) \tag{12.5}$$

Now we have to specify the membership functions $\mu_i(x)$. $\mu_i(x)$ should be 0 if the constraints (including objective function) are strongly violated, and

1 if they are very well satisfied (i.e., satisfied in the crisp sense); and $\mu_i(x)$ should increase monotonously from 0 to 1, that is:

$$\mu_i(x) = \begin{cases} 1 & \text{if } B_i x \leq d_i \\ \in [0, 1] & \text{if } d_i < B_i x \leq d_i + p_i \\ 0 & \text{if } B_i x > d_i + p_i \end{cases} \qquad i = 1, \ldots, m+1 \tag{12.6}$$

Using the simplest type of membership function we assume them to be linearly increasing over the "tolerance interval" p_i:

$$\mu_i(x) = \begin{cases} 1 & \text{if } B_i x \leq d_i \\ 1 - \dfrac{B_i x - d_i}{p_i} & \text{if } d_i < B_i x \leq d_i + p_i \\ 0 & \text{if } B_i x > d_i + p_i \end{cases} \qquad i = 1, \ldots, m+1 \tag{12.7}$$

The p_i are subjectively chosen constants of admissible violations of the constraints and the objective function. Substituting (12.7) into (12.5) yields, after some rearrangements [Zimmermann 1976] and with some additional assumptions,

$$\max_{x \geq 0} \min_i \left(1 - \frac{B_i x - d_i}{\mu_i}\right) \tag{12.8}$$

Introducing one new variable, λ, which corresponds essentially to (12.4), we arrive at

$$\begin{aligned} &\text{maximize} \quad \lambda \\ &\text{such that} \quad \lambda p_i + B_i x \leq d_i + p_i \qquad i = 1, \ldots, m+1 \\ &\qquad\qquad\qquad x \geq 0 \end{aligned} \tag{12.9}$$

If the optimal solution to (12.9) is the vector (λ, x_0), then x_0 is the maximizing solution (12.5) of model (12.2) assuming membership functions as specified in (12.7).

The reader should realize that this maximizing solution can be found by solving one standard (crisp) LP with only one more variable and one more constraint than model (12.3). This makes this approach computationally very efficient.

A slightly modified version of models (12.8) and (12.9), respectively, results if the membership functions are defined as follows: A variable t_i, $i = 1, \ldots, m+1$, $0 \leq t_i \leq p_i$, is defined which measures the degree of violation of the ith constraint: The membership function of the ith row is then

$$\mu_i(x) = 1 - \frac{t_i}{p_i} \tag{12.10}$$

The crisp equivalent model is then

$$\begin{aligned}
\text{maximize} \quad & \lambda \\
\text{such that} \quad & \lambda p_i + t_i \leq p_i \qquad i = 1, \ldots, m+1 \\
& B_i x - t_i \leq d_i \\
& t_i \leq p_i \\
& x, t \geq 0
\end{aligned} \tag{12.11}$$

This model is larger than model (12.9), even though the set of constraints $t_i \leq p_i$ is actually redundant. Model (12.11) has some advantages, however, in particular when performing sensitivity analysis, which will be discussed in the second volume on decisions in fuzzy environments.

Example 12–5

A company wanted to decide on the size and structure of its truck fleet. Four differently sized trucks (x_1 through x_4) were considered. The objective was to minimize cost and the constraints were to supply all customers (who have a strong seasonally fluctuating demand). That meant certain quantities had to be moved (quantity constraint) and a minimum number of customers per day had to be contacted (routing constraint). For other reasons, it was required that at least 6 of the smallest trucks be included in the fleet. The management wanted to use quantitative analysis and agreed to the following suggested linear programming approach:

minimize

$$41{,}400x_1 + 44{,}300x_2 + 48{,}100x_3 + 49{,}100x_4$$

subject to constraints

$$\begin{aligned}
0.84x_1 + 1.44x_2 + 2.16x_3 + 2.4x_4 &\geq 170 \\
16x_1 + 16x_2 + 16x_3 + 16x_4 &\geq 1{,}300 \\
x_1 \qquad\qquad\qquad\qquad\qquad &\geq 6 \\
x_2, x_3, x_4 &\geq 0
\end{aligned}$$

The solution was $x_1 = 6$, $x_2 = 16.29$, $x_3 = 0$, $x_4 = 58.96$. Min Cost = 3,864,975. When presenting the results to management it turned out that they were considered acceptable but that the management would rather have some "leeway" in the constraints. They felt that because demand forecasts had been used to formulate the constraints (and because forecasts never turn out to be correct!), there was a danger of not being able to meet higher demands by their customers.

When inquiring whether or not they really wanted to "minimize transportation cost" they answered: Now you are joking. A few months ago you told us that we have to minimize cost otherwise you could not model our problem. Nobody knows minimum cost anyway. The budget shows a cost figure of 4.2 millions, a figure that must not be exceeded. If you want to keep your contract, you better stay considerably below this figure.

Since management felt that it was forced into giving precise constraints (because of the model) in spite of the fact that it would rather have given some intervals, model (12.3) was selected to model the management's perceptions of the problem satisfactorily. The following parameters were estimated: Lower bounds of the tolerance interval:

$$d_1 = 3{,}700{,}000 \qquad d_2 = 170 \qquad d_3 = 1{,}300 \qquad d_4 = 6$$

Spreads of tolerance intervals:

$$p_1 = 500{,}000 \qquad p_2 = 10 \qquad p_3 = 100 \qquad p_4 = 6$$

After dividing all rows by their respective p_i's and rearranging in such a way that only λ remains on the left-hand side, our problem in the form of (12.9) became:

Maximize λ subject to constraints

$$\begin{aligned}
&0.083x_1 + 0.089x_2 + 0.096x_3 + 0.098x_4 + \lambda \leq 8.4\\
&0.084x_1 + 0.144x_2 + 0.216x_3 + 0.24x_4 \ \ - \lambda \geq 17\\
&0.16x_1 \ \ + 0.16x_2 \ \ + 0.16x_3 \ \ + 0.16x_4 \ \ - \lambda \geq 13\\
&0.167x_1 \qquad\qquad\qquad\qquad\qquad\qquad\qquad - \lambda \geq 1\\
&\qquad\qquad\qquad\qquad \lambda, x_1, x_2, x_3, x_4 \geq 0
\end{aligned}$$

Solution

Unfuzzy	Fuzzy
$x_1 = 6$	$x_1 = 17.414$
$x_2 = 16.29$	$x_2 = 0$
$x_4 = 58.96$	$x_4 = 66.54$
$Z = 3{,}864{,}975$	$Z = 3{,}988{,}250$
Constraints:	
1. 170	174.33
2. 1,300	1,343.328
3. 6	17.414

As can be seen from the solution, "leeway" has been provided with respect to all constraints and at additional cost of 3.2 percent.

The main advantage, compared to the unfuzzy problem formulation, is the fact that the decision maker is not forced into a precise formulation because of mathematical reasons even though he might only be able or willing to describe his problem in fuzzy terms. Linear membership functions are obviously only a very rough approximation. Membership functions which monotonically increase or decrease, respectively, in the interval of $[d_i, d_i + p_i]$ can also be handled quite easily, as will be shown later.

So far the objective function as well as all constraints were considered fuzzy. If some of the constraints are crisp, $Dx \leq b$, then these constraints can easily be added to formulations (12.9) or (12.11), respectively. Thus (12.9) would, for instance, become:

$$\begin{aligned} &\text{maximize} && \lambda \\ &\text{such that} && \lambda p_i + B_i x \leq d_i + p_i \qquad i = 1, \ldots, m+1 \\ &&& Dx \leq b \\ &&& x, \lambda \geq 0 \end{aligned} \qquad (12.12)$$

Let us now turn to the case in which the objective function is crisp and the solution space is fuzzy.

12.2.2 Fuzzy LP with Crisp Objective Function

A model in which the objective function is crisp, that is, has to be maximized or minimized and in which the constraints are all or partially fuzzy is no longer symmetrical. The roles of objective functions and constraints are different, the latter define the decision space in a crisp or fuzzy way and the former induces an order of the decision alternatives. Therefore the approach of models (12.3)–(12.5) is not applicable. The main problem is the scaling of the objective function (the domain of which is not normalized) when aggregating it with the (normalized) constraints. In very rare real cases a scaling factor can be found that has a real justification.

The problem we are faced with is the determination of an extremum of a crisp function over a fuzzy domain, which we have already discussed in section 7.2 of this book. In definition 7–3 we defined the notion of a maximizing set which we will specify here and use as a vehicle to solve our LP problem. Two approaches are conceivable:

1. The determination of the fuzzy set "decision."
2. The determination of a crisp "maximizing decision" by aggregating the objective function after appropriate transformations with the constraints.

1: The Determination of a Fuzzy Set "Decision". Orlovski [1977] suggests to compute, for all α-level sets of the solution space, the corresponding optimal values of the objective function, and to consider as the fuzzy set "decision" the optimal values of the objective functions with the degree of membership equal to the corresponding α-level of the solution space.

Definition 12–3 [Werners 1984]

Let $R_\alpha = \{x \mid x \in X, \mu_R(x) \geq \alpha\}$ be the α-level sets of the solution space and $N(\alpha) = \{x \mid x \in R_\alpha, f(x) = \sup_{x' \in R_\alpha} f(x')\}$ the set of optimal solutions for each α-level set.

The fuzzy set "*decision*" is then defined by the membership function

$$\mu_{\text{opt}}(x) = \begin{cases} \sup\limits_{x \in N(\alpha)} \alpha & \text{if} \quad x \in \bigcup\limits_{\alpha > 0} N(\alpha) \\ 0 & \text{else} \end{cases}$$

The fuzzy set "*optimal values of the objective function*" has the membership function

$$\mu_f(r) = \begin{cases} \sup\limits_{x \in f^{-1}(r)} \mu_{\text{opt}}(x) & \text{if} \quad r \in \mathbb{R}_1 \wedge f^{-1}(r) \neq \emptyset \\ 0 & \text{else} \end{cases}$$

$f(x)$ is the objective function with functional values r.

For the case of linear programming the determination of the r's and $\mu_{\text{opt}}(x)$ can be obtained by parametric programming [Chanas 1983]. For each α an LP of the following kind would have to be solved:

$$\begin{aligned} &\text{maximize} && f(x) \\ &\text{such that} && \alpha \leq \mu_i(x) \qquad i = 1, \ldots, m \\ & && x \in X \end{aligned} \tag{12.13}$$

The reader should realize, however, that the result is a fuzzy set and that the decision maker would have to decide which pair $(r, \mu_f\, f(r))$ he considers optimal if he wants to arrive at a crisp optimal solution.

Example 12–6 [Werners 1984]

Consider the LP-Model

$$\begin{array}{ll} \text{maximize} & z = 2x_1 + x_2 \\ \text{such that} & x_1 \lessapprox 3 \\ & x_1 + x_2 \lessapprox 4 \\ & 5x_1 + x_2 \lessapprox 3 \\ & x_1, x_2 \geq 0 \end{array}$$

The "tolerance intervals" of the constraints are $p_1 = 6$, $p_2 = 4$, $p_3 = 2$.

The parametric linear program for determining the relationships between $f(x) = r$ and degree of membership is then

$$\begin{array}{ll} \text{maximize} & z = 2x_1 + x_2 \\ \text{such that} & x_1 \leq 9 - 6\alpha \\ & x_1 + x_2 \leq 8 - 4\alpha \\ & 5x_1 + x_2 \leq 5 - 2\alpha \\ & x_1, x_2 \geq 0 \end{array}$$

Figure 12–4 shows the feasible regions for R_0 and R_1 for $\mu_{\tilde{R}}(x) = 0$ and $\mu_{\tilde{R}}(x) = 1$. Figure 12–5 shows the resulting membership function $\mu_f(r)$. Additionally fig. 12–5 shows the membership function of the goal and the fuzzy decision that will be discussed in the following.

Obviously the decision maker has to decide which combination $(r, \mu_f(r))$ he considers best.

Decision aids in this respect can either be derived from external sources or they may depend on the problem itself. In the following we shall consider an approach which suggests a crisp solution that depends on the solution space.

2: The Determination of a Crisp Maximizing Decision

Some authors [Kickert 1978; Nguyen 1979; Zadeh 1972] suggest approaches based on the notion of a maximizing set, which seem to have some disadvantages [see Werners 1984]. We shall therefore present a model that is particularly suitable for the type of Linear Programming model we are considering here. Werners [1984] suggests the following definition.

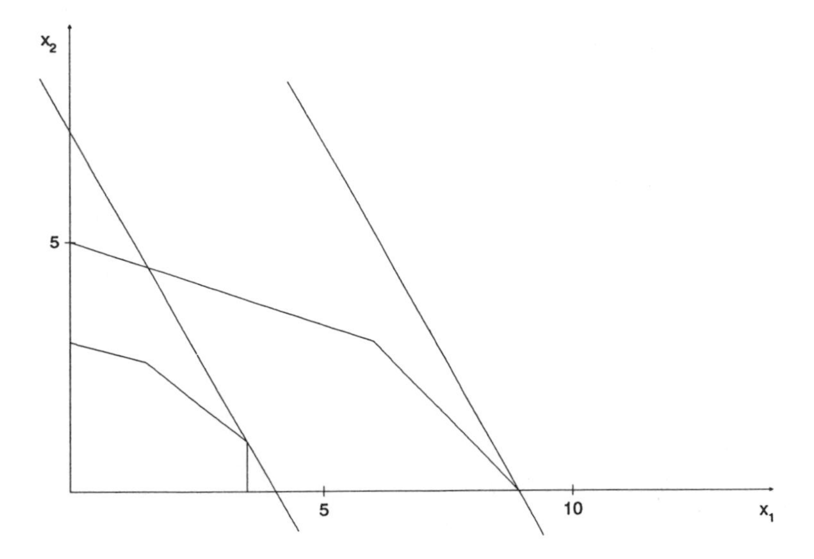

Figure 12–4. Feasible regions for $\mu_{\tilde{R}}(x) = 0$ and $\mu_{\tilde{R}}(x) = 1$.

Definition 12–4

Let $f: X \rightarrow \mathbb{R}^1$ be the objective function, $\widetilde{R}$ a fuzzy region (solution space) and $S(\widetilde{R})$ the support of this region. The *maximizing set over the fuzzy region*, $\widetilde{MR}(f)$, is then defined by its membership function

$$\mu_{\widetilde{MR}(f)}(x) = \begin{cases} 0 & \text{if } f(x) \leq \inf\limits_{S(\widetilde{R})} f \\ \dfrac{f(x) - \inf\limits_{S(\widetilde{R})} f}{\sup\limits_{S(\widetilde{R})} f - \inf\limits_{S(\widetilde{R})} f} & \text{if } \inf\limits_{S(\widetilde{R})} f < f(x) < \sup\limits_{S(\widetilde{R})} f \\ 1 & \text{if } \sup\limits_{S(\widetilde{R})} f \leq f(x) \end{cases}$$

The intersection of this maximizing set with the fuzzy set "decision" (figure 12–5) could then be used to compute a maximizing decision x_0 as the solution with the highest degree of membership in this fuzzy set. It does not seem reasonable that the judgment of the decision maker is calibrated by looking at the smallest value of f over the feasable region. A better benchmark would be the largest value for f that can be obtained at a degree of membership of 1 of the feasible region. This leads to the following definition.

Definition 12–5 (Werners 1984)

Let $f: X \to \mathbb{R}^1$ be the objective function, $\widetilde{R}$ = fuzzy feasible region, $S(\widetilde{R})$ = support of $\widetilde{R}$, and R_1 = α-level cut of $\widetilde{R}$ for $\alpha = 1$. The *membership function of the goal* (objective function) *given solution space* $\widetilde{R}$ is then defined as

$$\mu_{\tilde{G}}(x) = \begin{cases} 0 & \text{if } f(x) \leq \sup_{R_1} f \\ \dfrac{f(x) - \sup_{R_1} f}{\sup_{S(\widetilde{R})} f - \sup_{R_1} f} & \text{if } \sup_{R_1} f < f(x) < \sup_{S(\widetilde{R})} f \\ 1 & \text{if } \sup_{S(\widetilde{R})} f \leq f(x) \end{cases}$$

The corresponding membership function in functional space is then

$$\mu_{\tilde{G}}(r) := \begin{cases} \sup_{x \in f^{-1}(r)} \mu_{\tilde{G}}(x) & \text{if } r \in \mathbb{R}, f^{-1}(r) \neq 0 \\ 0 & \text{else} \end{cases}$$

Example 12–7

Consider the model of example 12–6. For this model R_1 is the region defined by

$$\begin{aligned} x_1 \quad\quad &\leq 3 \\ x_1 + x_2 &\leq 4 \\ 5x_1 + x_2 &\leq 3 \\ x &\geq 0 \end{aligned}$$

The supremum of f over this region is

$$\sup_{R_1} 2x_1 + x_2 = 7$$

Figure 12–5 shows the membership functions $\mu_f(r)$ and $\mu_{\tilde{G}}(r)$. Using the min-max-approach the resulting solution is $x_1^0 = 5.84$, $x_2^0 = .05$, $r_0 = 11.73$, and the attained degree of membership $\mu_{\tilde{R}}(x_0) = .53$.

Let us now return to model (12.2) and modify it by considering the

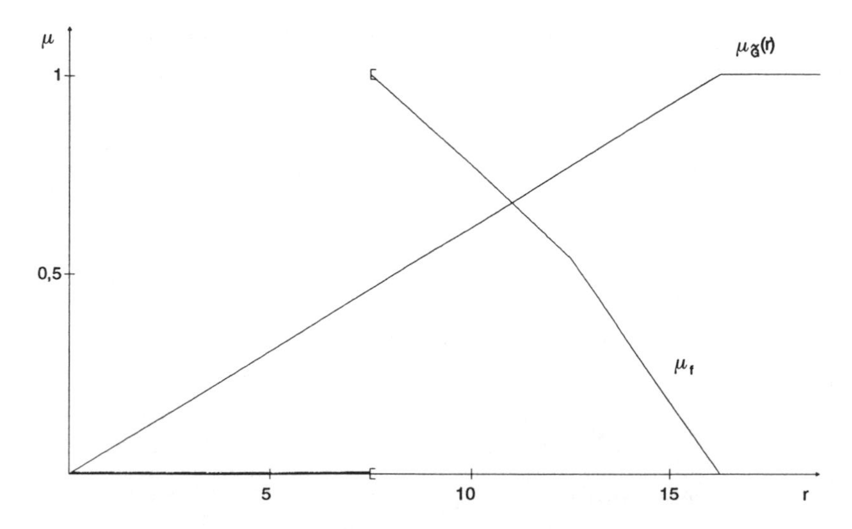

Figure 12–5. Fuzzy decision.

objective function to be crisp and by adding a set of crisp constraints $Dx \leq b'$:

$$\begin{aligned} &\text{maximize} && f(x) = c^T x \\ &\text{such that} && \left.\begin{aligned} Ax &\lessapprox b \\ Dx &\leq b' \\ x &\geq 0 \end{aligned}\right\} \widetilde{R} \end{aligned} \tag{12.14}$$

Let the membership functions of the fuzzy sets representing the fuzzy constraints be defined in analogy to (12.7) as

$$\mu_i(x) = \begin{cases} 1 & \text{if } A_i x \leq b_i \\ \dfrac{b_i + p_i - A_i x}{p_i} & \text{if } b_i < A_i x \leq b_i + p_i \\ 0 & \text{if } A_i x > b_i + p_i \end{cases} \tag{12.15}$$

The membership function of the objective function (12.5) can be determined by solving the following two LP's:

$$\begin{aligned} &\text{maximize} && f(x) = c^T x \\ &\text{such that} && Ax \leq b \\ & && Dx \leq b' \\ & && x \geq 0 \end{aligned} \tag{12.16}$$

yielding $\sup_{R_1} f = (c^T x)_{\text{opt}} = f_1$; and

$$\begin{array}{ll} \text{maximize} & f(x) = c^T x \\ \text{such that} & Ax \leq b + p \\ & Dx \leq b' \\ & x \geq 0 \end{array} \tag{12.17}$$

yielding $\sup_{S(\tilde{R})} f = (c^T x)_{\text{opt}} = f_0$

The membership function of the objective function is therefore

$$\mu_{\tilde{G}}(x) = \begin{cases} 1 & \text{if } f_0 \leq c^T x \\ \dfrac{c^T x - f_1}{f_0 - f_1} & \text{if } f_1 < c^T x < f_0 \\ 0 & \text{if } c^T x \leq f_1 \end{cases} \tag{12.18}$$

Now we have again achieved "symmetry" between constraints and the objective function and we can employ the approach we used to derive model (12.9) as an equivalent formulation of (12.2).

The equivalent model to (12.6) is

$$\begin{array}{ll} \text{maximize} & \lambda \\ \text{such that} & \lambda(f_0 - f_1) - c^T x \leq -f_1 \\ & \lambda p \qquad + Ax \leq b + p \\ & \qquad Dx \leq b' \\ & \lambda \qquad\qquad \leq 1 \\ & \qquad \lambda, x \geq 0 \end{array} \tag{12.19}$$

Example 12–8

We shall again consider the model in example 12–6. In example 12–7 we have computed $f_1 = 7$. By solving (12.17) we obtain $f_0 = 16$. Therefore (12.19) is

$$\begin{array}{ll} \text{maximize} & \lambda \\ \text{such that} & 9\lambda - 2x_1 - x_2 \leq -7 \\ & 6\lambda + x_1 \leq 9 \\ & 4\lambda + x_1 + x_2 \leq 8 \\ & 2\lambda + 5x_1 + x_2 \leq 5 \end{array}$$

$$\lambda \qquad\qquad \leq 1$$
$$\lambda, x_1, x_2 \geq 0$$

The solution to this problem is $x_1^0 = 5.84$, $x_2^0 = 0$, $\lambda_0 = .52$.

Before turning to fuzzy dynamic programming it should be mentioned that on the basis of the approach described so far, suggestions for a duality theory [Rödder and Zimmermann 1980], for sensitivity analysis in fuzzy linear programming [Hamacher, Leberling, and Zimmermann 1978],for integer fuzzy programming [Zimmermann and Pollatschek 1984], and for the use of other than linear membership functions and other operators have been published [Werners 1988]. These topics will not, however, be discussed here. They have been discussed in more detail in Zimmermann [1987]. Other approaches introducing fuzziness into mathematical programming have been published by a number of authors. Often these approaches have been developed in the context of multiobjective decision making. In order to avoid duplication these approaches will be mentioned at the end of the discussion of the vector-maximum problem in section 12.4.

12.3 Fuzzy Dynamic Programming

Traditional dynamic programming [Bellman 1957] is a technique well known in operations research and used to solve optimization problems that can be composed into subproblems of one variable (decision-variable) each. The idea underlying dynamic programming is to view the problem as a multistage decision process, the optimal policy to which can be determined recursively.

Generally the problem is formulated in terms of state variables, x_i, decision variables, d_i, stage rewards, $r_i(x_i, d_i)$, a reward function, $R_i(d_N, \ldots, d_{N-i}, x_N)$, and a transformation function, $t_i(d_i, x_i)$. Figure 12–6 illustrates the basic structure.

The problem is solved by solving recursively the following:

$$\max_{d_i} R_i(x_i, d_i) = \max_{d_i} r_i(x_i, d_i) \circ R_{i+1}(x_{i+1})$$

such that

$$x_{i+1} = t_i(x_i, d_i)$$
$$i = 1, \ldots, N-1$$

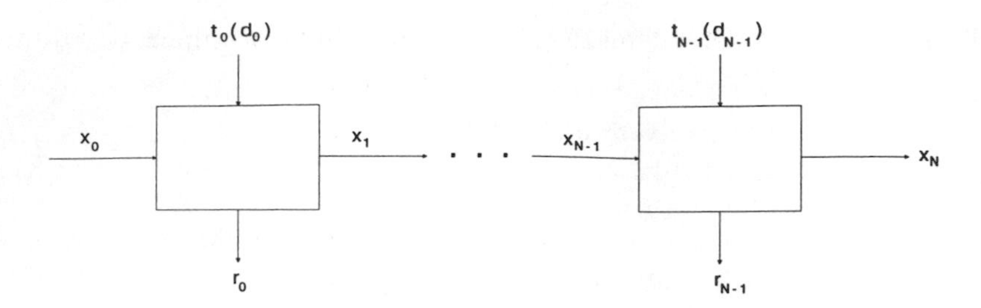

Figure 12–6. Basic structure of a dynamic programming model.

or

$$\max_{d_i} R_i(x_i, d_i) = \max_{d_i} \{r_i(x_i, d_i) \circ R_{i+1}(t_i(x_i, d_i))\}$$

All variables, rewards, and transformations are supposed to be crisp.

Fuzzy Dynamic Programming with Crisp State Transformation Function

In their famous paper, Bellman and Zadeh [1970] suggested for the first time a fuzzy approach to this type of problem. Conceivably they based their considerations on the symmetrical model of a decision as defined in definitions 12–1 and 12–2. The following terms will be used to define the fuzzy dynamic programming model [Bellman and Zadeh 1970, B-151]: $\widetilde{X}_i \in \widetilde{X}$, $i = 0, \ldots, N$: (crisp) state variable where $\widetilde{X} = \{\tau_1, \ldots, \tau_N\}$ is the set of values permitted for the state variables; $d_i \in \widetilde{D}$, $i = 1, \ldots, N$: (crisp) decision variable where $\widetilde{D} = \{\alpha_1, \ldots, \alpha_m\}$ is the set of possible decisions.

$$x_{i+1} = t(x_i, d_i)\text{: (crisp) transformation function}$$

For each stage t, $t = 0, \ldots, N - 1$, we define:

1. a fuzzy constraint $\widetilde{C}_t$ limiting the decision space and characterized by its membership function

$$\mu_{\tilde{C}_t}(d_t)$$

2. a fuzzy goal $\widetilde{G}_N$ characterized by the membership function

$$\mu_{\tilde{G}_N}(x_N)$$

The problem is to determine the maximizing decision

$$\widetilde{D}^0 = \{d_i^0\} \qquad i = 0, \ldots, N, \text{ for a given } x_0$$

The Model. According to definition 12–1 the fuzzy set decision is the "confluence" of the constraints and the goal(s), that is,

$$\widetilde{D} = \bigcap_{t=0}^{N-1} \widetilde{C}_t \cap \widetilde{G}_N$$

Using the min-operator for the aggregation of the fuzzy constraints and the goal, the membership function of the fuzzy set decision is

$$\mu_{\widetilde{D}}(d_0, \ldots, d_{N-1}) = \min\{\mu_{\widetilde{C}_0}(d_0), \ldots, \mu_{\widetilde{C}_{N-1}}(d_{N-1}), \mu_{\widetilde{G}_N}(x_N)\} \qquad (12.20)$$

The membership function of the maximizing decision is then

$$\mu_{\widetilde{D}^0}(d_0^0, \ldots, d_{N-1}^0) = \max_{d_0, \ldots, d_{N-2}} \max_{d_{N-1}} [\min\{\mu_{\widetilde{C}_0}(d_0), \ldots, \mu_{\widetilde{G}_N}(t_N(x_{N-1}, d_{N-1}))\}] \qquad (12.21)$$

where d_i^0 denotes the optimal decision on stage i. If K is a constant and g is any function of d_{N-1}, we can write

$$\max_{d_{N-1}} \min\{g(d_{N-1}), K\} = \min\{K, \max_{d_{N-1}} g(d_{N-1})\}$$

and (12.21) can be expressed as

$$\mu_{\widetilde{D}^0}(d_0^0, \ldots, d_{N-1}^0) = \max_{d_0, \ldots, d_{N-1}} \min\{\mu_{\widetilde{C}_0}(d_0), \ldots, \mu_{\widetilde{G}_{N-1}}(x_{N-1})\} \qquad (12.22)$$

with

$$\mu_{\widetilde{G}_{N-1}}(x_{N-1}) = \max_{d_{N-1}} \min\{\mu_{\widetilde{C}_{N-1}}(d_{N-1}), \mu_{\widetilde{G}_N}(t_N(x_{N-1}, d_{N-1}))\} \qquad (12.23)$$

We can thus determine $\widetilde{D}^0$ recursively.

Example 12–9 [Bellman and Zadeh 1970, B-153]

Let $\widetilde{d}_1$, $\widetilde{d}_2$ be the two decision variables the possible values of which can be α_2, α_2. The state variables are x_t, $t = 0, \ldots, 2$ with a finite range $X = \{\tau_1, \tau_2, \tau_3\}$.

The fuzzy constraints for $t = 0$ and $t = 1$ are

$$\widetilde{C}_0(\alpha_i) = \{(\alpha_1, .7), (\alpha_2, 1)\}$$
$$\widetilde{C}_1(\alpha_i) = \{(\alpha_1, 1), (\alpha_2, .6)\}$$

The fuzzy goal is specified as

$$\tilde{G}(x_2) = \{(\tau_1, .3), (\tau_2, 1), (\tau_3, .8)\}$$

and the crisp transformation function is defined by the following matrix:

	x_t		
d_t	τ_1	τ_2	τ_3
a_1	τ_1	τ_3	τ_1
a_2	τ_2	τ_1	τ_3

Solution. Using (12.23) we can compute the fuzzy goal induced at $t = 1$ as follows: We start at stage $t = 2$. The state-decision combinations that yield τ_i on state $t = 1$ are obtained from the above matrix.

So we can compute:

$$\begin{aligned}
\mu_{\tilde{G}_1}(\tau_1) &= \max_{d_1} \{\min [\mu_{\tilde{C}_1}(d_1), \mu_{\tilde{G}_2}(t(\tau_1, \alpha_1))], \\
&\qquad \min [\mu_{\tilde{C}_1}(d_1), \mu_{\tilde{G}_2}(t(\tau_1, \alpha_2))]\} \\
&= \max \{\min [1, .3], \min [.6, 1]\} \\
&= \max \{.3, .6\} = .6 \\
&\to d_1^0 = \alpha_2 \\
\mu_{\tilde{G}_1}(\tau_2) &= \max \{\min [1, .8], \min [.6, .3]\} \\
&= \max \{.8, .3\} = .8 \\
&\to d_1^0 = \alpha_1 \\
\mu_{\tilde{G}_1}(\tau_3) &= \max \{\min [1, .3], \min [.6, .8]\} \\
&= \max \{.3, .6\} = .6 \\
&\to d_1^0 = \alpha_2 \\
\mu_{\tilde{G}_0}(\tau_1) &= \max \{\min [.7, .6], \min [1, .8]\} \\
&= .8 \\
&\to d_0^0 = \alpha_2 \\
\mu_{\tilde{G}_0}(\tau_2) &= \max \{\min [.7, .6], \min [1, .6]\} \\
&= .6 \\
&\to d_0^0 = \alpha_1 \quad \text{or} \quad \alpha_2 \\
\mu_{\tilde{G}_0}(\tau_3) &= \max \{\min [.7, .6], \min [1, .6]\} \\
&= .6 \\
&\to d_0^0 = \alpha_1 \quad \text{or} \quad \alpha_2
\end{aligned}$$

Thus for

$$x_0 = \tau_1 : d_0^0 = \alpha_2, \quad d_1^0 = \alpha_1$$
$$\text{with } \mu_{\tilde{G}_2^0} = .8$$

$$x_0 = \tau_2: d_0^0 = \alpha_1, \quad d_1^0 = \alpha_2 \quad \text{or}$$
$$d_0^0 = \alpha_2, \quad d_1^0 = \alpha_2$$
$$\text{with } \mu_{\tilde{G}_2^0} = .6$$
$$x_0 = \tau_3: d_0^0 = \alpha_1, \quad d_1^0 = \alpha_2 \quad \text{or}$$
$$d_0^0 = \alpha_2, \quad d_1^0 = \alpha_2$$
$$\text{both with } \mu_{\tilde{G}_2^0} = .6$$

12.4 Fuzzy Multi Criteria Analysis

In the recent past it has become more and more obvious that comparing different ways of action as to their desirability, judging the suitability of products, or determining "optimal" solutions in decision problems can in many cases not be done by using a single criterion or a single objective function. This area, multi criteria decision making, has led to numerous evaluation schemes (e.g., in the areas of cost benefit analysis and marketing) and to the formulation of vector-maximum problems in mathematical programming.

Two major areas have evolved, both of which concentrate on decision making with several criteria: Multi Objective Decision Making (MODM) and Multi Attribute Decision Making (MADM). The main difference between these two directions is: The former concentrates on continuous decision spaces, primarily on mathematical programming with several objective functions; the latter focuses on problems with discrete decision spaces. There are some exceptions to this rule (e.g., integer programming with multiple objectives), but for our purposes this distinction seems to be appropriate.

The literature on multi criteria decision making has grown tremendously in the recent past. We shall only mention one survey reference for each of these two areas: Hwang and Yoon [1981] for MADM and Hwang and Masud [1979] for MODM. Fuzzy set theory has contributed to MODM as well as to MADM. We shall illustrate these contributions by describing one model in each of these areas. This topic has been treated in much more detail in the volume on Fuzzy Sets and Decision Analysis.

12.4.1 Multi Objective Decision Making

In mathematical programming the MODM-problem is often called the "vector-maximum" problem, and was first mentioned by Kuhn and Tucker [1951].

Definition 12–6

The *vector-maximum problem* is defined as

$$\text{“maximize”} \ \{Z(x) \mid x \in X\}$$

where $Z(x) = (z_1(x), \ldots, z_k(x))$ is a vector-valued function of $x \in \mathbb{R}^n$ into $\mathbb{R}^k$ and X is the “solution space.”

Two stages can generally be distinguished, at least categorically, in vector-maximum optimization:

1. The determination of efficient solutions.
2. The determination of an optimal compromise solution.

Definition 12–7

Let “max” $\{Z(x) \mid x \in X\}$ be a vector-maximum problem such as defined in definition 12–6. $\bar{x}$ is an *efficient solution* if there is no $\hat{x} \in X$ such that

$$z_i(\hat{x}) \geq z_i(\bar{x}) \qquad i = 1, \ldots, k$$

and

$$z_i(\hat{x}) > z_i(\bar{x}) \quad \text{for at least one } i = 1, \ldots, k$$

The set of all efficient solutions is generally called the “*complete solution*.”

Definition 12–8

An *optimal compromise solution* of a vector-maximum problem is a solution $x \in X$ which is preferred by the decision maker to all other solutions, taking into consideration all criteria contained in the vector-valued objective function. It is generally accepted, that an optimal compromise solution has to be an efficient solution according to definition 12–7.

In the following we shall restrict our considerations to the determination of optimal compromise solutions in linear programming problems with vector-valued objective functions.

Three major approaches are known to single out one specific solution from the set of efficient solutions which qualifies as an “optimal” compromise solution.

1. The utility approach [see, e.g., Keeney and Raiffa 1976]
2. Goal Programming [see, e.g., Charnes and Cooper 1961]
3. Interactive approaches [see, e.g., Dyer 1973]

The first two of these approaches assume that the decision maker can specify his "preference function" with respect to the combination of the individual objective functions in advance, either as "weights" (utilities) or as "distance functions" (concerning the distance from an "ideal solution," for example). Generally they assume that the combination of the individual objective functions that arrives at the compromise solution with the highest overall utility is achieved by linear combinations (i.e., adding the weighted individual objective functions). The third approach uses only local information in order to arrive at an acceptable compromise solution.

The following example illustrates a fuzzy approach to this problem.

Example 12–10

A company manufactures two products 1 and 2 on given capacities. Product 1 yields a profit of \$2 per piece and product 2 of \$1 per piece. Product 2 can be exported, yielding a revenue of \$2 per piece in foreign countries; product 1 needs imported raw materials of \$1 per piece. Two goals are established: (1) profit maximization and (2) maximum improvement of the balance of trade, that is, maximum difference of exports minus imports. This problem can be modeled as follows:

$$\text{"maximize" } Z(x) = \begin{pmatrix} -1 & 2 \\ 2 & 1 \end{pmatrix} \begin{pmatrix} x_1 \\ x_2 \end{pmatrix} \quad \begin{matrix} \text{(effect on balance of trade)} \\ \text{(profit)} \end{matrix}$$

such that

$$\begin{aligned} -x_1 + 3x_2 &\leq 21 \\ x_1 + 3x_2 &\leq 27 \\ 4x_1 + 3x_2 &\leq 45 \\ 3x_1 + x_2 &\leq 30 \\ x_1, x_2 &\geq 0 \end{aligned}$$

Figure 12–7 shows the solution space of this problem. The "complete solution" is the edge $x^1 - x^2 - x^3 - x^4$. x^1 is optimal with respect to objective function $z_1(x) = -x_1 + 2x_2$ (i.e., best improvement of balance of trade). x^4 is optimal with respect to objective function $z_2(x) = 2x_1 + x_2$ (profit). The "optimal" values are $z_1(x^1) = 14$ (maximum net export) and $z_2(x^4) = 21$ (maximum profit), respectively. For $x^1 = (7;\ 0)^T$ total profit is $z_2(x^1) = 7$ and $x^4 = (9;\ 3)^T$ yields $z_1(x^4) = -3$, that is, a net import of 3.

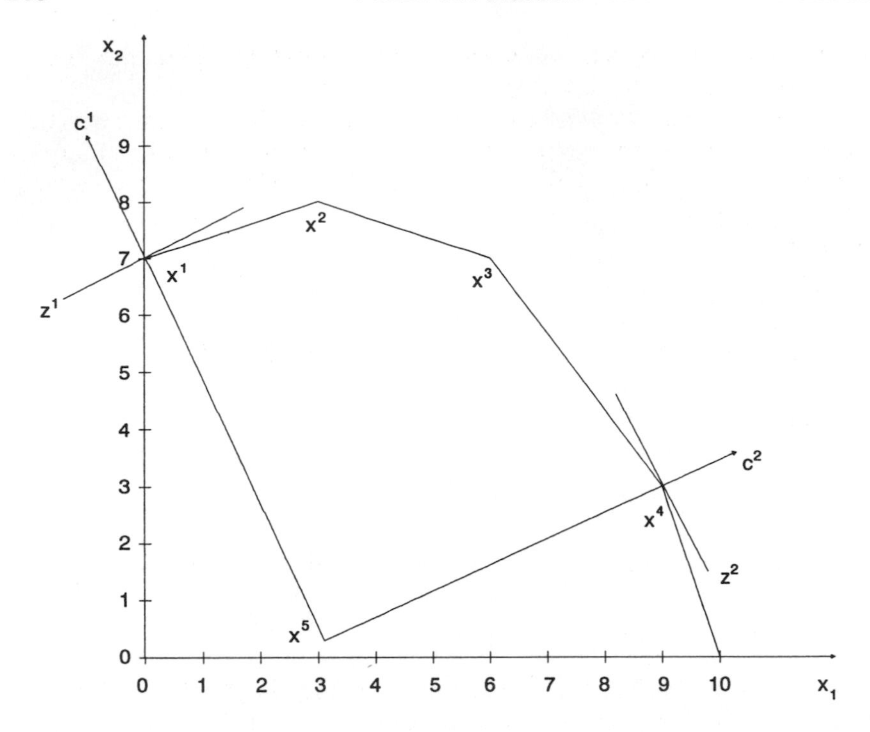

Figure 12–7. The vector-maximum problem.

Solution $x^5 = (3.4;\ 0.2)^T$ is the solution that yields $z_1(x^5) = -3$, $z_2(x^5) = 7$ which is the lowest "justifiable" value of the objective functions in the sense that a further decrease of the value of one objective function can not be balanced or even counteracted by an increase in the value of the other objective function.

To solve problems of the kind shown in example 12–10 we can use the following approach. We first assume that either the decision maker can specify aspiration levels for the objective functions, or we define properties of the solution space for "calibration" of the objective functions. Let us consider the objective functions as fuzzy sets of the type "solutions acceptable with respect to objective function 1." In example 12–10 we would have to construct two fuzzy sets: "Solutions acceptable with respect to objective function 1" and "solutions acceptable with respect to objective function 2." As calibration points we shall use the respective "individual optima" and the "least justifiable solution."

The membership functions $\mu_1(x)$ and $\mu_2(x)$ of the fuzzy sets characterizing the objective functions rise linearly from 0 to 1 at the highest achievable values of $z_1(x) = 14$ and $z_2(x) = 21$, respectively.

That means that we assume that the level of satisfaction with respect to the improvement of the balance of trade rises from 0 for imports of 3 units or more to 1 for exports of 14 and more; and the satisfaction level rises with respect to profit from 0 if the profit is 7 or less to 1 if total profit is 21 or more.

$$\mu_1(x) = \begin{cases} 0 & \text{for} \quad z_1(x) \leq -3 \\ \dfrac{z_1(x)+3}{17} & \text{for} \quad -3 < z_1(x) \leq 14 \\ 1 & \text{for} \quad 14 < z_1(x) \end{cases}$$

$$\mu_2(x) = \begin{cases} 0 & \text{for} \quad z_2(x) \leq 7 \\ \dfrac{z_2(x)-7}{14} & \text{for} \quad 7 < z_2(x) \leq 21 \\ 1 & \text{for} \quad 21 < z_2(x) \end{cases}$$

We are now faced with a problem of type (12.3) in which crisp constraints have been added (i.e., the problem consists of 2 rows representing our fuzzified objectives and 4 crisp constraints). We can now employ (12.12).

Example 12–10 (continuation)

In analogy to formulation (12.12) and including the crisp constraints we arrive at the following problem formulation:

$$\begin{aligned} &\text{maximize} && \lambda \\ &\text{such that} && \lambda \leq -0.05882x_1 + 0.117x_2 + 0.1764 \\ &&& \lambda \leq +0.1429\ x_1 + 0.714x_2 - 0.5 \\ &&& 21 \geq -x_1 + 3x_2, \\ &&& 27 \geq x_1 + 3x_2, \\ &&& 45 \geq 4x_1 + 3x_2, \\ &&& 30 \geq 3x_1 + x_2, \\ &&& x \geq 0, \end{aligned}$$

depicted in figure 12–8.

The maximum degree of "overall satisfaction" ($\lambda_{max} = 0.74$) is achieved for the solution $x_0 = (5.03;\ 7.32)^T$. This is the "maximizing solution" which, in our example yields a profit of \$17.38 and an export contribution of \$4.58. The basic solutions x^1 and x^4 yield $\lambda = 0$.

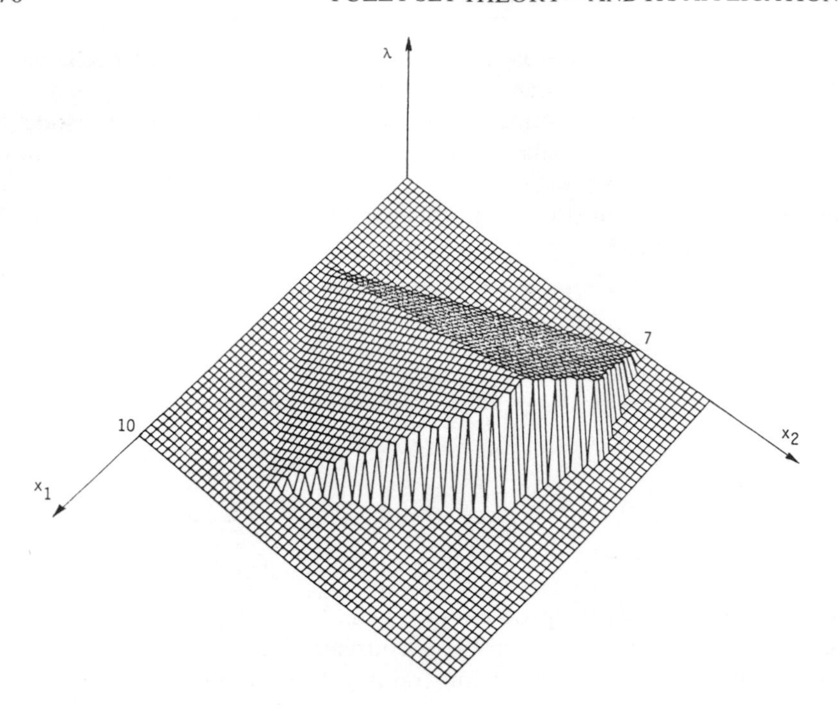

Figure 12–8. Fuzzy LP with min-operator.

By contrast to the usual vector-maximum models, the efficient solutions contained in the "complete solution" are ordered (distinguishable) by their degree of membership to the fuzzy set decision. It should be obvious that the approach described above can only be applied if the "symmetrical model" of a decision (definition 12–1) is accepted. Otherwise we will have to use approaches applicable to (12.13) which will, however, not be discussed in this volume.

At the beginning of section 12.2 many simplifying assumptions which are generally accepted in traditional linear programming models were pointed out. These assumptions concerned the use of real numbers rather than fuzzy numbers for the coefficients of linear programming as well as the use of crisp relations rather than fuzzy. One approach used in section 12.2 for the fuzzification of crisp mathematical programming problems seems to be computationally very efficient, well applicable in practice, and understandable by practioners. In the literature the reader will find numerous different approaches which, from a mathematical point of view, are quite interesting. It would certainly exceed the scope of this book to describe the majority of these suggestions. We shall, however, mention a few of them. The reader will find quite a number of references to other approaches in the bibliography at the end of the book.

Approaches that use fuzzy sets to describe the parameter of linear programming models can be traced, in particular, to the paper by Negoita, Minouiu, and Stan [1976]. They use fuzzy sets to describe the parameters of the matrix A and the capacity vector b and then formulate for each α the respective α-cuts. The resulting crisp problem can then be solved by the usual LP codes. If the membership functions have only a finite number of values, an optimal alternative and an objective function value can be determined for each case. This, however, is connected with a high computational effort. Afterwards the decision maker has to choose a desirable degree of membership and the associated solution. Kacprzyk and Orlovski [1987], in their review article, mention a number of additional references in which special representations of fuzzy parameters are used. Here we shall mention only the work of Tanaka and Asai [1984], who use triangular membership functions, and Ramik and Rimanek [1985, 1989], who use fuzzy parameters in LR representation and replace each resulting fuzzy relation by four strict relations.

Other authors consider nonlinear vector-maximum problems in which all parameters are defined fuzzily. Sakawa and Yano [1987], for instance, formulate a fuzzy nonlinear vector-maximum problem with fuzzy parameters $\tilde{a}_e$, $L = 1, \ldots, k$ in the k objective functions and $\tilde{b}_i$, $i = 1, \ldots, m$ in the m constraints. Here the fuzzy parameters are regarded as real-valued fuzzy numbers. For each α-degree a crisp equivalent model can be formulated for which the values of the fuzzy numbers can be considered as variables subject to the condition that they belong to the fuzzy number at least with the degree of membership α. Sakawa and Jano [1987] define the notion of an α-paretooptimal solution in generalizing the classical paretooptimality with respect to the crisp equivalent models. The authors suggest an interactive algorithm which leads the decision maker to a satisfying solution. The decision maker has to provide as starting values the desired α and the aspiration level for the objective function. The algorithm then solves an equivalent model which minimizes for a given α the deviation from the aspiration level and supplies additional trade-off information to the decision maker. This approach assumes that the decision maker can choose the states which are expressed in the fuzzy numbers. Therefore, this approach seems to be only suitable if the decision maker can really influence these values, that is, if they are not dependent on the environment. Due to the fact that it is assumed that the parameters are variables, the resulting α-model is at least quadratic even if the basic model is linear.

If the fuzzy coefficients are the result of insufficient information that can be improved by additional effort, an optimal context-dependent allocation of additional effort is of interest. Tanaka, Ishihashi, and Asai [1986] discuss the value of additional information and suggest a model for the allocation of information on the basis of sensitivity analysis. In the recent past, fuzzy models have also been suggested for fractional programming, integer programming, geometric programming, and other versions of mathematical programming problems. Of particular interest is the application of possibility theory to mathematical programming suggested by Buckley [1988a, 1988b].

12.4.2 Multi Attributive Decision Making

The general multi attributive decision-making model can be defined as follows.

Definition 12–9

Let $X = \{x_i \mid i = 1, \ldots, n\}$ be a (finite) set of decision alternatives and $G = \{g_j \mid j = 1, \ldots, m\}$ a (finite) set of goals according to which the desirability of an action is judged. Determine the optimal alternative x^0 with the highest degree of desirability with respect to all relevant goals g_j.

Most approaches in MADM consist of two stages:

1. The aggregation of the judgments with respect to all goals and per decision alternative.
2. The rank ordering of the decision alternatives according to the aggregated judgments.

In crisp MADM models it is usually assumed that the final judgments of the alternatives are expressed as real numbers. In this case the second stage does not pose any particular problems and suggested algorithms concentrate on the first stage. Fuzzy models are sometimes justified by the argument that the goals, g_j, or their attainment by the alternatives, x_i, respectively, cannot be defined or judged crisply but only as fuzzy sets. In this case the final judgments are also represented by fuzzy sets which have to be ordered to determine the optimal alternative. Then the second stage is, of course, by far not trivial.

In the following, we shall describe two fuzzy MADM models, the first one, by Yager, because it shows very clearly the general structure of the problem and the second, by Baas and Kwakernaak, because many of the publications refer to this model, which is one of the first models of this kind published.

Model 12–1 [Yager 1978]. Let $X = \{x_1, \ldots, x_n\}$ be a set of alternatives. The goals are represented by the fuzzy sets $\tilde{G}_j$, $j = 1, \ldots, m$. The "importance" (weight) of goal j is expressed by w_j. The "attainment" of goal $\tilde{G}_j$ by alternative x_i is expressed by the degree of membership $\mu_{\tilde{G}_{2j}}(x_i)$.

The decision is defined in line with definition 12–1 as the intersection of all fuzzy goals, that is,

$$\tilde{D} = \tilde{G}_1^{w_1} \cap \tilde{G}_2^{w_2} \cap \cdots \cap \tilde{G}_m^{w_m}$$

and the optimal alternative is defined as that achieving the highest degree of membership in $\tilde{D}$.

The rationale behind using the weights as exponents to express the importance of a goal can be found in definition 9–3: There the modifier "very" was defined as the squaring operation. Thus the higher the importance of a goal the larger should be the exponent of its representing fuzzy set, at least for normalized fuzzy sets and when using the min-operator for the intersection of the fuzzy goals. Yager concentrates on the problem of determining the weights of the goals. As a solution to that problem he suggests Saaty's hierarchical procedure for determining weights by computing the eigenvectors of the matrix M of relative weights of subjective estimates [Saaty 1978].

> The membership grade in all objectives having little importance ($w < 1$) becomes larger, and while those in objectives having more importance ($w > 1$) become smaller. This has the effect of making the membership function of the decision subset D, which is the min value of each X over all objectives, being more determined by the important objectives, which is as it should be. Furthermore, this operation (min) makes particularly small those alternatives that are bad in important objectives, therefore when we select the x_i that maximizes D, we will be very unlikely to pick one of these [Yager 1978, p. 90].

The solution procedure can now be described as follows: Given the set $X = \{x_1, \ldots, x_n\}$ and the degrees of membership $\mu_{\tilde{G}_j}(x_i)$ of all x_i in the fuzzy sets $\tilde{G}_j$ representing the goals,

1. establish by pairwise comparison the relative importance, α_i, of the goals among themselves. Arrange the α_i in a matrix M.

$$M = \begin{bmatrix} \frac{\alpha_1}{\alpha_1} & \frac{\alpha_1}{\alpha_2} & \cdots & \frac{\alpha_1}{\alpha_n} \\ \frac{\alpha_2}{\alpha_1} & & & \vdots \\ \frac{\alpha_n}{\alpha_1} & & & \frac{\alpha_n}{\alpha_n} \end{bmatrix}$$

2. Determine consistent weights w_j for each goal by employing Saaty's eigenvector method.
3. Weight the degrees of goal attainment, $\mu_{\tilde{G}_j}(x_i)$ exponentially by the respective w_j. The resulting fuzzy sets are $(\tilde{G}_j(x_i))^{w_j}$
4. Determine the intersection of all $(\tilde{G}_j(x_i))^{w_j}$:

$$\tilde{D} = \{(x_i, \min_j (\mu_{\tilde{G}_j}(x_i))^{w_j}) \mid i = 1, \ldots, n; j = 1, \ldots, m\}$$

5. Select the x_i with largest degree of membership in $\tilde{D}$ as the optimal alternative.

Example 12–11 [Yager 1978, p. 94]

Let $X = \{x_1, x_2, x_3\}$ and the goals be given as

$$\begin{aligned} \tilde{G}_1(x_i) &= \{(x_1, .7), (x_2, .5), (x_3, .4)\} \\ \tilde{G}_2(x_i) &= \{(x_1, .3), (x_2, .8), (x_3, .6)\} \\ \tilde{G}_3(x_i) &= \{(x_1, .2), (x_2, .3), (x_3, .8)\} \\ \tilde{G}_4(x_i) &= \{(x_1, .5), (x_2, .1), (x_3, .2)\} \end{aligned}$$

The subjective evaluations have resulted in the following matrix of weights:

$$M = \begin{array}{c} \\ \tilde{G}_1 \\ \tilde{G}_2 \\ \tilde{G}_3 \\ \tilde{G}_4 \end{array} \begin{array}{c} \begin{array}{cccc} \tilde{G}_1 & \tilde{G}_2 & \tilde{G}_3 & \tilde{G}_4 \end{array} \\ \begin{bmatrix} 1 & 3 & 7 & 9 \\ \frac{1}{3} & 1 & 6 & 7 \\ \frac{1}{7} & \frac{1}{6} & 1 & 3 \\ \frac{1}{9} & \frac{1}{7} & \frac{1}{3} & 1 \end{bmatrix} \end{array}$$

Via Saaty's method we obtain the vector

$$w = \{w_1, w_2, w_3, w_4\} \quad \text{as}$$
$$w = \{2.32, 1.2, .32, .16\}$$

Exponential weighting of $\widetilde{G}_j(x_i)$ by their respective weight yields

$$\begin{aligned}
\widetilde{G}_1(x_i)^{2.32} &= \{(x_1, .44), (x_2, .2), (x_3, .12)\} \\
\widetilde{G}_2(x_i)^{1.2} &= \{(x_1, .24), (x_2, .76), (x_3, .54)\} \\
\widetilde{G}_3(x_i)^{.32} &= \{(x_1, .6), (x_2, .68), (x_3, .93)\} \\
\widetilde{G}_4(x_i)^{.16} &= \{(x_1, .9), (x_2, .69), (x_3, .77)\}
\end{aligned}$$

The fuzzy set decision $\widetilde{D}$, as the intersection of the $\widetilde{G}_j^{\alpha_j}(x_i)$ becomes

$$\widetilde{D} = \{(x_1, .24)\ (x_2, .2), (x_3, .12)\}$$

and the optimal alternative is x_1 with a degree of membership in $\widetilde{D}$ of $\mu_{\widetilde{D}}(x_1) = .24$.

Model 12–2 [Baas and Kwakernaak 1977]. Let again $X = \{x_i \mid i = 1, \ldots, n\}$ be the set of alternatives and $G = \{g_j \mid j = 1, \ldots, m\}$ the set of goals. r_{ij} is the "rating" of alternative i with respect to goal j and $w_j \in \mathbb{R}^1$ is the weight (importance) of goal j. It is assumed that the rating of alternative i with respect to goal j is fuzzy and is represented by the membership function $\mu_{\widetilde{R}_{ij}}(r_{ij})$ on $\mathbb{R}^1$.

Similarly the weight (relative importance) of goal j is represented by a fuzzy set w_j with membership function $\mu_{w_j}(w_j)$. All fuzzy sets are assumed to be normalized (i.e., have finite supports and take on the value 1 at least once!).

Step 1. The evaluation of an alternative x_i is, by contrast to model 12.1, assumed to be a fuzzy set which is computed on the basis of the r_{ij} and w_j as follows: Consider a function g: $\mathbb{R}^{2m} \to \mathbb{R}$ defined by

$$g(z) = \frac{\sum_{j=1}^{m} w_j r_j}{\sum_{j=1}^{m} w_j} \tag{12.24}$$

with $z = (w_1, \ldots, w_m, r_1, \ldots, r_m)$

On the product space $\mathbb{R}^{2n}$ a membership function μ_{z_i} is defined as

$$\mu_{z_i}(z) = \min\{\min_{j=1,\ldots,m}(\mu_{w_j}(w_j), \min_{k=1,\ldots,m}(\mu_{\widetilde{R}_{ik}}(r_k))\} \tag{12.25}$$

Through the function g the fuzzy set $\widetilde{Z} = (\mathbb{R}^{2m}, \mu_{z_i})$ induces a fuzzy set $\widetilde{R}_i = (\mathbb{R}, \mu_{\widetilde{R}_i})$ with the membership function

$$\mu_{\widetilde{R}_i}(\bar{r}) = \sup_{\bar{z}: g(z) = \bar{r}} \mu_{z_i}(z) \qquad \bar{r} \in \mathbb{R} \tag{12.26}$$

$\mu_{\tilde{R}_i}(\bar{r})$ is the final rating of alternative x_i on the basis of which the "rank ordering" is performed in step 2.

Step 2. For the final ranking of the x_i Baas and Kwakernaak start from the observation that if the x_i had received crisp rating r_i then a reasonable procedure would select the x_i that have received the highest rating, that is, would determine the set of preferred alternatives as $\{i \in I \mid \bar{r}_i \geq r_i, \forall j \in I\}$, $I = \{1, \ldots, n\}$.

Since here the final ratings are fuzzy the problem is somewhat more complicated. The authors suggest in their model two different fuzzy sets in addition to $\tilde{R}_i$, which supply different kinds of information about the preferability of an alternative.

a. They first determine the conditional set $(I \mid \tilde{R})$ with the characteristic function

$$\mu_{(I|\tilde{R})}(i \mid \bar{r}_1, \ldots, \bar{r}_n) = \begin{cases} 1 & \text{if } \bar{r}_i \geq \bar{r}_j \quad \forall j \in I \\ 0 & \text{else} \end{cases} \tag{12.27}$$

This "membership function" expresses that a given alternative x_i belongs to the preferred set iff

$$\bar{r}_i \geq \bar{r}_j \qquad \forall j \in I$$

The final fuzzy ratings $\tilde{R}$ define on $\mathbb{R}^n$ a fuzzy set $\tilde{R} = (\mathbb{R}^n, \mu_{\tilde{R}})$ with the membership function

$$\mu_{\tilde{R}}(\bar{r}_1, \ldots, \bar{r}_n) = \min_{i=1,\ldots,n} \mu_{\tilde{R}_i}(\bar{r}_i) \tag{12.28}$$

This fuzzy set together with the conditional fuzzy set (12.27) induces a fuzzy set $\tilde{I} = (I, \mu_{\tilde{I}})$ with the membership function

$$\mu_{\tilde{I}}(i) = \sup_{\bar{r}_1, \ldots, \bar{r}_n} (\min \{\mu_{(I|\tilde{R})}(i \mid \bar{r}_1, \ldots, \bar{r}_n), \mu_{\tilde{R}}(\bar{r}_1, \ldots, \bar{r}_n)\}) \tag{12.29}$$

which can be interpreted as the degree to which alternative x_i is the best alternative. If there is a unique i, then x_i corresponds to the alternative that maximizes (12.29) if the w_j and r_{ij} are set to the values at which $\mu_{\tilde{w}_j}(w_j)$ and $\mu_{\tilde{R}_{ij}}(r_{ij})$, respectively, attain their supremum, namely 1.

b. This is, of course, not all the information that can be provided. x_i might not be the unique best alternative, but there might be some x_i attaining their maximum degree of membership at $r*$. They might, however, be represented by different fuzzy sets $\tilde{r}_{ij}$.

Baas and Kwakernaak therefore try to establish another criterion that might be able to distinguish such "preferable" alternatives from each other and rank them:

If the final ratings are crisp, $\bar{r}_1, \ldots, \bar{r}_n$, then

$$p_i = \bar{r}_i - \frac{1}{n-1}\sum_{\substack{j=1 \\ j \neq i}}^{n} \bar{r}_j$$

for fixed i, can be used as a measure of preferability of alternative x_i over all others.

If the ratings $\bar{r}_i$ are fuzzy, then the mapping h_i: $\mathbb{R}^m \rightarrow \mathbb{R}$ induces a fuzzy set $\widetilde{P}_i = (\mathbb{R}, \mu_{\tilde{P}_i})$ with the membership function

$$\mu_{\tilde{P}_i}(p) = \sup_{h_i(\bar{r}_i, \ldots, \bar{r}_n) = p} \mu_{\tilde{R}}(\bar{r}_1, \ldots, \bar{r}_n) \qquad p \in \mathbb{R}, \tag{12.30}$$

in which $\mu_{\tilde{R}}$ is defined by (12.28).

This fuzzy set can be used to judge the degree of preferability x_i over all other alternatives.

The computational aspects for determining all the fuzzy sets mentioned above shall not be discussed here; models 1 and 2 have been described because of their illustrative value. Baas and Kwakernaak mention and prove special conditions for the membership functions to make computations possible.

To summarize: 3 kinds of informations are provided:

1. $\mu_{\tilde{R}_i}(\bar{r})$ as the fuzzy rating of x_i,
2. $\mu_{\tilde{I}}(i)$ as the degree to which x_i is best alternative, and
3. $\mu_{\tilde{P}_i}(p)$ as the degree of preferability of x_i over all other alternatives.

Example 12–12 [Baas and Kwakernaak 1977, p. 54]

Let $X = \{x_1, x_2, x_3\}$ be the set of available alternatives and $G = \{g_1, g_2, g_3, g_4\}$ the set of goals. The weights and the ratings of the alternatives with respect to the goals are given as normalized fuzzy sets that resemble the terms of a linguistic variable (see definition 9–1). Figure 12–9 depicts the fuzzy sets representing weights and ratings. The table 12–1 gives the assumed ratings for all alternatives and goals and the respective weights:

Table 12–1. Ratings and weights of alternative goals

goal g_j	*weight* $\tilde{w}_j$	rating $\tilde{r}_{ij}$ for alternative x_i: *i = 1*	2	*3*
1	very important	good	very good	fair
2	moderately important	poor	poor	poor
3	moderately important	poor	fair to good	fair
4	rather unimportant	good	not clear	fair

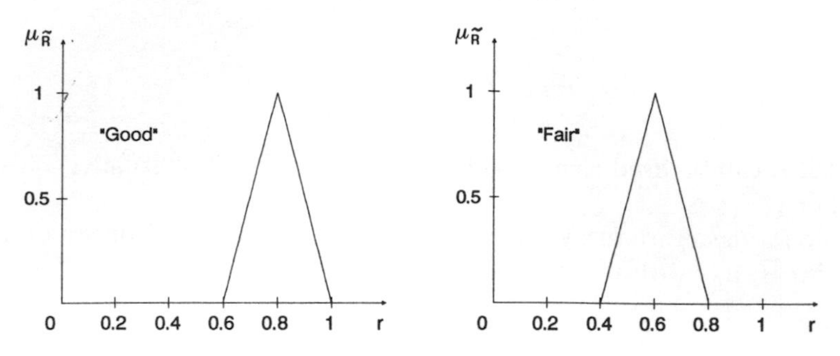

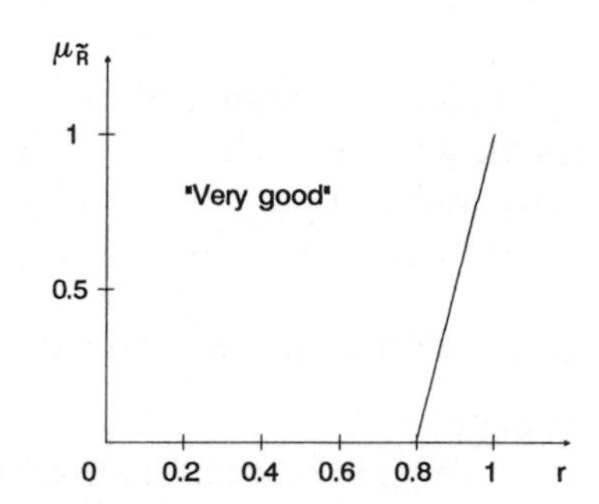

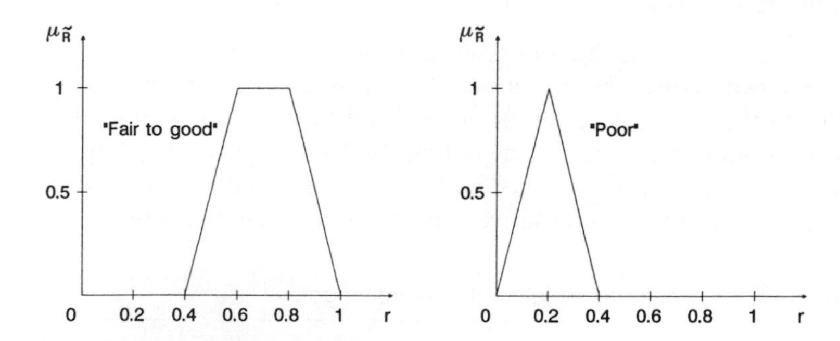

Figure 12–9. Fuzzy sets representing weights and ratings.

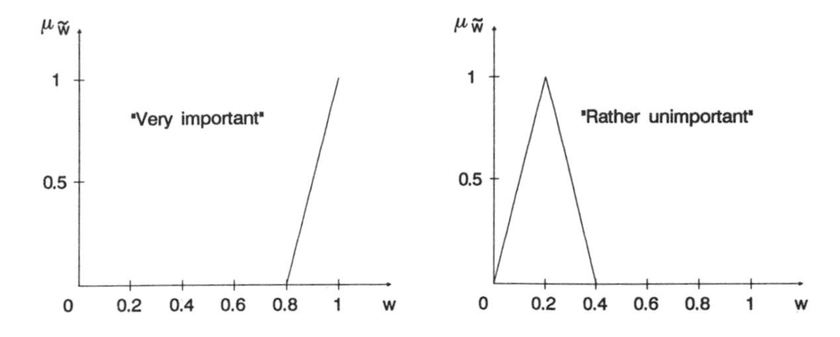

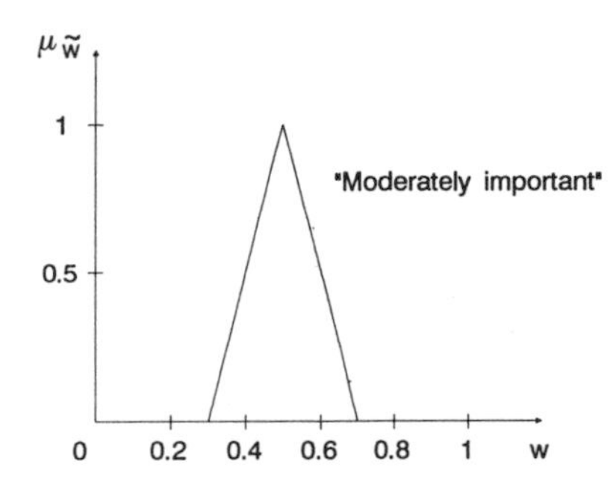

Figure 12–9b.

Figure 12–10 shows the $\mu_{\tilde{R}_i}(r_i)$ (final ratings for alternatives x_1, x_2, x_3).

The degrees of membership of the alternatives to the fuzzy set $(I, \mu_{\bar{I}})$, that is the degrees to which alternatives x_i are best are:

alternative	$\mu_{\bar{I}}(x_i)$
1	.95
2	1
3	.77

The fuzzy set $\widetilde{P}_2(p)$ indicating the degree to which alternative 2 is preferred to all others is shown in figure 12–11. p_2 is calculated as $p_2 = \bar{r}_2 - \frac{1}{2}(\bar{r}_1 + \bar{r}_3)$.

Many other fuzzy methods and models have been suggested to solve the MADM problem. They differ by their assumptions concerning the input data and by the measures used for aggregation and ranking. Also, they concentrate either on the first step (aggregation of ratings), or the second

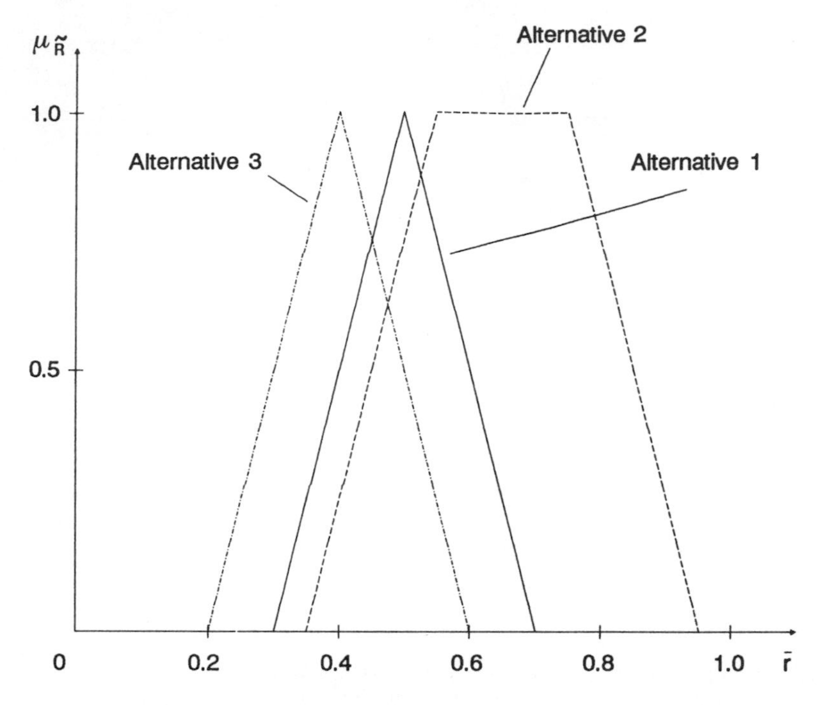

Figure 12–10. Final ratings of alternatives.

step (ranking), or both. Obviously all of them have advantages and disadvantages. They will, however, not be discussed here but will be in the second volume.

An interesting example of a more engineering-type application of multicriteria decision making using fuzzy sets is described by Muñoz-Rodriguez and Cattermole [1987].

Exercises

1. Explain the (mathematical) difference between the symmetric and nonsymmetric model of a decision in a fuzzy environment.
2. Consider example 12–4. What grade would the student get if the "and" was interpreted as the "bold-intersection" (definition 3–6), the "bounded difference" (definition 3–8), or the "bold union"?
3. Consider the following problem:

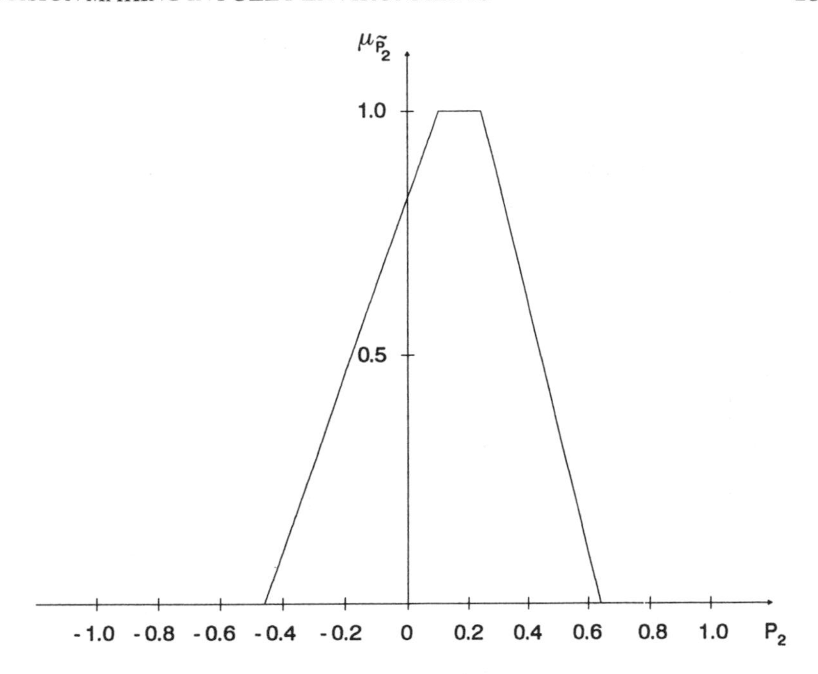

Figure 12–11. Preferability of alternative 2 over all others.

$$
\begin{array}{lr}
\text{Minimize} & z = 4x_1 + 5x_2 + 2x_3 \\
\text{such that} & 3x_1 + 2x_2 + 2x_3 \leq 60 \\
& 3x_1 + x_2 + x_3 \leq 30 \\
& 2x_2 + x_3 \geq 10 \\
& x_1, x_2, x_3 \geq 0
\end{array}
$$

Determine the optimal solution. Now assume that the decision maker has the following preferences:

a. He has a linear preference function for the objective function between the minimum and 1.5 the minimum.
b. The tolerance intervals can be established as

$$p_1 = 10, \quad p_2 = 12, \quad p_3 = 3$$

Now use model (12.9) to determine the optimal solution and compare it with the crisp optimal solution.

4. Solve the example of exercise 3 by assuming the objective function to be crisp and by using (12.18).

5. Consider the problem:

$$\text{“maximize”} \; Z(x) = \begin{Bmatrix} -x_1 - 3x_2 \\ 1.5x_1 + 2.5x_2 \end{Bmatrix}$$

$$\text{such that} \qquad \begin{aligned} -x_1 + 2x_2 &\le 18 \\ 4x_1 + 3x_2 &\le 40 \\ 3x_1 + x_2 &\le 25 \\ x_1, x_2 &\ge 0 \end{aligned}$$

Determine an optimal compromise solution by using the model from example 12–10 (continuation).

6. What is the optimal alternative in the following situation (use Yager's method!)?

Alternatives: $X = \{x_1, x_2, x_3, x_4\}$

Goals: $\tilde{G}_1(x_i) = \{(x_1, .8), (x_2, .6), (x_3, .4), (x_4, .2)\}$
$\tilde{G}_2(x_i) = \{(x_1, .4), (x_2, .6), (x_3, .6), (x_4, .8)\}$
$\tilde{G}_3(x_i) = \{(x_1, .6), (x_2, .8), (x_3, .8), (x_4, .6)\}$

The relative weights of the goals have been established as: $G_1{:}G_2{:}G_3 = 1{:}4{:}6$.

13 FUZZY SET MODELS IN OPERATIONS RESEARCH

13.1 Introduction

The contents and scope of operations research has been described and defined in many different ways. Most of the people working in operations research will agree, however, that the modeling of problem situations and the search for optimal solutions to these models are undoubtedly important parts of it. The latter activity is more algorithmic, mathematical, or formal in character. The former comprises many more disciplines than mathematics, has been more neglected than mathematical research in operations research, and, therefore, would probably need more new advances in theory and practice.

Considering that operations research is to a large extent applied in areas closely related to human evaluations, decisions, and perceptions the need for a modeling language geared to the social sciences (such as traditional mathematics) becomes apparent.

If the model does not consist of crisply defined mathematical statements and relations—if it is, for instance, a verbal model or a model containing fuzzy sets, fuzzy numbers, fuzzy statements, or fuzzy relations—then traditional mathematical methods cannot be applied directly. Either fuzzy algorithms—that is, algorithms that can deal with fuzzy entities or algorithms the procedure of which is "fuzzily" described—can be applied or one has to find crisp mathematical models that are in some specific sense equivalent to the original fuzzy model and to which available crisp algorithms can then be applied.

All cases in which fuzzy set theory is properly used as a modeling tool are characterized by three features:

1. Fuzzy phenomena, relations, or evaluations are modeled by a well-defined and founded theory. (There is nothing fuzzy about fuzzy theory!)
2. By doing so, a better approximation of real phenomena by formal models is achieved.
3. A better modeling of real phenomena normally requires more and more detailed information, more, in fact, than is needed for rather rough dichotomous modeling.

The theory of fuzzy sets, even though still very young, has already been applied to quite a number of operations research problems. As can be expected for a theory of this age, the majority of these "applications" are applications to "model problems" rather than to real-world problems. Exceptions are the areas of classification (structuring), control, logistics, and blending. For these areas there is already considerable software commercially available. The same is true for planning languages (decision support systems), for instance, in the area of financial planning. The reader should realize that the lack of real applications cannot necessarily be blamed on the theory. A real application of a certain theory normally requires that the practitioner who has the problem to be solved is also familiar with and understands, or at least accepts the theoretical framework of the theory before it can really be applied. That obviously takes some more time.

Real applications, particularly the commercially successful ones, very often are not published or with a long delay. This is partially due to competitive considerations, partly to the fact that practitioners normally do not consider publications as one of their prime concerns.

Table 13–1 surveys the major applications of fuzzy set theory in operations research (OR) so far. The table, of course, does not claim to be complete; rather, it wants to indicate major areas of applications. Furthermore, some areas of OR to which fuzzy set theory has been applied extensively—for instance, decision and game theory—have not been included (mainly because they are not too close to real applications). On the other hand, some of the areas included in the table, such as media planning and structuring, might not be considered to be part of OR.

Table 13–1. Applications of fuzzy set theory in operations research.

Methodological Approach / *Functional Area*	*Linear and NL-Programming*	*Combinatorial Programming*	*Dynamic Progr. + Branch + Bound*	*Control and Aprox-Reasoning*	*Graph Theory*	*Clustering*	*Other Heuristics*
Media Selection	X						
Blending	X						
Logistics	X						
Maintenance	X			X			
Production and Process Control	X			X	X		X
Project Management		X					
Inventory Control	X		X	X	X		
Assignment			X				X
Structuring			X			X	

We shall describe some of the uses of fuzzy set theory in order to show the scope of its real and potential applications. Selection criterion for the applications presented was didactical utility rather than coverage of that full scope. Also, for reasons of space economy, we will present only the parts of the applications relevant to the contents of this book. The remainder of this chapter is structured according to major areas and illustrates the use of different methodological approaches to these problem areas.

13.2 Fuzzy Set Models in Logistics

OR has been applied extensively to the area of logistics in the past. In the following, two applications of fuzzy set theory are presented. At first, we show the "fuzzification" of a standard problem in OR: the transportation problem. Second—as an example of existing projects—we show a decision support system based on a fuzzy model.

13.2.1 Fuzzy Approach to the Transportation Problem [Chanas et. al. 1984]

The analysis of "fuzzy counterparts" of linear programming problems of some special structure, for example problems of flows in networks, transportation problems, and so on, appears to be an interesting task. The following model considers a transportation problem with fuzzy supply values of the suppliers and with fuzzy demand values of the receivers. For the solution of the problem parametric programming is used.

Model 13–1

$$\text{minimize} \quad c = \sum_{i=1}^{m} \sum_{j=1}^{n} c_{ij} x_{ij}$$

$$\text{such that} \quad \sum_{j=1}^{n} x_{ij} \cong \tilde{a}_i \qquad i = 1, 2, \ldots, m$$

$$\sum_{i=1}^{m} x_{ij} \cong \tilde{b}_j \qquad j = 1, 2, \ldots, n$$

$$x_{ij} \geq 0 \qquad i = 1, 2, \ldots, m; j = 1, 2, \ldots, n$$

$\tilde{a}_i$ and $\tilde{b}_j$ denote non-negative fuzzy numbers of trapezoidal form. Note the slight difference between definition 5–3 and the definition shown in figure 13–1 which is only used for this section. The value of $\mu_{\tilde{a}}(\Sigma_j x_{ij})(\mu_{\tilde{b}}(\Sigma_i x_{ij}))$ is interpreted as a feasibility degree of the solution with respect to the i-th (j-th) constraint in model 13–1.

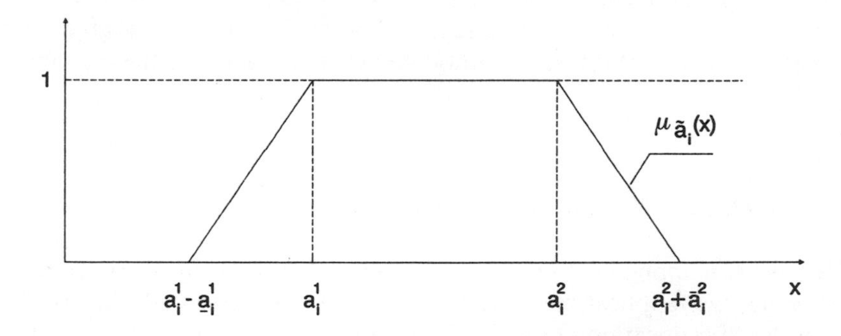

Figure 13–1. The trapezoidal form of a fuzzy number. $\tilde{a}_i = (a_i^1, \underline{a}_i^1, a_i^2, \bar{a}_i^2)$.

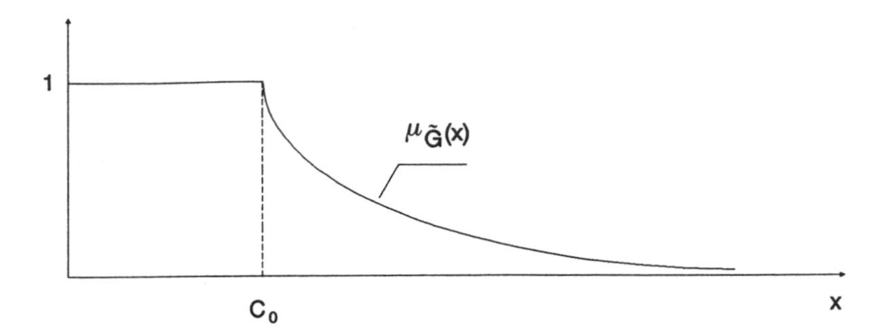

Figure 13–2. The membership function of the fuzzy goal $\tilde{G}$.

With the objective function of model 13–1, a fuzzy number $\tilde{G}$ is associated, expressing the "admissable" total transportation costs. The membership function, $\mu_{\tilde{G}}$, of the $\tilde{G}$ is assumed to be of the form

$$\mu_{\tilde{G}}(x) = \begin{cases} 1 & \text{for} \quad x < C_0 \\ f(x) & \text{for} \quad x \geq C_0 \end{cases}$$

where $f(x)$ is a continuous function, decreasing to zero and achieving the value 1 for $x = C_0$ (see figure 13–2). In particular, $f(x)$ may be a linear function. $\mu_{\tilde{G}}(x)$ determines the degree of the decision maker's satisfaction with the achieved level of the total transportation costs.

Model 13–1 now can be reduced to the symmetrical decision model 13–2, assuming goal and constraints are aggregated via the min-operator.

Model 13–2

$$\begin{aligned} &\text{maximize} && \lambda \\ &\text{such that} && \mu_{\tilde{G}}(c(x)) \geqslant \lambda \\ &&& \mu_{\bar{a}_i}\left(\sum_j x_{ij}\right) \geqslant \lambda \qquad i = 1, 2, \ldots, m \\ &&& \mu_{\bar{b}_j}\left(\sum_i x_{ij}\right) \geqslant \lambda \qquad j = 1, 2, \ldots, n \\ &&& \lambda \geqslant 0 \qquad x_{ij} \geqslant 0 \end{aligned}$$

Here this problem shall, however, be solved by parametric programming. For each level of a constraint's fulfilment λ, $\lambda \in [0, 1]$, one has to find the cheapest transportation plan. This plan satisfies the goal $\tilde{G}$ to the maximum degree for the respective λ. Hence in analogy to definition 12–5 and example 12–7 we shall determine

$$\max \{\mu_{\tilde{G}}(x) \wedge \mu_{\tilde{C}}(x)\},$$

where $\mu_{\tilde{C}}(x)$ will first be determined by an appropriate linear programming model. Here the min-operator is assumed to be acceptable. For the subsequent aggregation of $\mu_{\tilde{C}}(x)$ and $\mu_{\tilde{G}}(x)$ any nondecreasing operator and any decreasing function for $f(x)$ can be employed. Let us first turn to the determination of $\mu_{\tilde{C}}(x)$: The parameter of our parametric LP shall be denoted by r, $r \in [0, 1]$, and rather than determining λ-cuts we shall consider $(1-r)$-cuts. Using the definition given in figure 13–1 for the fuzzy numbers specifying supplies and demands the $(1-r)$-cuts are intervals of the form:

$$\tilde{a}_i^{1-r} = \{x \mid \mu_{\tilde{a}_i}(x) \geq 1-r\} = [a_i^1 - r\underline{a}_i^1,\ a_i^2 + r\bar{a}_i^2]$$
$$\tilde{b}_j^{1-r} = \{x \mid \mu_{\tilde{b}_j}(x) \geq 1-r\} = [b_j^1 - r\underline{b}_j^1,\ b_j^2 + r\bar{b}_j^2]$$

Our problem can then be modeled as follows:

Model 13–3

$$\text{minimize} \quad \sum_{i=1}^{m}\sum_{j=1}^{n} c_{ij}x_{ij}$$

$$\text{such that} \quad \sum_{j=1}^{n} x_{ij} \in [a_i^1 - r\underline{a}_i^1,\ a_i^2 + r\bar{a}_i^2] \qquad i = 1, 2, \ldots, m$$

$$\sum_{i=1}^{m} x_{ij} \in [b_j^1 - r\underline{b}_j^1,\ b_j^2 + r\bar{b}_j^2] \qquad j = 1, 2, \ldots, n$$

$$x_{ij} \geq 0 \qquad r \in [1-\bar{r}, 1]$$

Where $\bar{r} = \sup_X \mu_{\tilde{a}\cap\tilde{b}}(x)$, that is, the maximum value of $\mu_{\tilde{C}}(x)$ which can be achieved for a given r. Solving this model either as a parametric LP or with special algorithms for parametric transportation models we obtain $\mu_{\tilde{C}}(r)$ for $r \in [1-\bar{r}, 1]$. This can now be combined with $\mu_{\tilde{G}}(r)$ to define the membership function of the fuzzy set "decision."

Example 13–1 [Chanas et al. 1984]

There are two suppliers with supply values:

$\tilde{a}_1 = (10, 5, 10, 5)$ and $\tilde{a}_2 = (16, 5, 16, 5)$ (triangular fuzzy numbers);

and three receivers with demand values:

$$\tilde{b}_1 = (10, 5, 10, 5), \qquad \tilde{b}_2 = (9, 4, 9, 4); \qquad \tilde{b}_3 = (1, 1, 1, 1)$$

(also triangular fuzzy numbers) respectively. The unit transport costs are

$$c_{11} = 10 \qquad c_{12} = 20 \qquad c_{13} = 30$$
$$c_{21} = 20 \qquad c_{22} = 50 \qquad c_{23} = 60$$

The membership function of the fuzzy goal is linear:

$$\mu_{\tilde{G}}(x) = \begin{cases} 0 & \text{for } x \geq 800 \\ 1 & \text{for } x \leq 300 \\ \dfrac{800 - x}{500} & \text{for } x \in [300, 800] \end{cases}$$

Model 13–3 for this example becomes:

$$\begin{aligned}
\text{minimize} \quad & c = 10x_{11} + 20x_{12} + 30x_{13} + 20x_{21} + 50x_{22} + 60x_{23} \\
\text{such that} \quad & x_{11} + x_{12} + x_{13} \geq 10 - 5r \\
& x_{11} + x_{12} + x_{13} \leq 10 + 5r \\
& x_{21} + x_{22} + x_{23} \geq 16 - 5r \\
& x_{21} + x_{22} + x_{23} \leq 16 + 5r \\
& x_{11} + x_{21} \geq 10 - 5r \\
& x_{11} + x_{21} \leq 10 + 5r \\
& x_{12} + x_{22} \geq 9 - 4r \\
& x_{12} + x_{22} \leq 9 + 4r \\
& x_{13} + x_{23} \geq 1 - r \\
& x_{13} + x_{23} \leq 1 + r \\
& x_{ij} \geq 0 \qquad \forall\, i, j
\end{aligned}$$

Table 13–2 shows the parametric transportation problem table. Column FR denotes a "fictitious" receiver, row FD a "fictitious" supplier and M a large real number. The rows and columns without an asterisk correspond to the suppliers having supply values settled at minimum level. In this

Table 13–2. Table of the parametric transportation problem.

Receivers / *Suppliers*	1	2	3	1*	2*	3*	FR	Supply
1	10	20	30	10	20	30	M	10 − 5r
2	20	50	60	20	50	60	M	16 − 5r
1*	10	20	30	10	20	30	0	10r
2*	20	50	60	20	50	60	0	10r
FD	M	M	M	0	0	0	0	20r
Demand	10 − 5r	9 − 4r	1 − r	10r	8r	2r	6 + 20r	

Table 13–3. Solution to transportation problem.

	$.3 \leq r \leq 1/3$	$1/3 \leq r \leq .6$	$.6 \leq r \leq 1$
x_{12}	$3 + 14r$	$9 - 4r$	$9 - 4r$
x_{13}	$7 - 19r$	$1 - r$	$1 - r$
x_{21}	$10 + 5r$	$10 + 5r$	$16 - 5r$
x_{22}	$6 - 10r$	$6 - 10r$	

section the FD and FR are blocked by assigning a large transport cost M to their cells. The rows and columns with an asterisk correspond to the maximum surplus of the product that may be sent additionally (but it is not necessary and therefore the respective transport costs to the "fictious" receiver and suppliers are equal to zero) if the constraints are to be satisfied at least to the degree $1 - r$.

It should be observed that the joint supply value of all the suppliers is equal to $\tilde{a} = (26, 10, 26, 10)$ and the joint demand value of all the receivers is equal to $\tilde{b} = (20, 10, 20, 10)$. The maximum degree to which the constraints could be satisfied is equal to $\bar{r} = .7$. Therefore the relevant interval for analysis is $r \in [.3, 1]$.

The solution of this example is shown in table 13–3. The membership function $\mu_{\tilde{G}}(r)$ takes the form:

$$\mu_{\tilde{G}}(r) = \begin{cases} .06 + 1.38r & \text{for} \quad r \in [.3, \frac{1}{3}], \\ .18 + 1.02r & \text{for} \quad r \in [\frac{1}{3}, .6], \\ .54 + 0.42r & \text{for} \quad r \in [.6, 1]. \end{cases}$$

The maximizing solution is obtained for $r = .4059$ and $\mu_{\tilde{G}}(.4059) = .5941$. Figure 13–3 depicts this situation in analogy to figure 12–5.

13.2.2 Fuzzy Linear Programming in Logistics

Ernst [1982] suggests a fuzzy model for the determination of time schedules for containerships, which can be solved by branch and bound, and a model for the scheduling of containers on containerships, which results eventually in an LP. We shall only consider the last model (a real project).

The model contained in a realistic setting approximately 2,000 constraints and originally 21,000 variables, which could then be reduced to approximately 500 variables. Thus it could be handled adequately on a modern computer. It is obvious, however, that a description of this model in a textbook would not be possible. We shall, therefore, sketch the

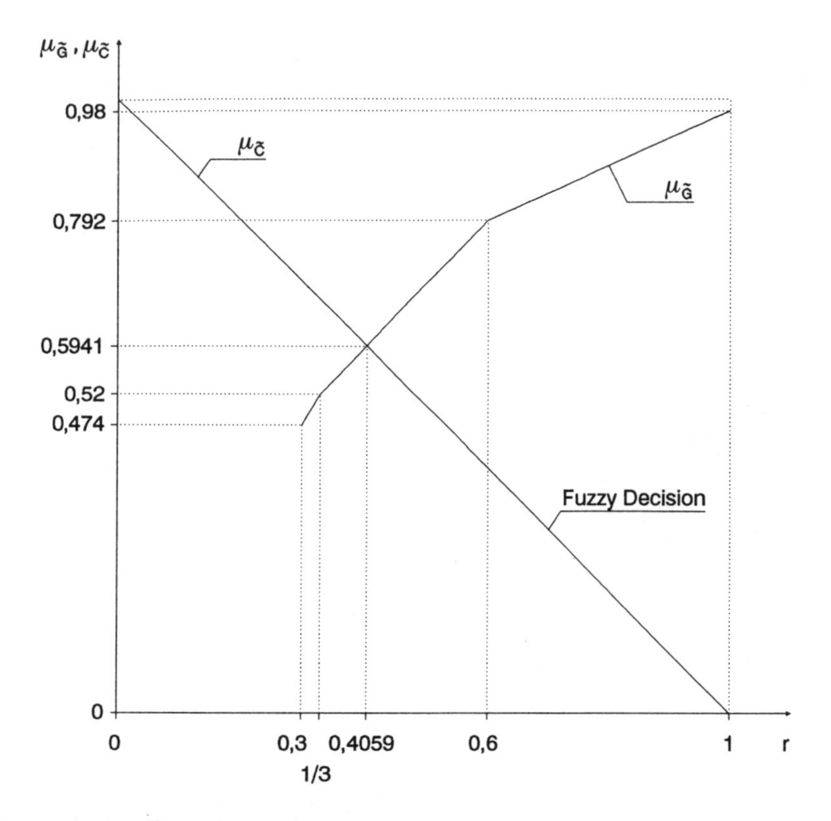

Figure 13–3. The solution of the numerical example.

contents of the modeling verbally and then concentrate on the aspects that included fuzziness.

The system is the core of a decision support system for the purpose of scheduling properly the inventory, movement, and availability of containers, especially empty containers, in and between 15 harbors. The containers were shipped according to known time schedules on approximately 10 big containerships worldwide on 40 routes. The demand for container space in those harbors was to a high extent stochastic. Thus the demand for empty containers in different harbors could either be satisfied by large inventories of empty containers in all harbors, causing high inventory costs, or they could be shipped from their locations to the locations where they were needed, causing high shipping costs and time delays.

Thus the system tries to control optimally primarily the movements and inventories of empty containers, given the demand in ports, the available number of containers, the capacities of the ships, and the predetermined time schedule of the ships.

This problem was formulated as a large LP model. The objective function maximized profit (from shipping full containers) minus cost of moving empty containers minus inventory cost of empty containers. When comparing data of past periods with the model it turned out, that very often ships had transported more containers than their specific maximum capacity. This, after further investigations, lead to a fuzzification of the ship's capacity constraints, which will be described in the next model.

Model 13–4. [Ernst 1982, p. 90]

Let

$z = c^T x$	the net profit to be maximized
$Bx \leq b$	the set of crisp constraints
$Ax \lessapprox d$	the set of capacity constraints for which a crisp formulation turned out to be inappropriate

Then the problem to be solved is:

$$\begin{aligned} \text{maximize} \quad & z = c^T x \\ \text{such that} \quad & Ax \lessapprox d \\ & Bx \leq b \\ & x \geq 0 \end{aligned} \tag{13.1}$$

This corresponds to (12.14). Rather than using (12.19) to arrive at a crisp equivalent LP model the following approach was used: Basing on (12.10) and (12.11) the following membership functions were defined for those constraints that were fuzzy:

$$\mu_i(t_i) = \frac{t_i}{p_i - d_i} \qquad 0 \leq t_i \leq p_i - d_i, \quad i \in I,$$

I = Index set of fuzzy constraints.

As the equivalent crisp model to (13.1) the following LP was used:

$$\begin{aligned} \text{maximize} \quad & z' = c^T x - \sum_{i \in I} s_i (p_i - b_i) \mu_i(t_i) \\ \text{such that} \quad & Ax \leq d + t \\ & Bx \leq b \end{aligned}$$

$$t \leq p - b$$

$$x, t \geq 0 \qquad (13.2)$$

where the s_i are problem-dependent scaling factors with penalty character.

Formulation (13-2) only makes sense if problem-dependent penalty terms s_i, which also have the required scaling property, can be found and justified.

In this case the following definitions performed successfully: First the crisp constraints $Bx \leq b$ were replaced by $Bx \leq .9b$, providing a 10% leeway of capacity, which was desirable for reasons of safety. Then "tolerance" variables t were introduced:

$$Bx - t \leq .9b$$
$$t \leq .1b$$

The objection function became

$$\text{maximize } z = c'x - s't$$

where s was defined to be

$$s = \frac{\text{average profit of shipping a full container}}{\text{average number of time periods which elapsed between departure and arrival of a container}}$$

Because of this definition more than 90% of the capacity of the ships was used only if and when very profitable full containers were available for shipping at the ports, a policy that seemed to be very desirable to the decision makers.

Before turning to another application area it should be mentioned that other applications of fuzzy set theory can be found in the literature [Oh Eigeartaigh 1982] and that the development of (12-9) was initially triggered by a real problem in logistics described by Zimmermann [1976].

13.3 Fuzzy Set Models in Production Control and Scheduling

Production control and scheduling is one of the main functional areas to which traditional OR has been extensively applied. Hence a lot of interesting cases are reported. We will focus on six applications in this area which cover the wide range from slight modifications of standard OR approaches to highly sophisticated modular systems that employ multiple fuzzy models.

13.3.1 A Fuzzy Set Decision Model as Optimization Criterion [v. Altrock 1990]

In the following case a fuzzy decision model is employed as optimization criterion for a search algorithm. It proved to better reflect human judgment in a scheduling problem when due dates were tight. Within an optimizing algorithm, decisions have to be made about which of a given set of schedules is to be prefered. Measures commonly used to rate schedules with respect to their degree of lateness proved to be poor approximations to human judgment. The described decision model can be used in both precise and heuristic algorithms. It either better directs the computation toward a solution that satisfies all due dates or allows the determination of a schedule with the smallest degree of lateness.

The problem, arisen in the chemical industry, is to schedule a given set of jobs on a single machine. Each job is associated with a due date (*DD*) set by the customer and a makespan estimated by the producer's decision maker. The set-up times are sequence-dependent and on an average of the same order of magnitude as the makespans. Prime objective of the scheduling is to maximize overall customer satisfaction. Customer satisfaction is achieved by job completion before or at its *DD*.

The standard OR approach is to model the tardiness of a schedule as the objective function to be minimized, defined by:

- the number of late jobs
- the sum of delays of all late jobs
- the delay of the latest job
- the weighted sum of late jobs (penalty cost approach)

All these rules aggregate the lateness of a given schedule to a real number. This characterizing number is then used to compare schedules and to direct the algorithm's search. Alas, those rules only depict a very rough image of a decision maker's rating. Therefore the resulting solution may not reflect the decision maker's concept of an optimal schedule. While discussing the problem with the decision maker some observations were made:

- *DD*s are not equally important. The importance of satisfying the customer-given *DD* hinges on the customer himself and on the product.
- Most of the aggregation rules presented above assume a linear relation between the lateness of a completion date (*CD*) and the associated loss of satisfaction. This implication could not be found in the decision maker's judgment.

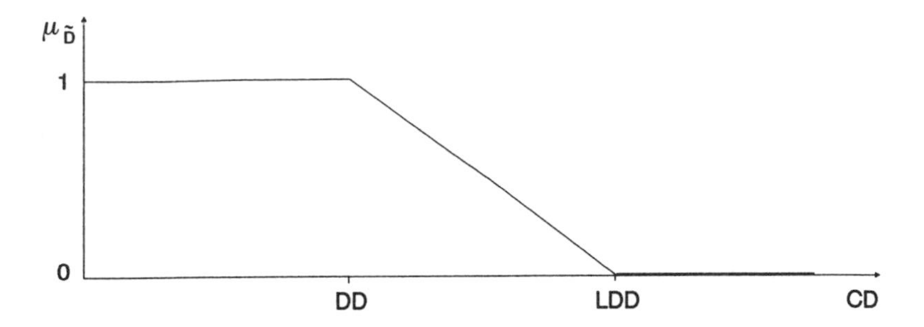

Figure 13–4. Membership function of CDs in the set of fully satisfying CDs.

- Rather than something like penalty costs the decision maker consideres a concept of a latest due date (*LDD*) in rating schedules. The *LDD* is an estimate of the ultimate latest *CD* by the decision maker or the marketing department.

Those observations led to the formulation of a fuzzy decision model. Two fuzzy sets were defined. One describes the membership of a given schedule to the fuzzy set of fully satisfying schedules, $\mu_{\tilde{S}}$. The schedule with the highest membership in this set is the one with the best overall customer satisfaction and will be selected by the algorithm for further evaluation.

For each job in the schedule the other fuzzy set $\mu_{\tilde{D}}$ describes the membership of a given *CD* to the fuzzy set $\tilde{D}$ of fully satisfying *CD*s. This fuzzy set is shown in figure 13–4. Every *CD* prior and equal to the *DD* belongs to the fuzzy set of fully satisfying completion dates with the membership $\mu_{\tilde{D}} = 1$. Every *CD* later than the *LDD* has the degree of membership $\mu_{\tilde{D}} = 0$. The degree of membership of *CD*s between *DD* and *LDD* is assumed to be linear.

To find the degree of membership for all schedules in the set of fully satisfying schedules $\mu_{\tilde{S}}$, for every schedule the $\mu_{\tilde{D}}$'s of every job's *CD* have to be aggregated. The linguistic formulation to maximize the overall customer satisfaction implies that EVERY customer has to be satisfied. The semantic concept of the decision maker's EVERY corresponds to a compensatory AND. For this reason, the γ-operator was chosen. The degree of compensation was determined by generating sets of different schedules with the same $\mu_{\tilde{S}}$ for different values of γ. The decision maker was asked to choose that set of schedules, which in his judgment only contained schedules that were equal with respect to their overall customer satisfaction. It was observed that the decision maker rated the set of different schedules with the same $\mu_{\tilde{S}}$ and $\gamma = .2$ as very much alike.

A greedy-type algorithm was employed. Starting from a tentative schedule, made up using a priority rule, it modifies this schedule by systematically exchanging jobs. The tentative schedule is then compared with the modified schedule by the fuzzy decision model. If the modified schedule is rated better than the tentative schedule, the tentative is replaced by the modified.

Once the algorithm finds a schedule with $\mu_{\tilde{S}} = 1$, the schedule meets all due dates, and is consequently optimal. From then on, the schedule can be further optimized with respect to secondary objectives.

13.3.2 Job-Shop Scheduling with Expert Systems [Bensana et al. 1988]

In the following, we will present a job shop scheduling approach where concepts of the field of artificial intelligence and concepts of fuzzy set theory enrich traditional OR.

Different kinds of knowledge cooperate in the determination of feasible schedules. One kind of knowledge is represented by rules. Relevances of rules with respect to facts and goals are expressed by concepts of fuzzy set theory. First, we will sketch the system. Second, we will focus on the application of fuzzy set theory within the system.

The scheduling problem in a workshop can be stated as follows: given a set of machines and technological constraints, and given production requirements expressed in terms of quantities, product quality, and time constraints expressed by means of earliest starting times and due-dates for jobs, find a feasible sequence of processing operations.

A set of K jobs must be performed by a set of M machines. Each job k is characterized by a set of Operations O_k assigned to machines on which they have to be performed. A schedule is described by means of a precedence graph, expressed by a set of pairs (O_i, O_j) denoting that O_i must precede O_j.

The system, implemented in LISP and named OPAL, consists of two planning modules—the "constraint-based" analysis module and the "decision-support" module—whose interaction is guided by a "supervisor" module. The supervisor module plays the role of the inference engine and guides the search process. The structure of the system is shown in figure 13–5.

The *constraint-based analysis module* (CBA) deals with a partial order of operations derived from the processing sequence of parts and the schedule in progress on one side and the time constraints for job processing

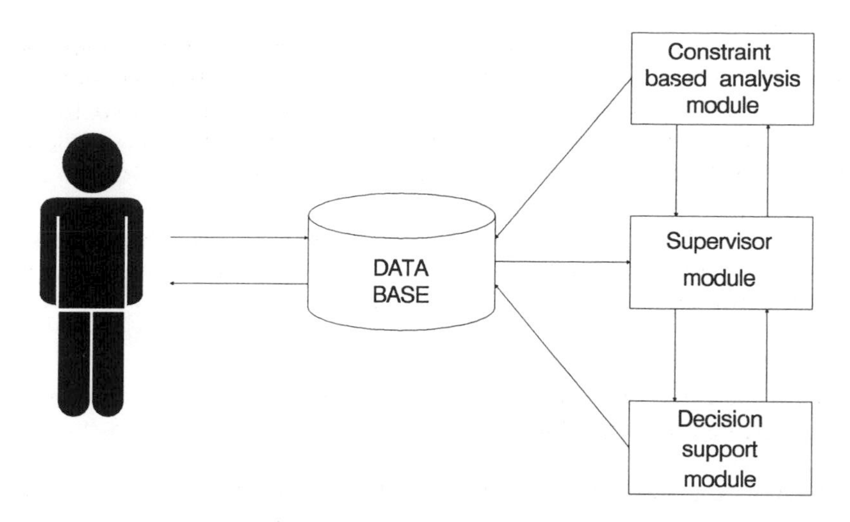

Figure 13–5. Structure of OPAL.

on the other. By subsequent systematical comparisons of the existing precedence constraints, new precedence constraints are generated. This procedure stops in one of the following states:

success: A feasible and complete schedule is derived.

failure: Due to conflicting precedence constraints a feasible schedule does not exist.

wait: The schedule in progress is incomplete (i.e., the set of precedence constraints does not form a complete order) and no more precedence constraints can be generated.

If the CBA module reaches a "wait" state, the decision pertaining to operation ranking is no longer dictated by feasibility considerations with respect to due dates. Such decisions can be made according to other kinds of criteria of a technological nature (e.g., it is better not to cut a workpiece made of metal M before a workpiece made of metal M'), or related to productivity (facilitate material flow, avoid filling up machine input buffers, avoid long set-up times ...).

According to these criteria, a **decision-support module** (DS) generates new precedence constraints. First, it selects a subset C of the set of all unordered pairs of operations and second, it choses one element of C and forms a new precedence constraint.

The selection can be based on criteria like specific machines, specific operations, temporal location, influence on the quality of the schedule, or influence on the resolution speed. The grades of membership of the unfixed pairs of operations in the sets defined by those criteria may be expressed fuzzily. If more than one criterion is used for selection, the corresponding fuzzy sets are intersected by the minimum-operator.

In the second step one element of this fuzzy set is chosen to be fixed, that is, to be the new precedence constraint. This step is carried out by using a collection of pieces of advice expressed as "if . . . then" rules. Rules differ by their origin and by their range of application (general or application-dedicated). Moreover, their efficiency is more or less well known and depends upon the prescribed goal, or the state of completion of the schedule. They can express antagonistic points of view. Lastly they are usually pervaded by imprecision and fuzziness, because their relevance in a given situation cannot be determined in an all-or-nothing manner.

To take these features into account, each rule r is assigned a grade of relevance $\pi_r(k)$ with respect to goal k. $\pi_r(k)$ can be viewed as the grade of membership of rule r to the fuzzy set of relevant rules for goal k. The aim of these coefficients is basically to create an order on the set of rules. For every pair of operations the "if" part of a rule is evaluated as to the extent to which O_i should precede O_j according to the attribute of the rule. Let v be the index qualifying this attribute, v_{ij} be the value of this index when O_i precedes O_j. The ratio $x_{ij} = \frac{v_{ij}}{v_{ij} + v_{ji}}$ is then evaluated. To avoid thresholding effects three fuzzy sets H = high ratio, M = medium ratio, and S = small ratio are defined (see figure 13–6). Hence the relation appearing in the rule is a fuzzy relation.

The "then" part of all rules is of the same format. It provides advice

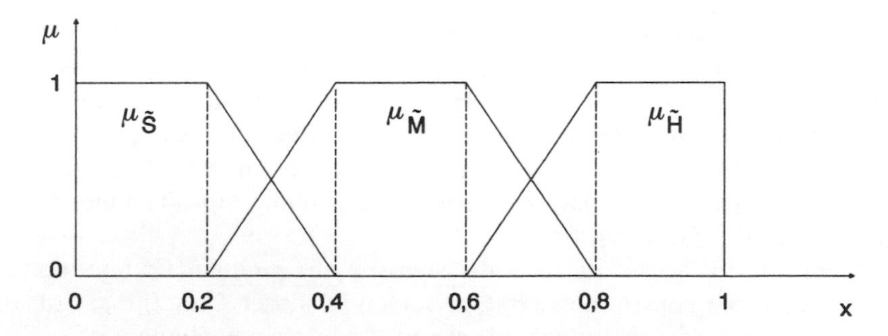

Figure 13–6. Fuzzy sets for the ratio in the "if" part of the rules.

about whether O_i should precede O_j $(i<j)$ or if the rule does not know $(i \sim j)$. This advice is expressed by three numbers:

$$\mu_r(i<j) = \min(\mu_{\tilde{S}}(x_{ij}), \pi_r(k))$$
$$\mu_r(j<i) = \min(\mu_{\tilde{M}}(x_{ij}), \pi_r(k))$$
$$\mu_r(j \sim i) = \min(\mu_{\tilde{H}}(x_{ij}), \pi_r(k))$$

The rules relevant for goal k are all triggered and applied to all facts in the set C. The proportions of relevant triggered rules preferring $i<j, j<i, i \sim j$ are obtained as relative cardinalities (see definition 2–5):

$$\rho(i<j) = \Sigma\mu_r(i<j)/\Sigma\pi_r(k)$$
$$\rho(j<i) = \Sigma\mu_r(j<i)/\Sigma\pi_r(k)$$
$$\rho(i \sim j) = \Sigma\mu_r(i \sim j)/\Sigma\pi_r(k)$$

When $p(i \sim j)$ is close to 1, one is unable to decide which of the two operations should precede the other because the rules are indifferent. In contrast, when $\rho(i \sim j)$ is close to 0, but $\rho(i<j)$ is close to $\rho(j<i)$, the set of rules is strongly conflicting. The preference index for decision $i<j$ is defined as $\min\{\rho(i<j), 1-\rho(i \sim j)\}$; in terms of fuzzy logic, it expresses to what extent most of the triggered rules prescribe $i<j$ and most are not indifferent about O_i preceding O_j.

The schedule is gradually built up by adding precedence constraints between operations. The search graph is developed as follows: each time the CBA module stops, a new node is generated and the current schedule is stored. The DS module then generates a new precedence constraint to the schedule graph and the CBA module checks for consequent precedence constraints. Back-tracking occurs if the explored path leads to a failure state. When no feasible schedule at all exists, the data must be modified in order to recover feasibility.

13.3.3 A Method to Control Flexible Manufacturing Systems [Hintz and Zimmermann 1989]

The following application shows the usage of multiple concepts of fuzzy set theory within a hybrid system for Production Planning and Control (PPC) in Flexible Manufacturing Systems (FMS). FMS are integrated manufacturing systems consisting of highly automated work stations linked by a computerized material-handling system making it possible for jobs to follow diverse routes through the system (see figure 13–7). They facilitate

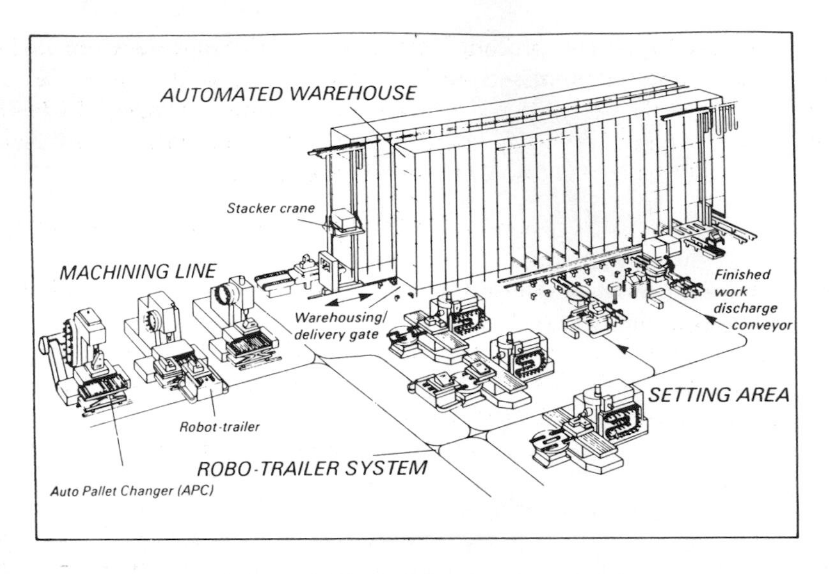

Figure 13–7. Example for an FMS. (Hartley, 1984, p. 194).

small batch sizes, high quality standards and efficiency of the production process at the same time.

Decentralized PPC-systems for each FMS are provided with schedules of complete orders by an aggregate planning system. They are responsible for meeting the due dates, minimizing flow times, and maximizing machine utilizations. Generally, these objectives are conflicting. The planning process is carried out by subsequently solving the subproblems:

1. Master scheduling
2. Tool loading
3. Releasing scheduling
4. Machine scheduling

Subprobem 1 is solved by using fuzzy linear programming (FLP), subproblem 2 is solved by a heuristic algorithm, and subproblems 3 and 4 are solved using approximate reasoning (AR). We will just sketch the master scheduling, omit the tool loading, and focus on the release and machine scheduling.

Master Scheduling. The objective of the master schedule is to determine a short-term production program with a well-balanced machine utilization which optimally meets all due dates. Its determination is a quite well-structured problem, although some important input data are rather uncertain. Since nearly the complete manufacturing of a part can be performed within an FMS, a simultaneous approach using FLP (as defined in section 12.2.1) has been employed for the master scheduling. Restrictions to be considered in the master schedule are:

(1) Parts can only be processed when they are released from earlier production stages.
(2) They have to meet given due dates in order to match the following operations and assembling.
(3) The capacity of the FMS must not be exceeded. Because the machines may partially be substituted by each other, they have to be classified into appropriate groups.
(4) There is only a limited number of (expensive) fixtures and pallets available.

In restrictions (1) and (2), release and due dates are often rough estimates which include safety buffers and unnecessary work-in-process inventories. In practice it is often possible to supply some parts earlier than initially planned (i.e., by overtime) or to violate the due dates only for a portion of an order (for instance by lot size splitting) without seriously disturbing processing or assembling. On the other hand, if release dates or due dates are chosen too stringently, there may be no feasible solution at all.

For these reasons, restrictions (1) and (2) are modeled as fuzzy constraints while (3) and (4) are modeled as crisp constraints. The solution of the FLP yields a solution

- that is feasible according to the restrictions (1)–(4), if possible, or
- that minimizes the deviations from given due dates and distributes them uniformly among the different orders. The value of the maximized variable then denotes the degree of membership of the optimal solution to the set of feasible and optimal solutions.

Release and Machine Scheduling. Decisions concerning the parts schedule for both releasing and machining are arrived at by AR. This is considered to be an appropriate way to model a very complex situation with many interdependencies. The decision criteria are formulated in terms of production rules, which have been shown to lead to quite stable decisions. It will be shown later that this approach also leads at least to a very good compromise of the three mentioned conflicting goals of scheduling. In addition, this method is very suitable for interactive deci-

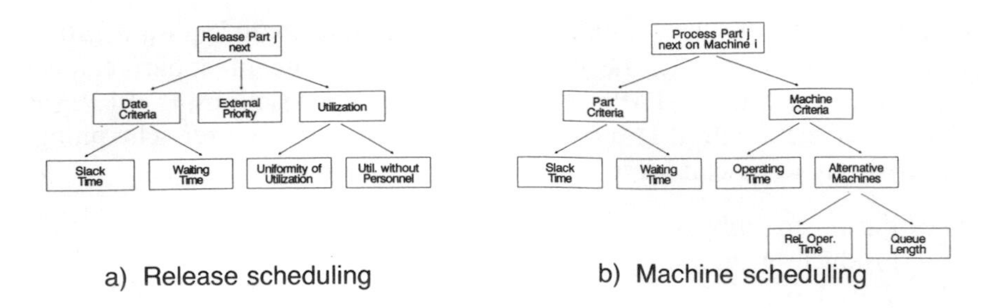

Figure 13–8. Criteria hierarchies.

sion making, where the decision maker can employ familiar linguistic descriptions of the situations.

The basic *release scheduling* procedure can be regarded as dispatching parts for a single capacity unit (the FMS) with several work stations: As long as there are working places unused and appropriate pallets with fixtures are available, new parts can be released into the FMS. Once the upper limit of parts has been reached, the remaining parts have to wait in a queue until one of the parts leaves the FMS. Then the decision of which part has to be released next will be made using an AR procedure.

The *machine scheduling* procedure is very similar to dispatching when using priority rules. That means that no machine is allowed to wait if there is a part that can be processed on that machine. If there are several parts at a time waiting for a machine then another AR procedure is used to choose a part from the waiting line.

For both AR procedures a *hierarchy of decision criteria* is defined (see figure 13–8). This hierarchy corresponds to stepwise operationalizing the decision criteria until they can easily be used by the decision maker. On the other hand, such a hierarchy can be considered as the combination of elementary local-priority rules in a more comprehensive global-priority or decision rule. The single elements or concepts of the hierarchy may in general consist of arithmetic or linguistic terms. Both the hierarchy and the ways to make the concepts operational are heuristic in nature. Hence no optimal solution can be guaranteed.

Let us further concentrate on the criteria hierarchy depicted in figure 13–8a for the release scheduling. The decision of which part to release next mainly depends on date criteria of the parts under consideration or the impact of parts on machine utilization or it may be some kind of external priority. For the date criteria we furthermore distinguish between the slack time of a part and the time the part has already waited for processing.

The impact on the effect on machine utilization can be twofold. First, we have to take care that the machines are used as uniformly as possible thus trying to avoid bottlenecks. For this purpose we define a criterion "uniformity of utilization." On the other hand, we want to ensure a good utilization in the shift with reduced personnel, during which no parts can be fixed on pallets. On the contrary, parts can only be processed as long as they do not need any manual operation, be it for changing a pallet or in any case of failure. We shall take this into consideration by the concept "processing time until the next fixturing." The external priority can be given by the plant manager or some other person responsible.

To illustrate the AR process we will look at the definitions of the concepts of the hierarchy and the aggregation of concepts by the rule set. We will focus on the derivation of the date criterion of the slack time and the waiting time criteria. Slack time and waiting time are considered linguistic variables as defined in section 9.1:

Linguistic variable:	Term set:
slack time	critically_short, short
waiting time	short, medium, long
date criterion	urgent, not_urgent

The base variable is defined for all possible values for the indicator, that is, in general, all real numbers within a reasonable interval. The meaning of the terms can be defined by giving the degree of membership as a function of the above-defined indicator as base variable. As membership functions, piecewise linear functions are used. The parameters were obtained by extensive simulation studies for a specific structure of orders to be processed in a specific FMS.

An essential task before aggregating these two criteria to the date criteria is the assignment of degrees of sensibleness to each element (rule) of the cartesian product defined by the assumptions and the conclusion: {long, medium, short} $\otimes$ {critically_short, short} $\otimes$ {urgent, not_urgent}. This can be done by an expert (scheduler) and results in the "degrees of sensibleness" shown in parentheses for each rule above.

An example rule set might be (degree of sensibleness given in parentheses):

(1) **IF** waiting time is long **AND** slack time is critically_short **THEN** date criterion is urgent (1.0)
(2) **IF** waiting time is medium **AND** slack time is critically_short **THEN** date criterion is urgent (0.8)
(3) **IF** waiting time is short **AND** slack time is critically_short **THEN** date criterion is urgent (0.6)

(4) **IF** waiting time is long **AND** slack time is short **THEN** date criterion is urgent (0.5)
(5) **IF** waiting time is medium **AND** slack time is short **THEN** date criterion is urgent (0.2)
(6) **IF** waiting time is medium **AND** slack time is short **THEN** date criterion is not_urgent (0.7)

Each of these rules can now be interpreted as one possible aggregation of the two criteria "slack time" and "waiting time" to the "date criteria" (see figure 13–8a). Only rules with a nonzero degree of sensibleness are considered. The AR procedure applied is depicted in figure 13–9. That is, first the conditional parts of the rules which are connected by "AND" or "OR" are aggregated by using the γ operator. The "THEN" of the rule is then interpreted as "the conditions hold *and* the rule is valid," where this "AND" is also modeled by the γ operator. In this case, however, γ is taken

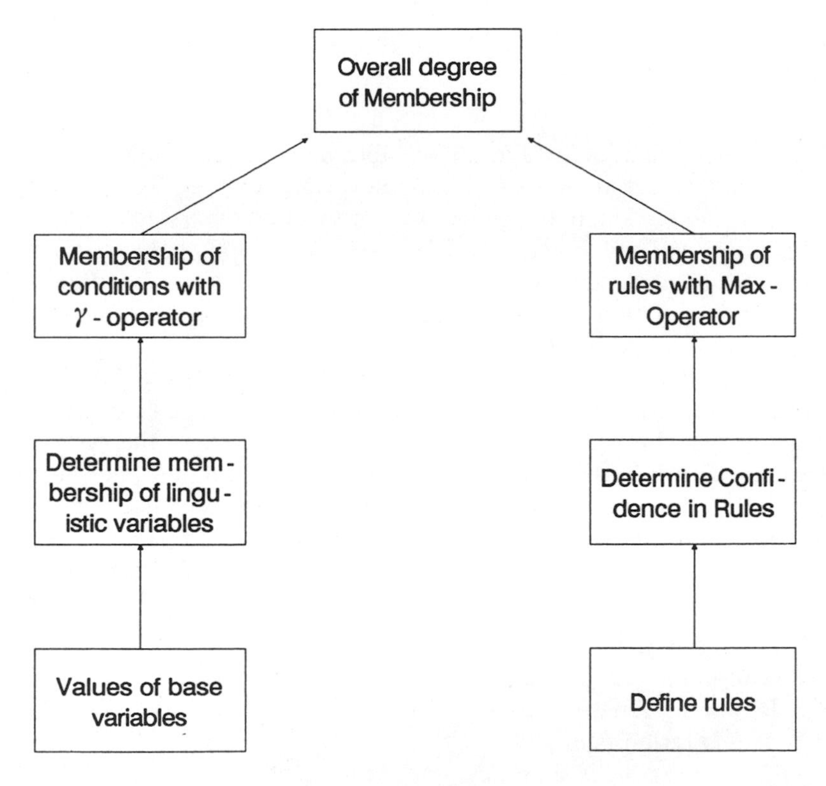

Figure 13–9. Principle of approximate reasoning.

to be zero since no compensation is assumed between the truth of the rule and the validity of its conditions. If more than one rule leads to a certain condition, the maximum of the respective degrees of membership determines the final result.

Example 13–2

We want to compute the values (degrees of membership) of the terms of the "date criterion" in figure 13–8a. Consider three parts, whose slack time and waiting time are linguistic variables as described above. The grades of membership to the terms of the linguistic variables are given in table 13–4.

In the first step the conditional parts of the rules are aggregated by using the γ operator. In this example $\gamma = .5$ is used. The results are depicted in table 13–5. In the second step the rules are evaluated. The use of the γ operator with $\gamma = 0$ is equivalent to the multiplication of the degree of membership of the condition and the degree of sensibleness. The results

Table 13–4. Membership grades for slack time and waiting time.

		Membership grade of part		
		1	*2*	*3*
Waiting time:	long	0.7	0	0.7
	medium	0.2	0.8	0.3
	short	0	0.4	0
Slack time:	critically_short	0.4	0.8	0.7
	short	0.6	0.2	0.3

Table 13–5. Membership grades for conditional parts of the rules.

		Part	
	1	*2*	*3*
Condition (1)	0.58	0.00	0.67
Condition (2)	0.20	0.78	0.41
Condition (3)	0.00	0.53	0.00
Condition (4)	0.72	0.00	0.41
Condition (5)	0.29	0.37	0.21
Condition (6)	0.29	0.37	0.21

Table 13–6. Membership grades for the rules.

	Part		
	1	*2*	*3*
Date criterion is urgent:			
conclusion (1)	**0.58**	0.00	**0.67**
conclusion (2)	0.16	**0.62**	0.33
conclusion (3)	0.00	0.32	0.00
conclusion (4)	0.36	0.00	0.21
conclusion (5)	0.06	0.07	0.04
Date criterion is not urgent:			
conclusion (6)	**0.20**	**0.26**	**0.15**

Table 13–7. Results.

Criteria	*Suggested Approach*	*Priority Rule Approach*
Mean in-process waiting time [min]	2884	3369
Part of lots that have met their due dates [%]	97	28
Mean machine utilization [%]	80	79

are summarized in table 13–6 where the maxima of the respective degrees of membership for the two terms (urgent, not_urgent) of the linguistic variable "date criteria" are printed in bold. Part 3 shows the highest degree of membership in the fuzzy set of parts with urgent date criteria and the lowest degree of membership in the fuzzy set of parts with not_urgent date criteria. The results are used for further aggregation as shown in figure 13–8a.

Results. The approach described above has been programmed and its performance has been compared to systems with no master scheduling and employing only simple priority rules for release and machine scheduling using a general simulation program for FMS. The results are shown in table 13–7. The suggested approach dominated the classical priority scheduling with respect to all three objectives.

13.3.4 Aggregate Production and Inventory Planning [Rinks 1982a,b]

The "HMMS-model" [Holt et al. 1960] is one of the best-known classical models in aggregate production planning. It assumes that the main objective of the production planner is to minimize total cost which is assumed to consist of costs of regular payroll, overtime and lay-offs, inventory, stock-outs, and machine set up. The model assumes quadratic cost functions and then derives linear decision rules for the production level and the work-force level. The following terminology is used:

FS_t = sales forecast for period t
W_{t-1} = work force level in period $t-1$
I_{t-1} = inventory level at the end of period $t-1$
ΔW_t = change in work force level in period t
P_t = production level in period t

In general, the decision variables are related to the cue variables as

$$P_t = f(FS_t, W_{t-1}, I_{t-1})$$
$$\Delta W_t = g(FS_t, I_{t-1})$$

By contrast to most other models the HMMS-model was tested empirically for a paint factory. The cost coefficients were derived in different ways (statistically, heuristically, etc.) and the performance of the decision rules was compared to the actual performance of the paint factory managers [Holt et al. 1960].

The following model resulted for the paint factory.

Model 13–5

$$\text{minimize } C_N = \text{minimize} \sum_{t=1}^{N} C_t$$

where

$$\begin{aligned} C_t = {} & [340W_t] && \text{Regular payroll costs} \\ & + [64.3(W_t - W_{t-1})^2] && \text{Hiring and layoff costs} \\ & + [0.20(P_t - 5.67W_t)^2 + 51.2P_t - 281W_t] && \text{Overtime costs} \\ & + [0.0825(I_t - 320)^2] && \text{Inventory-connected costs} \end{aligned}$$

and subject to restraints

$$I_{t-1} + P_t - S_t = I_t \qquad t = 1, 2, \ldots, N$$

Even though the HMMS-model performed quite well and is used as a common benchmark for later models it was rarely used in practice. The main objection was generally that managers would not use it, roughly speaking, because too much mathematics was involved.

Rinks tries to avoid this lack of acceptance by suggesting a model basing on the concepts described in chapters 9 and 10 of this book. He developed one production and one work-force algorithm which consist of a series of relational assignment statements (rules) of the form

If FS_t is ... and I_{t-1} is ...
and W_{t-1} is ... then P_t is ...
Else ...

and

If FS_t is ... and I_{t-1} is ...
and W_{t-1} is ... then ΔW_t is ...
Else ...

respectively.

He uses the definition (given in table 13–8) of the terms of linguistic variables. Figure 13–10 sketches the membership functions of the terms of the linguistic variables used. Forty decision rules were suggested (see table

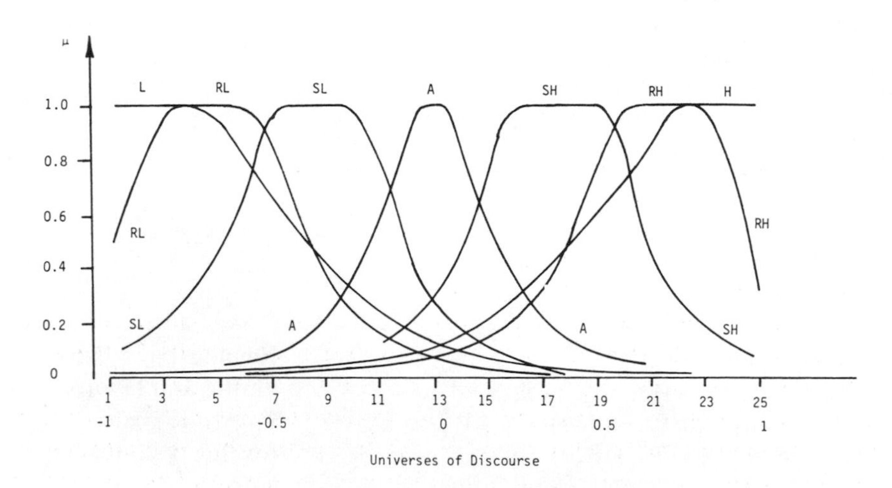

Figure 13–10. Membership functions for several linguistic terms.

Table 13–8. [Rinks 1982] Definition of linguistic variables.

Linguistic Terms	*Acronym*	*Base variable*[1]	*Membership Function Expression*[2,3]	
VERY HIGH (POSITIVE, VERY BIG)	VH (PVB)	x (dx)	HIGH x * HIGH x	
HIGH (POSITIVE BIG)	H (PB)	x (dx)	$1 - e^{[-(0.5/\|1-x\|)^{2.5}]}$	
RATHER HIGH (POSITIVE, RATHER BIG)	RH (PRB)	x (dx)	$1 - e^{[-(0.25/\|0.7-x\|)^{2.5}]}$	
SORTOF HIGH (POSITIVE, SORTOF BIG)	SH (PSB)	x (dx)	$1 - e^{[-(0.25/\|0.4-x\|)^{2.5}]}$	
AVERAGE (ZERO)	A (Z)	x (dx)	$1 - e^{[-5\|x\|]}$	
SORTOF LOW (NEGATIVE, SORTOF BIG)	SL (NSB)	x (dx)	$1 - e^{[-(0.25/\|-0.4-x\|)^{2.5}]}$	
RATHER LOW (NEGATIVE, RATHER BIG)	RL (NRB)	x (dx)	$1 - e^{[-(0.25/\|-0.7-x\|)^{2.5}]}$	
LOW (NEGATIVE BIG)	L (NB)	x (dx)	$1 - e^{[-(0.5/\|1-x\|)^{2.5}]}$	
VERY LOW (NEGATIVE, VERY BIG)	VL (NVB)	x (dx)	LOW x * LOW x	
AT LEAST AVERAGE	ALA	x	$1 - e^{[-5\|x\|]}$ 1	$-1 \leqslant x \leqslant 0$ $0 < x \leqslant 1$
AT MOST AVERAGE	AMA	x	1 $1 - e^{[-5\|x\|]}$	$-1 \leqslant x \leqslant 0$ $0 < x \leqslant 1$

[1] x is any one of the following variables: W_{t-1}, FS_t, W_t, and P_t. dx is ΔW_t.
[2] All variables are scaled to be placed in the $[-1, 1]$ interval.
[3] dx replaces x in the membership function expression for use with ΔW_t.

13–9); these were not claimed to be optimal but rather heuristic in character and acceptable to the manager.

To test the performance of the suggested approach the data of the paint factory of the HMMS-model were used. In order to apply Rinks decision rules the membership functions of the terms, as shown in figure 13–10, had to be calibrated. In fact the range $[-1, 1]$ on the horizontal axis of this figure had to be calibrated to the data. For test purposes, lower and upper bounds as shown in the following tabulation were derived from available historical data (HMMS).

Table 13–9. Membership functions.

	Cue variables			Decision variables	
Rule no.	FS_t	I_{t-1}	W_{t-1}	P_t	ΔW_t
1	H	AMA	H	H	Z
2	H	AMA	A	RH	PRB
3	H	AMA	L	SH	PVB
4	SH	L	H	H	Z
5	SH	L	A	RH	PRB
6	SH	L	L	SH	PVB
7	SH	SH	H	SH	NRB
8	SH	SH	A	A	Z
9	SH	SH	L	A	PRB
10	A	A	H	SH	NRB
11	A	A	A	A	Z
12	A	A	L	A	PRB
13	SL	SL	H	SH	NRB
14	SL	SL	A	A	Z
15	SL	SL	L	SL	PRB
16	RL	L	H	SH	NRB
17	RL	L	A	A	Z
18	RL	L	L	A	PRB
19	L	ALA	H	SL	NVB
20	L	ALA	A	RL	NRB
21	L	ALA	L	L	Z
22	SL	H	H	SL	NVB
23	SL	H	A	RL	NRB
24	SL	H	L	RL	Z
25	H	AMA	SH	H	PSB
26	H	AMA	SL	SH	PB
27	SH	L	SH	H	PSB
28	SH	L	SL	SH	PB
29	SH	SH	SH	A	Z
30	SH	SH	SL	A	PSB
31	A	A	SH	A	NSB
32	A	A	SL	A	PSB
33	SL	SL	SH	A	NSB
34	SL	SL	SL	A	Z
35	RL	L	SH	A	NSB
36	RL	L	SL	A	Z
37	L	ALA	SH	RL	NB
38	L	ALA	SL	L	NSB
39	SL	H	SH	RL	NB
40	SL	H	SL	RL	Z

1. Acronyms for the values of the linguistic variables are defined in table 13–8.
2. Each production rule is a fuzzy relational assignment statement of the form "IF FS_t is ______ AND I_{t-1} is ______ AND W_{t-1} is ______ THEN P_t is ______."
3. Each work force rule is a fuzzy relational assignment statement of the form "IF FS_t is ______ AND I_{t-1} is ______ AND W_{t-1} is ______ THEN ΔW_t is ______."

Variable	Lower Bound	Upper Bound
W_{t-1}	60	115
ΔW_t	-10	10
P_t	250	750
I_{t-1}	150	490
FS_t	250	750

In the absence of historical data the manager would use his judgment to make the determinations. For computations the max-min compositions were used, resulting in fuzzy sets as representing the "conclusion" or "decision." Since, however, a decision concerning the workforce, production, or inventory of next period should be a crisp decision, Rinks used

Table 13–10. Cost results.

Costs (1000\$)	*Linear DR HMMS (optimal)*	*Fuzzy algorithm*
Regular payroll	1879	1814
Hiring and lay-off	20	22
Overtime	129	251
Inventory	25	43
Total cost	2053	2130

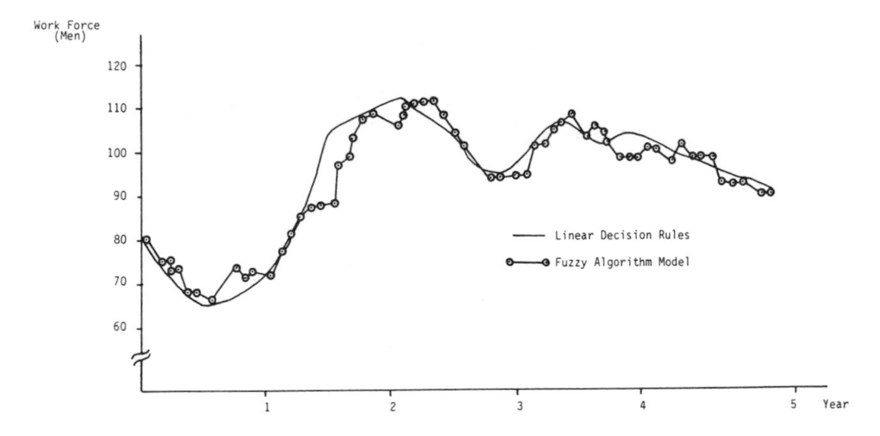

Figure 13–11. Comparison of work force algorithms.

the maximum rule if possible. If the membership function did not have a unique maximum he used other, heuristic, rules to choose the crisp decision to be implemented.

For the 60 month's data of the HMMS-model (1949–1953) the results of the work-force algorithm are shown in figure 13–11. The cost results are shown in table 13–10.

Rinks' own evaluation of the simulation results reads as follows:

> While the 5.0 per cent cost penalty evidenced by the production scheduling fuzzy algorithms is somewhat greater than that reported by other heuristics—Search Decision Rule [Taubert 1967] and Parametric Production Planning [Jones 1967] reported cost penalties of less than one percent for the paint factory—it must be remembered that the fuzzy algorithms do not even require an explicit cost function. For situations where restrictive assumptions cannot be rationalized and sufficient data is not available to construct a cost function, approximate reasoning based models would seem to offer an appealing alternative. [Rinks 1982b, p. 579]

If Rinks had compared his results to other benchmarks, he would probably have been more optimistic. Table 13–11 is from Bowman [1963, p. 104] and shows the real performance and the performance of another heuristic, the management coefficient approach, in the case of the HMMS paint factory and three other plants. Compared to the 139% and 124.7% performance of these two approaches the 105% performance of the fuzzy algorithm would look even better.

Table 13–11. Comparison of performances.

	Ice cream	*Chocolate*	*Candy*	*Paint*
Decision rule (perfect)	100%	100%	100%	100%
Decision rule (moving average)	104.9%	102.0%	103.3%	110%
Company performance	105.3%	105.3%	114.4%	139.5%
Management coefficients	102.3%	100.0%	124.1%	124.7%
Correlation	$W^{t,r} = .78$ $P^{t,r} = .97$	$W^{t,r} = .57$ $P^{t,r} = .93$	$W^{t,r} = .73$ $P^{t,r} = .86$	$W^{t,r} = .40$ $P^{t,r} = .66$

13.3.5 Fuzzy Mathematical Programming for Maintenance Scheduling

The following application, basing on a master thesis in Zittau, Germany is of interest because the effects of different operators were investigated and because parametrized membership functions were used.

Model 13–6 [Holtz and Desonki 1981]

The problem objective here is to determine optimal maintenance cycles in electrical power plants. Stochastic models had been used before, but because of the very low frequency of breakdowns it seemed that a model basing on frequentistic arguments was not appropriate.

T_j : Cycle times of maintenance operations for $j=1, \ldots, N$ maintenance crews (Decision Variable)
x_{ij} : Coefficients of the crisp cost function, $i = 1, 2, 3; j = 1, \ldots, N$
y_{ij} : Coefficients of the manpower requirement function, $i = 1, 2, 3;$ $j=1, \ldots, N$
z_{ij} : Coefficients of the breakdown function, $i=1, 2, 3;$ $j=1, \ldots, N$
Mh : Number of manhours available for maintenance per year
B : Number of breakdowns per year
$B_{\max}$: Maximum of acceptable breakdowns per year

Crisp mathematical model. For $N=2$ and $C=$ total cost the following crisp model was the point of departure:

$$\text{minimize} \quad C = C(T_1, \ldots, T_N) = \sum_{j=1}^{N} \left(x_{1j}T_j + x_{2j} + \frac{x_{3j}}{T_j} \right)$$

$$\text{such that} \quad \sum_{j=1}^{N} \left(y_{1j}T_j + y_{2j} + \frac{y_{3j}}{T_j} \right) \leq Mh$$

$$\sum_{j=1}^{N} \left(z_{1j}T_j + z_{2j} + \frac{z_{3j}}{T_j} \right) \leq B_{\max}$$

$$T_j \geq 0$$

The requirements were:

1. Cost should not exceed 500 considerably, in no case an upper bound which could be varied.
2. Manpower Mh should generally not exceed 1100, by no means 1200.
3. The number of breakdowns can exceed 50 but never 300 ($B_{\max}$).

Fuzzy Mathematical Model. The symmetrical concept of a decision (definition 12–1) was used and the optimal decision was defined to be

$$T_{j_0} = T_j \circ \mu_{\tilde{T}_j} = \max_j \min_i \mu_i(T_j)$$

Two types of membership functions were investigated: a linear membership function and a nonlinear two-parameter membership function.

Type 1 Membership Functions

$$\mu_{\tilde{C}}(T_j) = \frac{1}{2}\left\{[1 + \operatorname{sgn}(C_L - C)] + [1 + \operatorname{sgn}(C - C_L)] \cdot \left(\frac{C_U - C}{C_L - C_U}\right)\right\}$$

where C_L and C_U represent the lower and upper bounds for total cost.

$$\mu_{\tilde{M}h}(T_j) = \frac{1}{2}\left\{[1 + \operatorname{sgn}(Mh_L - Mh)] + [1 + \operatorname{sgn}(Mh - Mh_L)] \cdot \left(\frac{Mh_U - Mh}{Mh_L - Mh_U}\right)\right\}$$

with Mh_L and Mh_U the lower and upper bounds.

$$\mu_{\tilde{B}}(T_j) = \frac{1}{2}\left\{[1 + \operatorname{sgn}(B_L - B)] + [1 + \operatorname{sgn}(B_U - B_L)] \cdot \left(\frac{B_U - B}{B_L - B_U}\right)\right\}$$

with B_L and B_U the lower and upper bounds.

Type 2 Membership Function. We shall only show the membership function for the objective function. The others are defined accordingly:

$$\mu_{\tilde{C}}(T_j) = \frac{1}{2}\left\{[1 + \operatorname{sgn}(C_L - C)] + \frac{1 + \operatorname{sgn}(C - C_L)}{1 + \left(\frac{1}{b_1} - 1\right)\left|\frac{C - C_L}{c_1}\right|}\right\}$$

b_1 and c_1 serve as means of better fitting the membership function to the real situation. On the other hand, they obviously increase the computational effort.

Detailed numerical results, as well as a comparison of the performance of the min-operator versus the product operator as a model for the intersection, can be found in Holtz [1981].

13.3.6 Scheduling Courses, Instructors, and Classrooms

It is well known that the determination of time schedules in which several resources have to be combined belongs to the most difficult combinatorial problems in Operations Research. Rarely does one ever try to determine optimal schedules. The determination of feasible schedules is very often what one can hope for. The difficulty of obtaining such schedules by formal algorithms might partly be due to the fact that constraints are treated as crisp requirements even though in reality they often are flexible. The following case indicates how a combination of fuzzy set theory and heuristics can lead to quite acceptable results.

Model 13–7 [Prade 1979]

Problem Description: A quarter schedule in a french university is to be determined. There are N (here $N=4$) instruction programs, each of them lasts one year and a student can only attend one of these programs. Each instruction program I consists of $M(I)$ courses (here, $10 \leq M(I) \leq 14$). Each course contains lectures, lab work, and a final examination.

A course is taught by one instructor, supported by several teaching assistants. An instructor may teach several courses in one or several instruction programs. The availability of an instructor differs from person to person. An instructor may be present for only some predetermined days of a week, another may be available for only some weeks during the quarter. Information about the availability of instructors is only known approximately beforehand.

A schedule has to satisfy seven "global" constraints:

1. Each instruction program must be completely planned for the entire school year.
2. There are precedence constraints between courses (or sometimes parts of courses) which are elements of the same instruction program.
3. It is not desirable that more than four weeks elapse between the first lecture of a course and its final examination.
4. It is not desirable that any course that has already begun is interrupted for more than a week.
5. Some courses can be in common in several instruction programs.
6. An instructor is not always available.
7. It is very desirable that several courses (three or four) are planned during the course of the same week.

Constraints 1, 2, 5, and 6 are considered as "hard," 3, 4, and 7 as "soft" constraints. More local constraints will be considered later.

Solution: The flow time of a course is considered as a fuzzy number with a membership function similar to figure 13–12. These fuzzy numbers in L-R

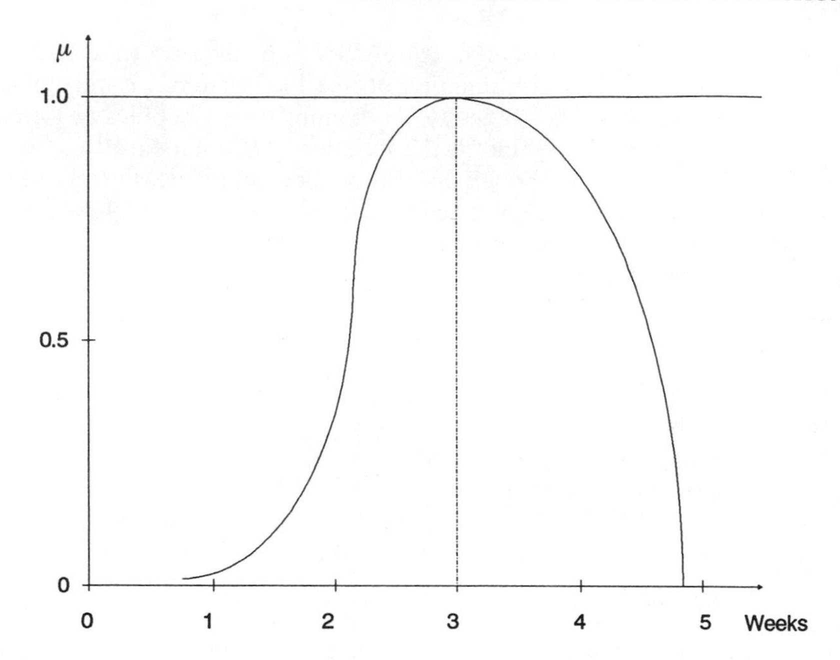

Figure 13–12. Flowtime of a course.

representation (see definition 5–6) are used to compute via fuzzy PERT a fuzzy early starting date $\tilde{r}_i$ and a late ending date, $\tilde{d}_i$. If x denotes time then the interval $\tilde{T}_i$ in which the course i will be taught is a fuzzily bounded interval (see figure 7–5), bounded by $\tilde{r}_i$ and $\tilde{d}_i$, respectively.

The membership function of these intervals $\tilde{T}_i$ is

$$\mu_{\tilde{T}_i}(x) = \begin{cases} \mu_{\tilde{r}_i}(x) & \text{for} \quad x \leq r_i \\ 1 & \text{for} \quad x \in [r_i, d_i] \\ \mu_{\tilde{d}_i}(x) & \text{for} \quad x \geq d_i \end{cases}$$

where r_i, d_i are the mean values of $\tilde{r}_i$, and $\tilde{d}_i$, respectively.

The "global" constraints are taken into consideration successively: Constraints (1) and (2) are used as a basis for PERT; (4) is used to compute whole programs from single courses. And if constraint (5) is relevant the intersection of the different possibility intervals for all relevant courses in all effected instruction programs is computed. Constraint (6) is taken care of similarly.

So far, the slack time for each course, the work load of each instructor,

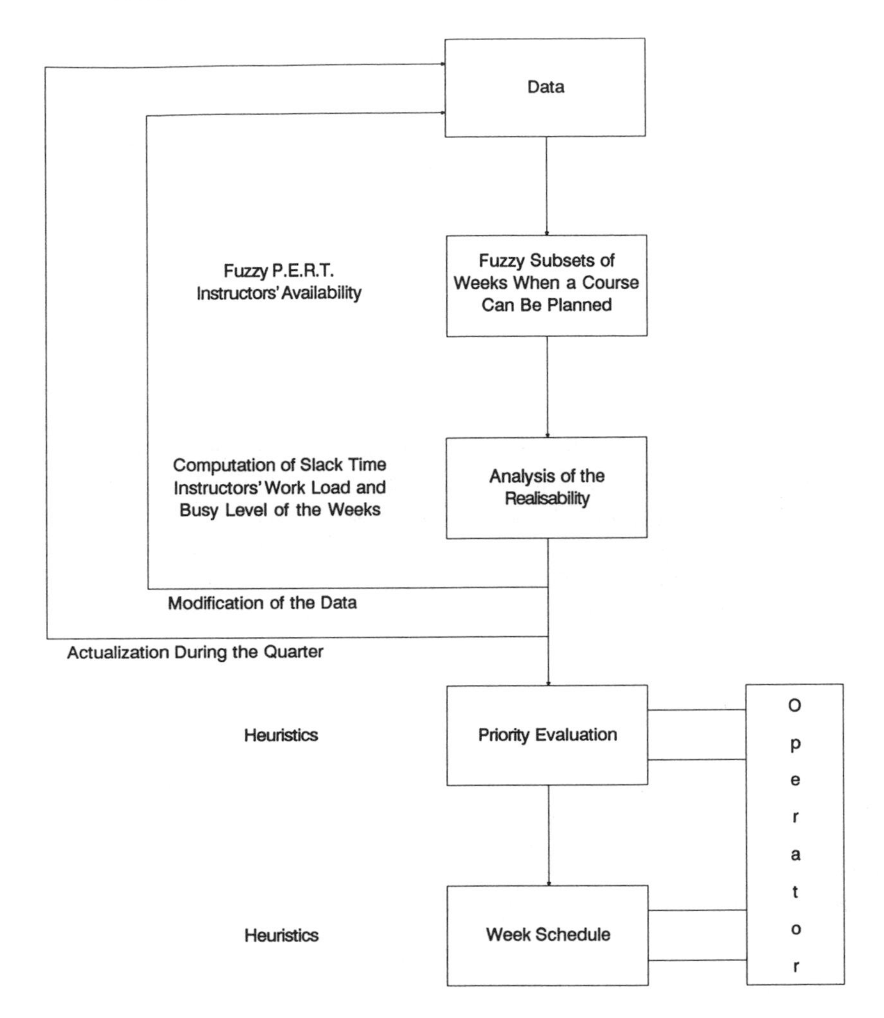

Figure 13–13. The scheduling process.

and the number of courses per week for each instruction program have been determined. Modifications of this schedule due to the availability of the instructors can now be made and the following "local" constraints are considered by interactively changing schedules which have been generated automatically via heuristic priority assignment. Figure 13–13 summarizes the entire process.

"Local" constraints are as follows:

1. There exist precedence constraints between lectures and lab work inside a course (the graph of these constraints is not the same for all the courses).
2. An instructor can teach only one lecture at a given moment.
3. It is generally desirable to plan two lecture of the same course in succession, but not three.
4. It is not desirable that an instructor teach more than two lectures of different courses in the same morning.
5. It is desirable to give priority to lectures in the morning and lab work in the afternoon.

Example 13–2

The following tables and figures can only serve to visualize the process. Details can be found in Prade [1977]. Figure 13–14 presents the data of one of the four instruction programs that were considered. All courses had to be scheduled within one quarter of 11 weeks. Table 13–12 gives the node number, name of courses, instructor number and category (1 to 4 indicate different availabilities of the instructor). p, α, and β are the mean

Table 13–12. Structure of instruction program.

N	*Name*	*Instructor number*	*category*	α	*Processing time* p	β
1	A231(L1 to 6)	9	3	0.5	2	1
2	A231(L7 to 10)	9	3	0.5	1	1
3	A141(L1 to 6)	12	3	0.5	2	1
4	A141(L7 to 10)	12	3	0.5	1	1
5	A121	12	3	1	3	1.5
6	A241	12	3	1	3	1.5
7	A510	9	3	1	3	1.5
8	M317(L1 to 8)	17	4	1	3	1.5
9	M317(L9 to 10)	17	4	1	3	1.5
10	PS16	8	1	0	4	0
11	V231	21	2	0	1	0
12	V211	1	4	1	3	1.5
13	E541	23	4	1	3	1.5
14	E551	23	4	1	3	1.5
15	M361	13	3	1	3	1.5
16	E531	11	1	0	4	0
17	E532	11	1	0	4	0

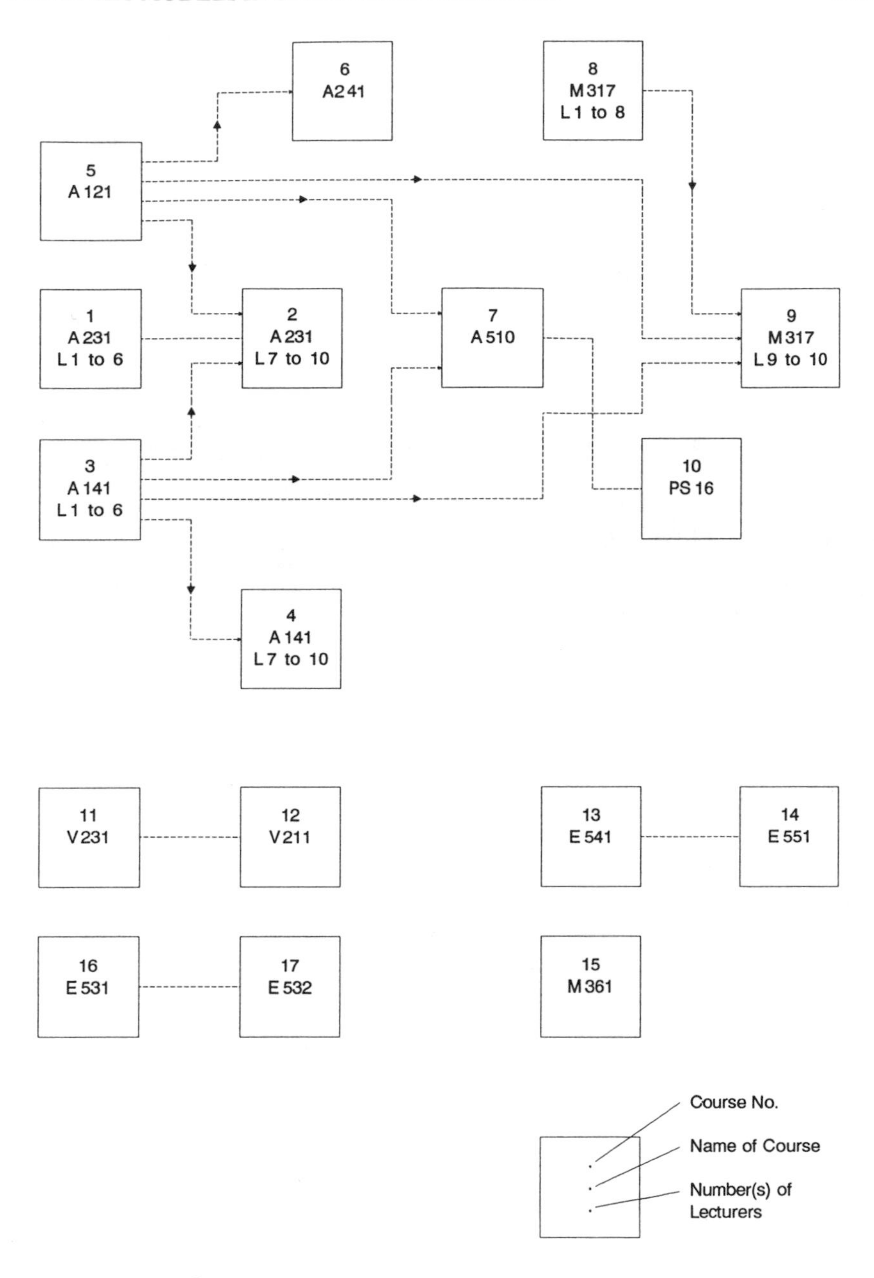

Figure 13–14. Courses of one instruction program.

Table 13–13. Availability of instructors.

Instructor number	*Weeks* 1	2	3	4	5	6	7	8	9	10	11
1	0	0	0	0	0.5	1	1	1	1	1	0.5
8	0	0	0	0	0.5	1	1	1	1	0.5	0.5
9	1	1	1	1	1	1	1	1	1	0	1
11	0	0	0	0.5	1	1	1	1	1	1	1
12	1	1	0.5	0	1	1	1	1	1	1	1
13	1	1	1	0.5	0	1	1	1	1	0.5	0.5
17	1	1	1	1	1	1	0.5	0	0.5	1	1
21	1	1	0.5	0	0	0	0	0	0	0	0
23	1	1	1	1	1	1	0.5	0.5	0.5	0	0

Table 13–14. PERT output.

Name	α	$\bar{p}$	β	α	d	β
A231	0	1	0	0	11	0
A141	0	1	0	2	6	2
A121	0	1	0	1.5	4	1
A241	1	4	1.5	0	11	0
A510	1	4	1.5	0	7	0
M317	0	1	0	0	11	0
PS16	2	7	3	0	11	0
V231	0	1	0	0	3	0
V211	0	5	0	0	11	0
E541	0	1	0	1.5	6	1
E551	1	4	1.5	0	9	0
M361	0	1	0	0	11	0
E531	0	4	0	0	7	0
E532	0	8	0	0	11	0

values, left and right spreads of the processing time for each course. The availability for each instructor is given in Table 13–13. Table 13–14 gives course numbers, initialized by the name of the instructor and early start and late finish times.

As reference (membership) functions for the fuzzy numbers in L-R-representation representing the "flowlines" of the course Prade used $L(x) = \exp[-x^2]$ and $R(x) = \max[0, 1 - x^2]$.

Table 13–15. Availability of weeks for courses.

Name	Weeks 1	2	3	4	5	6	7	8	9	10	11
A231	1	1	1	1	1	1	1	1	0	1	1
A141	1	1	0.5	0	1	1	0.8	0.4	0	0	0
A121	1	1	0.5	0	0.4	0	0	0	0	0	0
A241	0	0	0.4	0	1	1	1	1	1	1	1
M510	0	0	0.4	1	1	1	1	0	0	0	0
M317	1	1	1	1	1	1	0.5	0	0.5	1	1
PS16	0	0	0	0	0.4	0.8	1	1	1	0.5	0.5
V231	1	1	0.5	0	0	0	0	0	0	0	0
V211	0	0	0	0	0.5	1	1	1	1	1	0.5
E541	1	1	1	1	1	1	0.4	0	0	0	0
E551	0	0	0.4	1	1	1	0.5	0.5	0.5	0	0
M361	1	1	1	0.5	0	1	1	1	1	0.5	0.5
E531	0	0	0	0.5	1	1	1	0	0	0	0
E532	0	0	0	0	0	0	0	1	1	1	1

Table 13–16 First week's final schedule.

		Morning		*Afternoon*
Monday	A141 L.1	A141 L.2	A121 L.1	—
Tuesday	A231 L.1	A231 L.2	M317 L.1	—
Wednesday	A231 L.3	A231 L.4	M317 L.2	A231 L.W.1
Thursday	A141	A141	A121	Sports
Friday	A121	A121	M317	A141

The intersection of the availability schedule (Table 13–13) and the PERT schedule yields the possibility schedule of weeks in which courses can be scheduled (table 13–15). Table 13–16 shows an example of the final schedule for the first week.

13.4 Fuzzy Set Models in Inventory Control

There exist a large number of inventory models in operations research using a great variety of methods for their solution. For inventory models using linear or integer linear models the approach of section 12.2 or an algorithm described in Zimmermann and Pollatschek [1984] may be used. For solutions basing on differential calculus the models in chapter 7 might be useful. Kacprzyk and Staniewksi [1982] present a very interesting approach for aggregate inventory planning, using primarily the concept presented in chapters 3 and 5 of this book. We shall present a model that uses Bellman and Zadeh's approach to fuzzy dynamic programming discussed in section 12.3.

Model 13–8 [Sommer 1981]

The management of a company wants to close down a certain plant within a definite time interval. Therefore production levels should decrease to zero as steadily as possible and the stock level at the end of the planning horizon should be as low as possible. The demand is assumed to be deterministic.

Mathematical model

Let

$d_i \in D$, $i = 1, \ldots, N$ be the decision variable representing the production level in period i,

where

$D = \{\alpha_1, \ldots, \alpha_n\}$ is the set of values permitted for the decisions.

$\tilde{x}_i \in X$, $i = 1, \ldots, N+1$ be the state variable representing the inventory level at the beginning of period i,

where

$X = \{\tau_1, \ldots, \tau_m\}$ is the set of possible state values,

$a_{i,i} = 1, \ldots, N$ is the deterministic demand in period i,

$x_{i+1} = x_i + d_i - a_i$ is the crisp transformation function,

$\tilde{C}_i(d_i) = \{(d_i, \mu_{\tilde{C}_i}(d_i))\}$ are fuzzy constraints on the decision variables representing the goal "production should decrease as steadily as possible,"

$i = 1, \ldots, N$ and

$\tilde{G}_i(x_{N+1}) = \{x_{N+1}, \mu_{\tilde{G}_i}(x_{N+1}))\}$ is the fuzzy goal, representing the decision to have as low a stock level as possible at the end of the planning horizon.

Then, using (12.20), the membership function of the decision on stage i is

$$\mu_{\tilde{D}}(d_i) = \min \{\mu_{\tilde{C}_i}(d_i),\ \mu_{\tilde{G}}(x_{N+1})\}$$

and the membership function of the maximizing decision on stage i is

$$\mu^0_{\tilde{D}}(d_i) = \max_{d_i \in D}\{\min \{\mu_{\tilde{C}_i}(d_i,\ \mu_{\tilde{G}}(x_{N+1})\}\}$$

which can be determined recursively using (12.23).

As will be shown in the following numerical example, the state spaces can sometimes be reduced even further by introducing bound on the basis of heuristic considerations.

Example 13–3

Let

$$\mu_{\tilde{C}_i}(d_i) = \begin{cases} 0 & \text{if} \quad 0 \leq d_i \leq 60 - 10i \\ -3 + .5i + d_i/20 & \text{if} \quad 60 - 10i < d_i \leq 80 - 10i \\ 5 - .5i - d_i/20 & \text{if} \quad 80 - 10i < d_i \leq 100 - 10i \\ 0 & \text{if} \quad 100 - 10i < d_i \end{cases}$$

and

$$\mu_{\tilde{G}_{N+1}}(x_{N+1}) = \begin{cases} 1 - x_{N+1}/20 & \text{if} \quad 0 \leq x_{N+1} \leq 20 \\ 0 & \text{else} \end{cases}$$

$$a_1 = 45, \quad a_2 = 50, \quad a_3 = 45, \quad a_4 = 60, \quad \text{and} \quad N = 4$$

x the stock level at the beginning is supposed to be 0.

$$\tau_j = \{0, 5, 10, \ldots\}$$
$$\alpha_h = \{0, 5, 10, \ldots\}$$

Only $\{d_i \mid \mu_{\tilde{C}_i}(d_i) > 0\}$ are of interest. Hence we can bound the decision variables as follows:

i	d_i^l	d_i^u
1	55	85
2	45	75
3	35	65
4	25	55

Also $0 < x_5 \leq 20$.

Using the transformation function, we can also find upper and lower bounds for the state variables on the different intermediate stages. We proceed in three steps: First we determine upper bounds $x_i^{u'}$ and lower bounds $x_i^{l'}$ from the forward calculation. The according bounds $x_i^{u''}$ and $x_i^{l''}$ from backward calculation are computed in the second step. Then we can obtain the final bounds by

$$x_i^u = \min\{x_i^{u'}, x_i^{u''}\},$$
$$x_i^l = \max\{x_i^{l'}, x_i^{l''}\}.$$

The lower bound for the state variable x_i can be calculated as

$$x_i^{l'} = \max\{0, x_{i-1}^{l'} + d_{i-1}^l - a_{i-1}\} \qquad i = 2, \ldots, 4$$

The appropriate upper bound is

$$x_i^{u'} = x_{i-1}^{u'} + d_{i-1}^u - a_{i-1} \qquad i = 2, \ldots, 4$$

For the different stages we obtain, for $x_1 = 0$,

i	$x_i^{l'}$	$x_i^{u'}$
1	—	—
2	10	40
3	5	65
4	0	85
5	—	—

Starting with x_5 and assuming $x_5^{1''} = 0$ and $x_5^{4''} = 20$ we obtain recursively the following upper and lower bounds:

i	$x_i^{l''}$	$x_i^{u''}$
1	—	—
2	0	65
3	0	60
4	5	50
5	—	—

The final upper and lower bounds can be determined by

$$x_i^l = \max\{x_i^{l'}, x_i^{l''}\}$$
$$x_i^u = \min\{x_i^{u'}, x_i^{u''}\}$$

Hence:

i	x_i^l	x_i^u
1	0	0
2	10	40
3	5	60
4	5	50
5	0	15

Now we can determine the optimal d_i and x_i within the lower and upper bounds computed above:

Stage 1: Using (12.23) we obtain

$$\begin{aligned} \mu_{\tilde{G}_4}(x_4) &= \max_{d_4} \{\min [\mu_{\tilde{C}}(d_4),\ \mu_{\tilde{G}}(x_4,\ d_4)]\} \\ &= \max_{d_4} \{\min [\mu_{\tilde{G}}(d_4,\ \mu_{\tilde{G}}(x_4 + d_4 - a_4)]\} \end{aligned}$$

d_4 / x_4	25	30	35	40	45	50	55	$\mu_{\tilde{G}_4}(x_4)$
5							1/4	1/4
10						1/2	1/4	1/2
15					3/4	1/2	1/4	3/4
20				1	3/4	1/2	1/4	1
25			3/4	3/4	1/2	1/4		3/4
30		1/2	3/4	1/2	1/4			3/4
35	1/4	1/2	1/2	1/4				1/2
40	1/4	1/2	1/4					1/2
45	1/4	1/4						1/4
50	1/4							1/4

Stage 2: $\mu_{\tilde{D}}(x_3) = \max_{d_3} \{\min [\mu_{\tilde{C}}(d_3), \mu_{\tilde{G}}(x_3 + d_3 - a_3)]\}$

d_3 / x_3	35	40	45	50	55	60	65	$\mu_{\tilde{G}_3}(x_3)$
5			1/4	1/2	3/4	1/2	1/4	3/4
10		1/4	1/2	3/4	3/4	1/2	1/4	3/4
15	1/4	1/2	3/4	1	3/4	1/2	1/4	1
20	1/4	1/2	3/4	3/4	3/4	1/2	1/4	3/4
25	1/4	1/2	3/4	3/4	1/2	1/2	1/4	3/4
30	1/4	1/2	3/4	1/2	1/2	1/4	1/4	3/4
35	1/4	1/2	1/2	1/2	1/4	1/4		1/2
40	1/4	1/2	1/2	1/4	1/4			1/2
45	1/4	1/2	1/4	1/4				1/2
50	1/4	1/4	1/4					1/4
55	1/4	1/4						1/4
60	1/4							1/4

Stage 3: $\mu_{\tilde{D}}(x_2) = \max_{d_2} \{\min [\mu_{\tilde{C}}(d_2), \mu_{\tilde{D}}(x_2 + d_2 - a_2)]\}$

d_2 / x_2	45	50	55	60	65	70	75	$\mu_{\tilde{G}_2}(x_2)$
10	1/4	1/2	3/4	3/4	3/4	1/2	1/4	3/4
15	1/4	1/2	3/4	3/4	3/4	1/2	1/4	3/4
20	1/4	1/2	3/4	3/4	1/2	1/2	1/4	3/4
25	1/4	1/2	3/4	1/2	1/2	1/2	1/4	3/4
30	1/4	1/2	1/2	1/2	1/2	1/4	1/4	1/2
35	1/4	1/2	1/2	1/2	1/4	1/4	1/4	1/2
40	1/4	1/2	1/2	1/4	1/4	1/4		1/2

Stage 4: $\mu_{\tilde{D}}(x_1) = \max_{d_1} \{\min [\mu_{\tilde{C}}(x_1), \mu_{\tilde{D}}(x_1 + d_1 - a_1)]\}$

d_1 / x_1	55	60	65	70	75	80	85
0	1/4	1/2	3/4	3/4	1/2	1/2	1/4

13.5 A Discrete Location Model [Darzentas 1987]

For quite a number of years there has been a widespread interest in location models. For specific types of these problems there exist exellent review papers. One of the most popular models is the "simple plant location model" (SPLP) for which, for instance, Krarup and Pruzan [1983] summarizes the existing literature until the middle of the 80's. In this paper the authors also establish some relationships between SPLP, other location problems, set-covering problems, and integer programming. One of the problems, the discrete location problem (DLP), can be formulated as a set-covering problem and principally solved by pure zero-one programming algorithms. In this type of problem, a number of facilities are to be located at specific points within an area, according to precisely quantified criteria. This results in a districting, that is, a plan that shows where the facilities have to be located and what locations they serve. However, in many location problems, especially those associated with social policies, noncrisply defined criteria are used such as, how "near" or "accessible" a facility is, or how "important" certain issues are, etc. In these cases a fuzzy sets approach is more appropriate.

In such a problem, the decision maker's main task is the identification and evaluation of criteria on the basis of which an optimum will be obtained. The choice of specific locations can only be based on questions like:

- How "far" should people travel to reach a service point?
- How "important" are "bad" and "good" roads and public transport?
- Is "homogeneity" of social class and income within a subset important?
- Is it "very unfair" to locate two major facilities in one point?

The fuzzy nature of the problem can be accepted and introduced at various stages in the analysis.

There are two major obstacles to finding "optimal" solutions to DLP's: It is necessary but difficult to define all possible covers, that is, subsets of locations, which have to enter even the crisp DLP-model. For readers who are not aquainted with this type of problem the above mentioned paper by Krarup and Pruzan or the work of Darzentas [1987, pp. 330–337] are recommended. The second problem is the "evaluation" of the covers in order to select the best one.

The aim of a location project is easy to state: find the "best" districting; which means that the objective itself is a fuzzy set. There may also be a number of restrictions such as "the budget allows for approximately M facilities" or "it is preferable that village i serves village m," and vice-versa, or "it is very important that i and j belong to the same district" and so on. Hence constraints can be formulated as fuzzy sets.

In a crisp model the determination of the optimal districting can be performed by using integer programming algorithms. If the problem is of reasonable size, heuristic versions have to be used.

In fuzzy DLP's, possibly even with a multiple criteria this approach is not possible. One could then use either fuzzy integer programming (see, for example, Fabian and Stoica [1984] or Zimmermann and Pollatschek [1984], or one could try to reduce the number of possible districtings to a reasonable size by eliminating nonfeasible and dominated covers. The remaining covers could be evaluated with respect to relevant criteria (yielding a fuzzy set for each criterion) and then ordered in analogy to methods described in section 12.4.

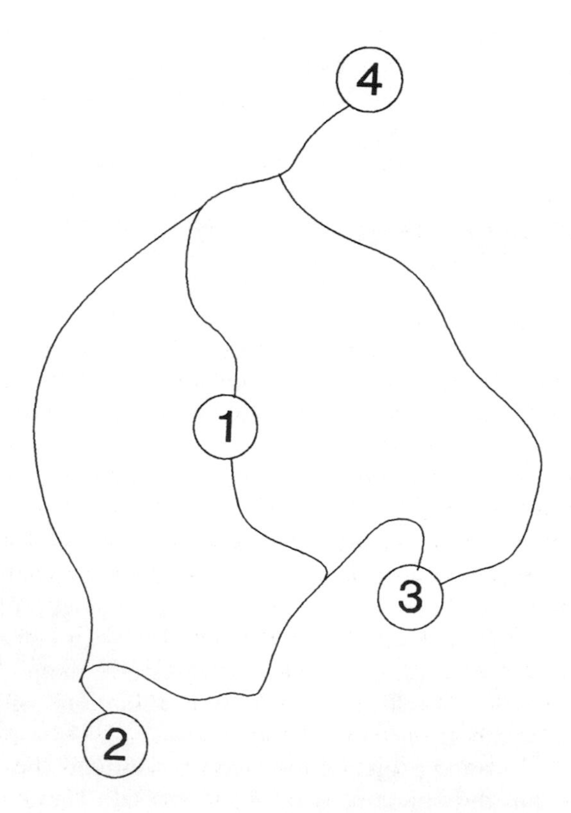

Figure 13–15. Road network.

Table 13–17a. Populations.

Village	*Population*
1	1,100
2	650
3	1,350
4	730

Table 13–17b. Distances between villages.

	(miles) 1	2	3	4
1	–	11	7	9
2	11	–	–	14
3	7	–	–	–
4	9	14	–	–

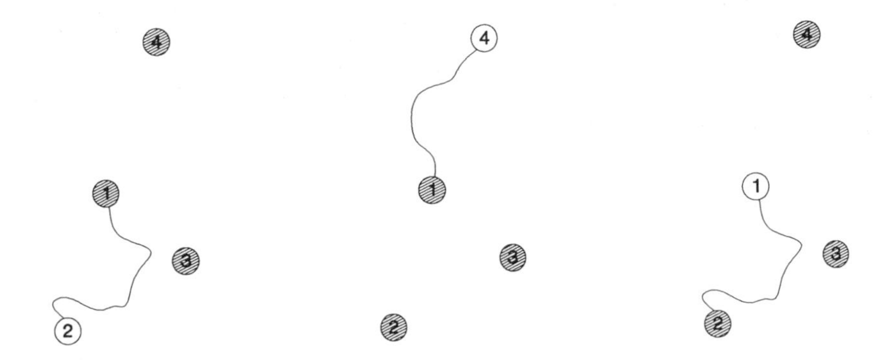

Figure 13–16. Feasible covers.

Example 13–4

Consider the road network shown in figure 13–15, which is part of a real road network. The points 1, . . . , 4 represent villages whose populations are given in table 13–17a. The distances between the villages are given in table 13–17b. The problem is to optimally locate three facilities in order to serve (cover) each village by only one facility. This problem in its nonfuzzy form can be formulated as a set-partitioning problem. The fuzzy version of the problem can be formulated as a symmetric fuzzy-decision model (see definition 12–1).

Suppose the three covers shown in figure 13–16 are the only covers feasible due to crisp constraints, which are omitted here. In figure 13–16 the villages hosting a facility are hatched. For the determination of the "best" cover the grades of membership of all three covers to every fuzzy criterion is rated. These ratings and the fuzzy criteria are given in table 13–18. In this example, the degrees of membership of the covers in the fuzzy set "decision" are obtained using the min-operator. These degrees

Table 13–18. Determination of the fuzzy set decision.

	Covers		
	c_1	c_2	c_2
It is a better policy to locate this type of facility in villages with high population:	.9	.8	.7
The facilities should not be located in polluted areas:	.6	.5	.2
The distance between a village without a facility and a facility should not exceed 8 miles considerably:	.6	.9	.6
Membership values of the decision:	.6	.5	.2

imply an order on the set of covers. If a crisp decision has to be made, the cover with the maximum degree of membership (c_1, $\mu_{\tilde{D}}(c_1) = .6$) is chosen.

Exercises

1. In what ways, and for what purposes can fuzzy sets be used in operations research?
2. Explain why in model 13–3 every nondecreasing operator can be used to combine the goal with all of the constraints.
3. Could approaches (12.9) or (12.18) have been used in model 13–2. If so, what would have been the consequences?
4. In section 13.3.1 a fuzzy decision model has been employed as an optimization criterion. Can this approach be used for both precise and heuristic algorithms?
5. In the system presented in section 13.3.2 the decision support module picks one precedence constraint out of the subset C of the set of all unordered pairs of operations. Consider multiple criteria for the selection of the subset C. Discuss possible fuzzy aggregation models for the derivation of the subset C.

6. Assume in model 13–7 that the instructors' availability is given by the following table:

Instructor Number	Weeks 1	2	3	4	5	6	7	8	9	10	11
1	0	0	0	.5	0.5	1	1	1	.5	0	0
8	0	0	.5	1	1	1	1	.5	.5	0	0
9	.5	1	1	1	1	1	1	.5	.5	0	0
11	0	0	0	.5	.5	1	1	1	.5	.5	0
12	1	1	.5	.5	1	1	1	1	1	1	1
13	0	1	1	.5	.5	1	1	1	.5	0	0
17	1	1	1	1	1	1	1	1	1	1	1
21	.5	.5	1	1	.5	0	0	0	0	0	0
23	1	1	1	1	.5	.5	.5	0	0	0	0

Determine a new table 13–15 of available weeks for courses and try to determine heuristically a first week's final schedule.

7. Discuss approaches, and their advantages and disadvantages, for PERT networks in which activity times are fuzzy and stochastically uncertain.
8. Determine the critical path for the network shown in table 13–14 by substituting for the addition of activity time in the normal critical path method the extended addition (section 5.3.1).
9. Determine an optimal policy for model 13–8 modified as follows: the demands are: $a_i = \{40, 40, 45, 50\}$; the fuzzy set goal is characterized by the membership function

$$\mu_{\tilde{G}_{N+1}} = \begin{cases} 1 - \dfrac{x_{N+1}}{10} & \text{if } \; 0 \leq x_{N+1} < 10 \\ 0 & \text{else} \end{cases}$$

10. In the example shown in table 13–18, the membership degrees of the districtings were evaluated subjectively by the decision maker. Consider fuzzy accessibility measures for the "nearness" or "accessibility" of a service point to every other point in a district for a location problem greater than shown in figure 13–15. Develop a model in which these accessibility measures are aggregated to a fuzzy measure for the "acceptability" of every district and are further aggregated to a fuzzy measure for the membership degree of every districting to the fuzzy set of "best districtings."

Discuss the sensitivity of such an approach to the choice of the intersection operator.

11. Discuss the possible use of expert systems and FLC models in operations research. Do those approaches satisfy sound OR principles?

14 EMPIRICAL RESEARCH IN FUZZY SET THEORY

14.1 Formal Theories vs. Factual Theories vs. Decision Technologies

The terms *model*, *theory*, and *law* have been used with a variety of meanings, for a number of purposes, and in many different areas of our life. It is therefore necessary to define more accurately what we mean by models, theories, and laws to describe their interrelationships and to indicate their use before we can specify the requirements they have to satisfy and the purposes for which they can be used. To facilitate our task we shall distinguish between definitions given and used in the scientific area and definitions and interpretations as they can be found in more application-oriented areas, which we will call "technologies" by contrast to "scientific disciplines." By technologies we mean areas such as operations research, decision analysis, and information processing, even though these areas call themselves sometimes theories (i.e., decision theory) or science (i.e., computer science, management science, etc.). This is by no means a value statement. We only want to indicate that the main goals of these areas are different. While the main purpose of a scientific discipline is to generate knowledge and to come closer to truth without making any value statements, technologies normally try to generate tools for solving problems better and very often by either accepting or basing on given value schemes.

Let us first turn to the area of scientific inquiry and consider the following quotation concerning the definition of the term *model*: "A possible realization in which all valid sentences of a theory T are satisfied is called a model of T."

Harré [1967, p. 86] calls: "A model, a, of a thing, A, is in one of many possible ways a replica or an analogue of A." And a few years later, "In certain formal sciences such as logic and mathematics a model for or of a theory is a set of sentences, which can be matched with the sentences in which the theory is expressed, according to some matching rule. . . . The other meaning of 'model' is that of some real or imagined thing or process, which behave similarly to some other thing or process, or in some other way than in its behavior is similar to it" [Harré, 1972, p. 173]. He sees two major purposes of models in science: (1) logical: to enable certain inferences, which would not otherwise be possible to be made; and (2) epistemological: that is, to express and enable us to extend our knowledge of the world. Models according to Harré are either used as a heuristic to simplify a phenomenon or to make it more readily handable and explanatorily where a model is a model of the real causal mechanism.

Leo Apostel [1961, p. 4] provides us with a very good example for various definitions of models as tuples of a number of components when defining: "Let then R (S, P, M, T) indicate the main variables of the modelling relationship. The subject S takes, in view of the purpose P, the entity M as a model of the prototype T." For the four components of the definition he gives a number of examples which are quite informative concerning the use of models in science and which can be summarized as follows:

Subjects (S) and purposes (P):

1. For a certain domain of facts let no theory be known. If we replace our study of this domain by the study of another set of facts for which a theory is well known, and that has certain important characteristics in common with the field under investigation then we use a model to develop our knowledge from a zero (or near zero) starting point.
2. For a domain D of facts, we do have a full-fledged theory but one too difficult mathematically to yield solutions, given our present techniques. We then interpret the fundamental notions of the theory in a model, in such a way that simplifying assumptions can express this assignment.
3. If two theories are without contact with each other we can try to use the one as model for the other or introduce a common model interpreting both and thus relating both languages to each other.

4. If a theory is well confirmed but incomplete we can assign a model in the hope of achieving completeness through the study of this model.
5. Conversely, if new information is obtained about a domain, to assure ourselves that the new and more general theory still concerns our earlier domain, we construct the earlier domain as a model of the later theory and show that all models of this theory are related to the initial domain, constructed as model, in a specific way.
6. Even if we have a theory about a set of facts, this does not mean that we have explained those facts. Models can yield such explanations.
7. Let a theory be needed about an object that is too big or too small, too far away, or too dangerous to be observed or experimented upon. Systems are then constructed that can be used as practical models, experiments which can be taken as sufficiently representative of the first system to yield the desired information.
8. Often we need to have a theory present to our mind as a whole for practical or theoretical purposes. A model realizes this globalization through either visualization or realization of a closed formal structure.

Thus, models can be used for theory formation, simplification, reduction, extension, adequation, explanation, concretization, globalization, action, or experimentation.

Entity (M) and model type (T):

M and *T* are both images or both perceptions or both drawings or both formalisms (calculi) or both languages or both physical systems. *M* can also be a calculus and *T* a theory or language, or vice versa.

Apostel believes that all models which can be constructed by varying the contents of the four components form a systematic whole: "Models are used for system restructuration because of their relations with the system (partial discrepancy); because of their relationship among each other (partial inconsistency at least multiplicity); because of their relationship with themselves (locally inconsistent or locally vague)."

By now two things should have become obvious:

1. There is a very large variety of types of models, which can be classified according to a number of criteria. For our deliberation one classification seems to be particularly important: The interpretation of a model as a "formal model" and the interpretation as a "factual, descriptive model." This corresponds to Rudolph Carnap's distinction between a logical and a descriptive interpretation of a calculus [Carnap 1946]. For him, a logically true interpretation of a model exists if, whenever a sentence is true, the second is equally true and whenever a sentence is refutable in the calculus, it is also false in the model. An inter-

pretation is a factual interpretation if it is not a logical interpretation, which means that whether a model is true or false does not depend only on its logical consistency but also on the (empirical) relationship of the sentences (axioms of the model) to the properties of the factual system of which the model is supposed to be an image. The second interpretation of a model is the one that is quite common in the empirical sciences and it is the one we will primarily be referring to in the following.

2. There is certainly a relationship between a model and a theory. This relationship, however, is seen differently by different scientists and by different scientific disciplines. We will now try to specify this relationship because theories, to our mind, are the focal point of all scientific activities.

For Harré [1972, p. 174] "A theory is often nothing but the description and exploitation of some model," or "Development of a theory on the other involves the superimposing of one model on another" [1967, p. 99].

White [1975] eventually simply points out that "There is a need to logically separate a model and a theory and that they play supporting roles in decision analysis, *viz.*, some theory is needed so that aspects of models can be tested and that some model is needed so that the affects of some changes can be examined. In particular validation of a model needs a theory." Thus, there seems to be a very intimate relationship between a model and a theory in scientific inquiry. Both, probably to varying degrees, base on hypotheses and these hypotheses can either be formal axioms or scientific laws. These scientific laws seem to us to fundamentally distinguish models and theories in scientific disciplines from the type of models (sometimes also called theories) in the more applied areas: "An experimental law, unlike a theoretical statement invariably possesses a determinate empirical content which in principal can always be controlled by observational evidence obtained by those procedures" [Nagel 1969, p. 83].

These laws as scientific laws assert invariance with respect to time and space. The tests to which such hypotheses have to be put before they can assert to be a law depend on the philosophical direction of the scientist. Karl Popper, as probably the most prominent representative of "critical rationalism," believes that laws are only testable by the fact of their falsifiability. Popper holds further that a hypothesis is "corroborated" (rather than confirmed) to the degree of severity of such tests. Such a corroborated hypothesis may be said to have stood up to the test thus far without being eliminated. But the test does not confirm its truth. A good hypothesis in science, therefore, is one that lends itself to the severest test, that is, one that generates the widest range of falsifiable consequences [Popper 1959].

14.1.1 Models in Operations Research and Management Science

The area of operations research will be considered as an example of a more application-oriented discipline, which I called "technology," in which modeling plays a predominant role. Even though one might dispute, whether operations research is a science or a technology, I will follow Symonds, who, as the President of the Institute of Management Science, stated: "Operations Research is the development of general scientific knowledge" [Symonds 1965, p. 385].

What, now, is a model in operations research? Most authors using the term *model* take it for granted that the reader knows what a model is and what it means. Arrow, for instance, uses the term *model* as a specific part of a theory, when he says: "Thus the model of rational choice as built up from pairwise comparisons does not seem to suit well the case of rational behaviour in the described game situation" [Arrow 1951]. He presumably refers to the model of rational choice, because the theory he has in mind does not give a very adequate description of the phenomena with which it is concerned, but only provides a highly simplified schema. In the social and behavioral sciences as well as in the technologies it is very common that a certain theory is stated in rather broad and general terms while models, which are sometimes required to perform experiments in order to test the theory, have to be more specific than the theories themselves. "In the language of logicians it would be more appropriate to say that rather than constructing a model they are interested in constructing a quantitative theory to match the intuitive ideas of the original theory" [Suppes 1961]. Rivett, in his book *Principles of Model Building* [Rivett 1972], offers three different kinds of classifications of models; when enumerating the models he suggests be put into the different classes, he no longer uses the term *model* but talks of "problems in this area" and "the theory of this area" as a not-too-well-defined entity of knowledge. Ackoff suggests as a model of decision making a six-phases-process that is supposed to be a good picture (model) of the real decision-making process [Ackoff 1962]. This is only one example of quite a number of very similar models of decision making.

If we consider the size of some of the models used in operations research, containing more than 10,000 variables and thousands of constraints, we can easily see what does not distinguish a theory from a model: It is not the complexity, it is not the size, it is not the language, and it is not even the purpose. In fact, there seems to be only a gradual distinction between theory and model. While a theory normally denotes an entire area or type of problem, it is more comprehensive but less specific than a model (e.g., decision theory, inventory theory, queueing theory, etc.), a model most often refers to a specific context or situation and is meant to be a mapping of a problem, a system, or a process. By contrast to a scientific theory, containing scientific laws as hypotheses, a model normally does not assert invariance with respect to time and space but requires modifications whenever the specific context, for which a model was constructed, changes.

In the following we will concentrate on models rather than on theories. Realizing that there is quite a variety of types of models, we do not think that it is important and necessary for our purposes to distinguish models by their language (mathematics or logic is considered to be a modeling language), by area, by problem type, by size, and so on. One classification, however, seems to be important: the distinction of models by their character. Scientific theories were already divided into formal theories and factual theories. For models, particularly in the area of the technology in which values and preferences enter our considerations, we will have to distinguish among the following:

1. *Formal models.* These are models that are purely axiomatic systems from which we can derive if-then statements and the hypotheses of which are purely fictitious. These models can only be checked for consistency, they can neither be verified nor falsified by empirical arguments.
2. *Factual models.* These models include in their basic hypotheses falsifiable assumptions about the object system, that is, conclusions drawn from these models have a bearing on reality and they, or their basic hypotheses, have to be verified or can be falsified by empirical evidence.
3. *Prescriptive models.* These are models that postulate rules according to which processes have to be performed or people have to behave. This type of model will not be found in science but it is a common type of model in practice.

The distinction between these three different kinds of models is particularly important when using models: All three kinds of models can look exactly the same but the "value" of their outputs is quite different. It is therefore rather dangerous not to realize which type of model is being used because we might take a formal model to be a factual model or a prescriptive model to be a factual model and that can have quite severe consequences for the resulting decision.

As an example, let us look at the above-mentioned Ackoff model of decision making. Is that a formal, a factual, or a prescriptive model? If it is a formal model, we can not derive from it any conclusion for real decision making. If it is a factual model then it would have to be verified or falsified before we can take it as a description of real decision making. The assertion, however, that decision making proceeds in phases was already empirically falsified in 1966 [Witte 1968]. Still, a number of authors stick to this type of model. Do they want to interpret their model as a prescriptive model? This would only be justified, if they could show that, for instance, decision making can be performed more efficiently when done in phases. This, however, has never been shown empirically. Therefore, we can only conclude that authors suggesting a multiphase scheme as a model for decision making take their suggestion as a formal model and do not want to make any statement about reality, or that they are using a falsified, that is, invalid and false, factual model.

14.1.2 Testing Factual Models

The quality of a model depends on the properties of the model and the functions the model is designed for. In general, models will have to have at least the following three major properties: formal consistency, usefulness, and efficiency. By *logical consistency* we mean that all operations and transformations have been performed properly and that all conclusions follow from the hypothesis. This consistency has to be demanded of all types of models, whether they are formal, factual, or prescriptive. By *usefulness* we mean that the model has to be helpful for the function for which it has been designed. By *efficiency* we mean that the model, as the tool to achieve an end, has to fulfill the desired function at a minimum of effort, time, and cost.

In decision making and problem solving factual models will be needed to describe, to explain, and to predict phenomena and consequences. For "conditional predictions" formal models will also be useful in order to obtain "if-then statements," for instance, in the framework of simulation. Formal models will also be useful and necessary for the area of communication within the decision-making process and for relaying the resolutions or conclusions of the decision or problem-solving process to the "actors." One should assume that prescriptive models are the most common in decision making. This, however, is only true if one calls all "decision models," that is, models which contain an objective function, by which an optimal solution can be determined, prescriptive models. To our mind this is not quite appropriate because these kinds of models only prepare suggestions for possible decisions and the normative or prescriptive character is acquired only after the "solution" has been declared a decision by the authorized decision maker. A much more important feature of these models, it seems to us, is that they have to describe or define properly the conditions that limit the action space (such as capacities, financial resources, legal restrictions, etc.).

We can now restate the notion of the quality of a model more precisely: we already mentioned that consistency is one of the necessary conditions for quality. Usefulness of a model will have to be defined for each of the three different types of models differently:

1. While a factual model can be called useful, if it is "factually true" (by contrast to logically true), that is, if it maps the object system with an appropriate precision (which can only be tested empirically), the model also has to generate knowledge, that is, the user of a model should gain knowledge he would not have gained without using the model or which he did not have available before using the model.
2. Formal models can be neither verified nor falsified empirically. Such a model will be considered useful, if activities, such as teaching, explaining, communication, become more efficient by using the model than without it.
3. Prescriptive models can also not be verified or falsified. They are the more useful, the more effectively they help to enforce the desired behavior, to control predefined performance measures, and to define ranges within which decision makers have freedom to decide.

Two prime factors in modeling are the modeling language and the quality of input data. The type of modeling language appropriate for models in decision making was already discussed in chapter 1. Here we shall elaborate some more on the quality of input data.

The saying "garbage in—garbage out" is well known and speaks for itself. The following quotation from Josiah Stamp [1975, p. 236] points in the same direction: "Governments are very keen in amassing statistics. They collect them, add them, raise them to the *n*th power, take the cube root and make wonderful diagrams. But what you must never forget is that every one of these figures comes in the first instance from the village watchman who just puts down what he damn pleases."

It must, however, be born in mind that the effort put into deriving and obtaining numerical values or relations must be geared to the value of the model, and that when data is scarce it may still be useful to draw conclusions from not fully satisfactory input data. In this case a tentative look at the dependence of the solution from the quality of the input data may be very advisable.

The quality of the input data is closely related to the question of operational definitions for the relevant variables. The processes of defining variables and their operational indicators and measurement are intertwined. To quote White [1975, p. 102] "We take 'measurement' to be a special aspect of a 'definition'." One might take the view that measurement is the actual procedure for assigning the real numbers that constitute the measure. However, as pointed out in a previous section, this is the quantification process and in itself does not constitute a measure unless it is a homomorphism. The homomorphism then defines the measure. Very often when modeling in the area of social sciences one will find that relations, data, or values are stated in very vague ways. Goals, for instance, may be stated as "trying to achieve satisfactory profits," data such as "the South of the country is much poorer than the North," and relations such as "his investment strategies were much more risky than those of his competitors." Very often these variables are measured subjectively and point scales are used to transform the "measurements" into numerical values. Even though it is necessary to include in the model variables that are considered important but very hard to operationalize and measure, the quality of the input data might have very limiting effects on their transformation which can be permitted in the model. Rather than neglecting these kinds of data one should consciously determine which scale quality these data have and then make sure that only admissible transformations are being used when processing these data in the model. Table 14–1 sketches the hierarchy of scale levels including the permissible transformations for each of the levels.

The testability of the components of a model—in the scientific and in the practical context—depends largely on the operational definition of the hypotheses. In this sense, observation and formal analysis prior to model building can very often improve the testability of hypotheses. Let us illustrate this with the following example. In decision analysis one normally distinguishes among decision making under certainty, decision making under risk, and decision making under uncertainty. One assumes that in decision making under risk the decision maker is able to store and process probability distribution functions. Here probabilities ought to be interpreted as Koopman-type probabilities—that is, probabilities as expressions of belief and not in the frequentistic sense. This hypothesis is hardly testable because a situation of decision making under risk is not homogenous with respect to the available information at all. An improvement in

Table 14–1. Hierarchy of scale levels.

	Permissible transformation			
Type of scale	*Verbal*	*Formal*	*Invariance*	*Example*
Nominal scale	One-to-one function	$x_i \neq x_j \rightarrow x_i' \neq x_j'$	Uniqueness of values	License plates
Ordinal scale	Monotonic increasing function	$x_i \leq x_j \rightarrow x_i' \leq x_j'$	Rankorder of values	Marks
Interval scale	Affine function	$x' = a \cdot x + b$	Ratio of differences	Temperature (C°, F°)
Ratio scale	Similarity function	$x' = a \cdot x$	Ratio of values	Length (cm, inch)
Absolute	Identity	$x' = x$	Values	Frequency

the testability of hypotheses could be achieved if one would distinguish, for instance, among:

1. Decision making when quantitative probabilities are known (interval-scale).
2. Decisions when interval probabilities are known (hyperordinal scale).
3. Decisions when qualitative probabilities are known (ordinal scale).
4. Decisions when partially ordered nominal probabilities are known (ordinal scale).
5. Decisions when nominal probabilities are known (states are known but not truth ratable).
6. Decisions when only some of the nominal probabilities are known.

It is obvious that the information storage and processing requirements a human would need in order to decide "rationally" are quite different in the above cases and that the permissible operations in the model will also be different depending on the type of probability that can be assumed to exist.

If the testing is done on the basis of the outputs of the analysis the decision maker might already be able to indicate that the output of the analysis is not satisfactory, probably because important relations or variables have been omitted. If the decision maker or expert rates the output of the model as satisfactory, it gains the status of face-validity, sometimes in practice the most we can hope for.

Ideally a model should now be tested by implementation, that is, by comparing actual with predicted results. This, however, in many instances is impossible for several reasons.

1. *Changes of environment:* Factors such as sales, price levels, and so on might have changed while the model was built and implemented and therefore the observed results after implementation of the model can no longer be compared with the predicted results.
2. *Changes in performance:* If, for instance, the model is tested after implementation by running the old procedure parallel to the model and if the old procedure included human activities, the performance of these activities might be improved by the persons because they know that the "new" model is being compared with their performance, which would probably drop again, if and when the operation of the new procedures would be terminated.
3. *Risk and uncertainty:* It is obvious that if procedures have been designed to optimally decide in situations of risk or uncertainty, the "real" results cannot meaningfully be compared with the probabilistic prediction.
4. *Optimality:* If only one solution is actually implemented there is, of course, no way to compare this with other alternatives. In many cases the optimal solution with which the model solution could be compared is not known at all because it is not computable or because optimality was defined subjectively in a way that is not objectively reproducible.

It has already been pointed out that all kinds of theories and models can be and ought to be tested for consistency. In formal analysis it might even be possible to prove consistency, which does not mean that models and theories, the consistency for which has not yet been proven, are not formally correct. For "factual" or "substantial" theories and models empirical testing of basic hypotheses, relations, and resulting outputs is absolutely necessary in order to achieve a certain degree of confirmation of the theory or the model. This fact is often neglected when working with theories and models. If, for instance, the hypothesis of "rationality" in decision-making models is "justified" by defining rationality by more basic axioms such as transitivity, reflexibility, existence of an ordering, and so on, which seem quite plausible and natural, then the model or the theory might become more testable but certainly not better confirmed. To confirm the model would require empirically testing either the main hypothesis or the presumably more operational basic axioms. This, of course, still does not determine uniquely the methods that can be used for testing hypotheses. These methods will depend on the area in which the model is being used (physics, engineering, management) and the purposes for which the model has been built. Thus, in scientific inquiry probabilistic tests might not be acceptable because scientific laws assert deterministic invariance. These methods, however, might be the only available ones for testing models in areas such as management, sociology, and political decision making.

In the following we shall report on empirical research concerning two main components of fuzzy set theory: Membership functions and operators (connectives, aggregators).

14.2 Empirical Research on Membership Functions

Measurement means assigning numbers to objects, such that certain relations between numbers reflect analogous relations between objects. With other words, measurement is the mapping of object relations into numerical relations of the same type.

If it is possible to prove that there is a homomorphic mapping $f: E \rightarrow N$ from an empirical relational structure $\langle E, P_1, \ldots, P_n \rangle$ with a set of objects E and an n-tuple of relations P_i into a numerical relational structure $\langle N, Q_1, \ldots, Q_n \rangle$ with a set of numbers N and relations Q_i, then a scale $\langle\langle E, N, f \rangle\rangle$ exists. By specifying the admissible transformations the grade of uniqueness is determined.

Therefore measurement starts by formulating the properties of the empirical structure; implicitly the intended object space is modeled on a non-numerical level. Strictly speaking, at the very beginning there should be a semantic definition of the central concepts. This would considerably facilitate the consistent use of the relevant principles. Unfortunately this has not yet been possible for the concept of membership. Membership has a clear cut formal definition. However, explicit requirements for its empirical/experimental measurement are still missing. Under these circumstances it is not surprising that apart from first steps by Norwich and Turksen [1981] genuine measurement structures have not yet been developed.

Under these circumstances one could wait and see, until a satisfactory definition is available. However, one should remember that up to the beginning of the twentieth century even in the "hard sciences" measures were used without being equipped with adequate measurement theories. Usually the measurement tools used were based on not much more but plausible reasons. Nevertheless, the success of the natural sciences is undoubted. Hence, for the purpose of empirical research it may be tolerable to use plausible techniques.

Firstly such a scale can serve as an operational definition of membership.

Secondly a specific concept can be criticized and hence may help to obtain useful improvements. We shall present two models for membership functions. Let us call the first "Type A-model" and the second "Type B-model."

14.2.1 Type A-Membership Model

Of prime importance is the determination of the lowest necessary scale level of membership for a specific application. The purpose of the model A-membership was to empirically investigate aggregation operators. In this instance it was sufficient to determine degrees of membership for a predefined set of objects rather than continuous membership functions. The requested scale level should be as low as possible in order to facilitate data acquisition, which usually affords the participation of human beings. On the other hand a suitable numerical handling is desirable in order to insure mathematically appropriate operating. Regarding the five classical scale types—nominal, ordinal, interval, ratio, and absolute scale—the interval scale level seems to be most adequate. In this respect we can not follow Sticha, Weiss, and Donnell [1979] who assert that membership has to be measured on an ordinal scale. Usually the intended mathematical operations require at least interval-scale quality.

The easiest way to obtain data is to ask some subjects directly for membership values. However, it is well known that scales developed by using the so called direct methods may be distorted by a number of response biases [Cronbach 1950]. On the other hand, indirect methods work on the basis of much weaker assumptions using ordinal judgments only. Their advantages are simplicity and robustness with respect to response biases.

Their disadvantage is that many judgments are needed since the ordinal judgment provides relatively little information. This seemed acceptable in order to avoid distortions of the data. Thus we decided to use a method that yields an interval scale on the basis of ordinal ratings: After a set of suitable objects has been established, subjects are asked for the grades of membership on a percentage scale. People are accustomed to this type of judgment and division by 100 provides the normalized 0–1 values. The obtained data are interpreted as ranks. The subsequent scaling procedure refers mainly to a method suggested by Diederich, Messick, and Tucker [1957] based on Thurstone's "Law of Categorical Judgement" [Thurstone 1927].

A detailed description of the method can be found in Thole, Zimmer-

Table 14–2. Empirically determined grades of membership.

Stimulus x		$\mu_{\tilde{M}}(x)$	$\mu_{\tilde{C}}(x)$	$\mu_{\tilde{M} \cap \tilde{C}}(x)$
1.	bag	0.000	0.985	0.007
2.	baking-tin	0.908	0.419	0.517
3.	ball-point-pen	0.215	0.149	0.170
4.	bathing-tub	0.552	0.804	0.674
5.	book wrapper	0.023	0.454	0.007
6.	car	0.501	0.437	0.493
7.	cash register	0.692	0.400	0.537
8.	container	0.847	1.000	1.000
9.	fridge	0.424	0.623	0.460
10.	hollywood-swing	0.318	0.212	0.142
11.	kerosene lamp	0.481	0.310	0.401
12.	nail	1.000	0.000	0.000
13.	parkometer	0.663	0.335	0.437
14.	pram	0.283	0.448	0.239
15.	press	0.130	0.512	0.101
16.	shovel	0.325	0.239	0.301
17.	silver-spoon	0.969	0.256	0.330
18.	sledge-hammer	0.480	0.012	0.023
19.	water-bottle	0.564	0.961	0.714
20.	wine-barrel	0.127	0.980	0.185

mann, and Zysno [1979]. Table 14–2 illustrates the type of membership information that was obtained and the type of objects used for experimentation. The transformation of the observed information to degrees of membership was performed by a computer program written for this purpose.

14.2.2 Type B-Membership Model

Often a certain concept can be considered as a context-specific version of a more general feature. For instance, the set of young men is a subset of all objects with the feature age. We shall call this general feature "base variable." This coincides with the definition of a base variable in definition 9–1. The scale of the base variable which is normally generally accepted (here age in years) will be called a "judgmental scale." By contrast to the scale of the base variable the scale of the "specific version" is context-dependent. Thus a term in definition 9–1 does not necessarily correspond to "the specific version" of the base variable, because "terms" did not explicitly assume a specific context. If the term *young* refers to the age of men (by contrast to the age of flies or cars, houses or dinosaurs), then we can assume that the observer has some idea about what "young" means with respect to men. He has a "standard" with respect to which he evaluates age in terms of "young," "old," et cetera. We shall, therefore, call this specific context-dependent scale an "evaluational scale." If there exist a judgmental scale and an evaluational scale, both referring to the same empirical relational structure, then a mapping from one numerical relative into the other which reflects the differences of the basic empirical relational structure with respect to the same set of elements would be possible. If, on the other hand, the scale of, for instance, the base variable and the mapping (function) was known, then the scale of the special feature could be determined. The mapping (function) can be considered as the membership function, which has to be determined. Concerning the required scale level of the membership function, essentially the same holds as for type A model. By contrast to model A we used, however, direct scaling methods. This involves less effort and is justified by the existence of the base variable which provides extra control with respect to judgmental errors of the subjects. The judgmental (valuation) of membership can be regarded as the comparison of object x with a standard (ideal) which results in a distance $d(x)$. If the object corresponds fully with the standard, the distance shall be zero; if no similarity between standard and object exists, the distance shall be "∞." If the evaluation concept is represented formally by a fuzzy set $\widetilde{P} \subset X$, then a certain degree of membership $\mu_{\widetilde{P}}(x)$ is assigned to each element x. We shall assume that this degree of membership is a function of the "distance," d, between the two above-mentioned scales ($\widetilde{P}$ representing a context-dependently defined fuzzy set as a subset of the universe X).

Thus we define

$$\mu_{\widetilde{P}}(x) = \frac{1}{1 + d(x)} \tag{14.1}$$

where $d(x)$ is the "distance" of the two scales for the element $x \in X$. The distance function now has to be specified. A specific monotonic function of the similarity with the ideal could as a first approximation be $d'(x) = 1/x$.

Experience shows, however, that ideals are very rarely ever fully realized. As an aid to determine the relative position, very often a context-dependent standard b is created. It facilitates a fast and rough preevaluation such as "rather positive," "rather negative," and so on. As another context-dependent parameter we can use the evaluation unit a, similar to unit of length such as feet, meters, yards, and so on. If one realizes furthermore that the relationship between physical unit and perceptions is generally exponential [Helson 1964], then the following distance function seems appropriate:

$$d(x) = \frac{1}{e^{a(x-b)}} \tag{14.2}$$

Substituting (14.2) into (14.1) yields the logistic function

$$\mu_{\tilde{P}}(x) = \frac{1}{1 + e^{-a(x-b)}} \tag{14.3}$$

It is S-shaped such as demanded by several authors [Goguen 1969; Zadeh 1971]. Formally b is the inflexion point and a is the slope of the function.

From the point of view of linear programming (14.3) has the additional advantage, that it can easily be linearized by the following transformation:

$$-\ln \frac{1-\mu}{\mu} = \ln \frac{\mu}{1-\mu} = a(x-b) \tag{14.4}$$

where μ stands for $\mu_{\tilde{P}}(x)$.

The parameters a and b will have to be interpreted differently depending on the situation which is modelled. From a linguistic point of view a and b can be considered as semantic parameters.

Model (14.3) is still too general to fit subjective models of different persons. Frequently only a certain part of the logistic function is needed to represent a perceived situation. This is also true for measuring devices such as scales, thermometers, and so on, which are designed for specific measuring intervals only.

In order to allow for such a calibration of our model we assume that only a certain interval of the physical scale is mapped into the open interval (0, 1) (see figure 14–1). Whenever stimuli are smaller or equal to the lower bound or larger or equal to the upper bound the grade of membership of 0 or 1 respectively is assigned to them. This is achieved by changing the range by legitimate scale transformations such that the desired interval is mapped into [0, 1].

Since we reqested an interval scale the interval of the degrees of membership may be transformed linearly. On this scale level the ratios of two distances are invariant. Let $\bar{\mu}$ and $\underline{\mu}$, respectively, be the upper and lower bounds of the normalized membership scale and μ_i a degree of membership between these bounds, $\underline{\mu} < \mu_i < \bar{\mu}$, and let $\underline{\mu}'$, μ_i', $\bar{\mu}'$ be the corresponding values on the transformed scale. Then

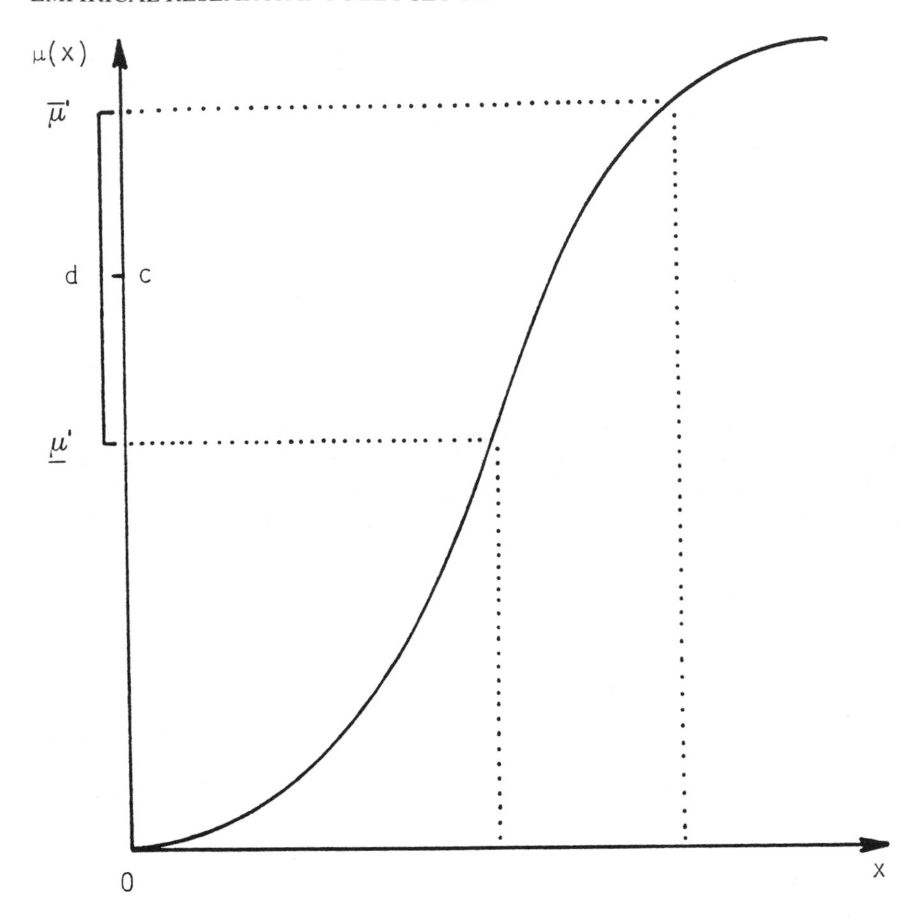

Figure 14–1. Calibration of the interval for measurement.

$$\frac{\mu_i - \underline{\mu}}{\overline{\mu} - \underline{\mu}} = \frac{\mu_i' - \underline{\mu}'}{\overline{\mu}' - \underline{\mu}'} \tag{14.5}$$

For the normalized membership function we have $\underline{\mu} = 0$ and $\overline{\mu} = 1$. Hence

$$\mu_i' = \mu_i(\overline{\mu}' - \underline{\mu}') + \underline{\mu}' \tag{14.6}$$

Generally it is preferable to define the range of validity by specifying the interval d with the center c as shown in figure 14–1.

Hence

$$\overline{\mu}' = d + \underline{\mu}' \tag{14.7}$$

and

$$\underline{\mu}' = 2c - \overline{\mu}' \tag{14.8}$$

Substituting (14.7) into (14.8) yields

$$\underline{\mu}' = 2c - d - \underline{\mu}' \tag{14.9}$$

Solving (14.9) for $\underline{\mu}'$ gives

$$\underline{\mu}' = c - d/2 \tag{14.10}$$

and inserting (14.10) and (14.7) into (14.6) yields

$$\mu_i' = d(\mu_i - 1/2) + c \tag{14.11}$$

The general model of membership (14.3) is specified by two parameters of calibration, if μ_i is replaced by μ_i'. Solving this equality for μ_i leads to the complete model of membership:

$$\mu_i = \overset{0}{\Bigg\lceil} \left(\frac{1}{1 + e^{-a(x-b)}} - c \right) \frac{1}{d} + \frac{1}{2} \overset{1}{\Bigg\rceil} \tag{14.12}$$

$\overset{0}{\lceil} \cdot \overset{1}{\rceil}$ indicates that values outside of the interval [0, 1] have no real meaning. The measurement instrument does not differentiate there. Hence

$$x < \underline{x} \rightarrow \mu(x) = 0$$
$$x > \overline{x} \rightarrow \mu(x) = 1 \tag{14.13}$$

The determination of the parameters from empirical data bases does not pose any difficulties in the general model (14.3). It should be mentioned that not only monotonic functions, such as discussed so far, can be described, but so can unimodal functions—by representing them by an increasing (S_I) and a decreasing (S_D) part. Formally they can be represented as the minimum or maximum, respectively, of two monotonic membership functions each:

$$\mu_{S_I S_D}(x) = \min \overset{0}{\lceil} \mu_{S_I}(x), \mu_{S_D}(x) \overset{1}{\rceil}$$
$$\mu_{S_I S_D}(x) = \max \overset{0}{\lceil} \mu_{S_I}(x), \mu_{S_D}(x) \overset{1}{\rceil}$$

A computer program was written to process the observed data.

Type B-model for membership functions, providing a membership function rather than degrees of membership for single elements of a fuzzy set (as Type A does), was also empirically tested.

We shall present results concerning a very common fuzzy set, "young men," "old men," and so on. Having available membership functions we could also test models of modifiers such as "very."

The evaluation of the data showed a good fit of the model. Figures 14–2 through 14–7 show the membership functions given by six different persons. As can be seen, the concepts "vym" and "ym" are realized in the monotonic type as well as in the unimodal. The detailed data and results can be found in a major report of the authors [Zimmermann and Zysno 1982].

One may ask whether a general membership function for each of the four sets can be established. Though the variety of conceptual comprehension is rather remarkable, there should be an overall membership function at least in order to have a standard of comparison for the individuals. This is achieved by determining the common parameter values a, b, c, and d for each set. Obviously the general membership functions of "old man" and "very old man" are rather similar (see figures 14–8 and 14–9). Practically, they differ only with respect to their inflection points,

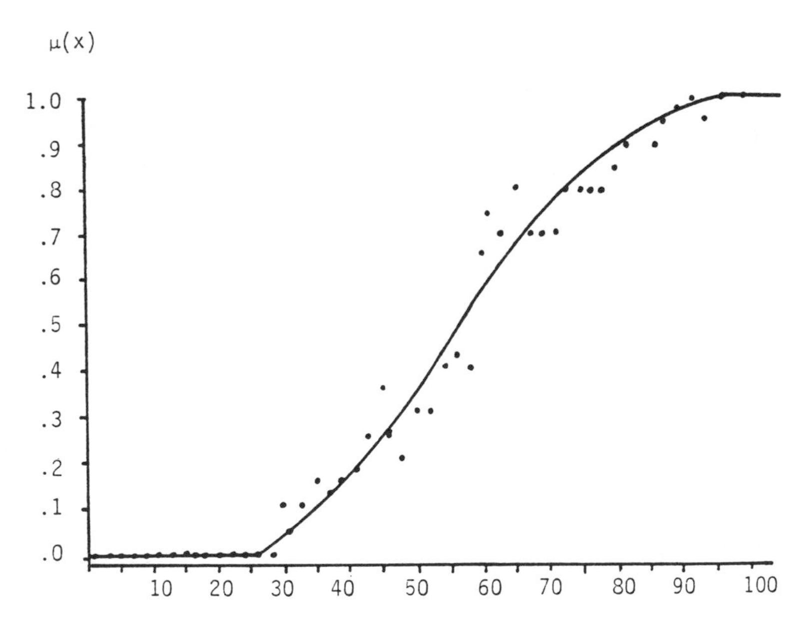

Figure 14–2. Subject 34, "Old Man."

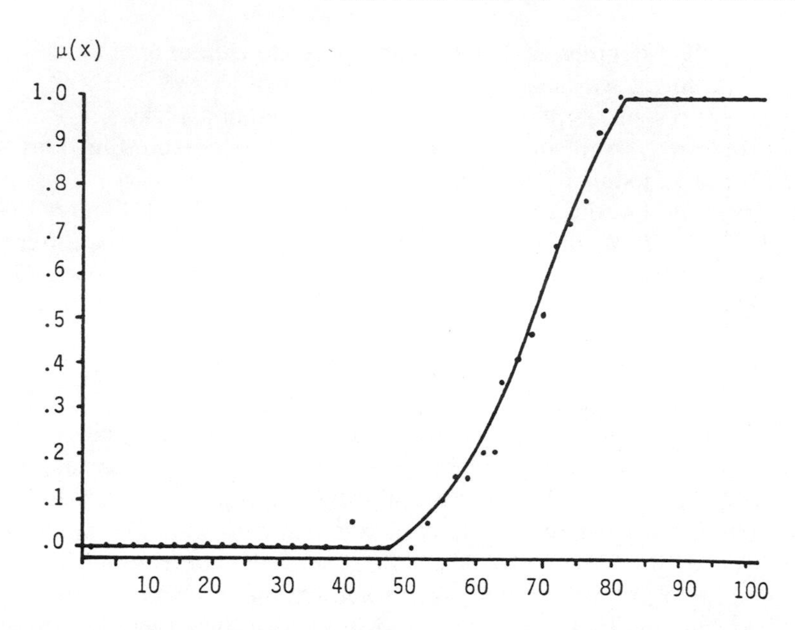

Figure 14–3. Subject 58, "Very Old Man."

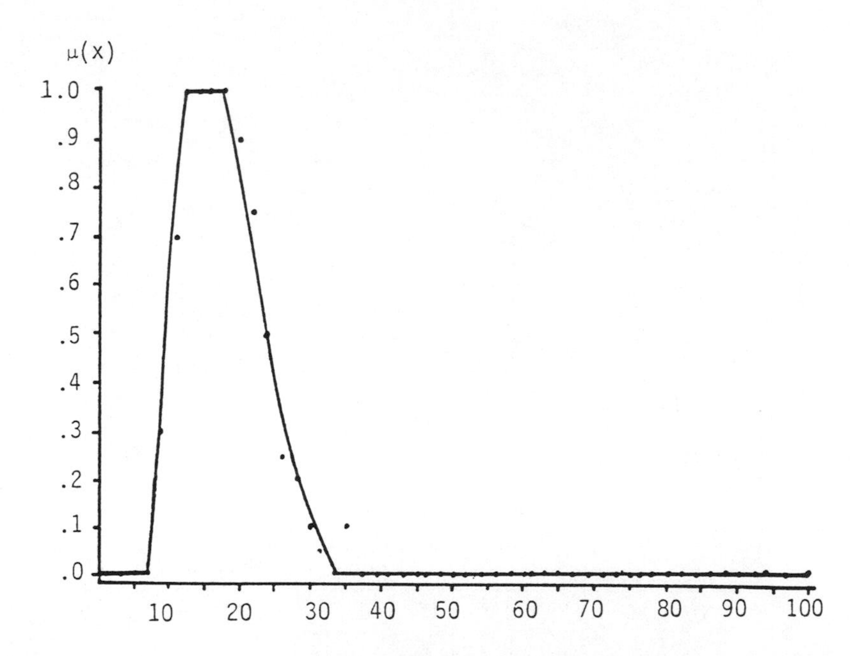

Figure 14–4. Subject 5, "Very Young Man."

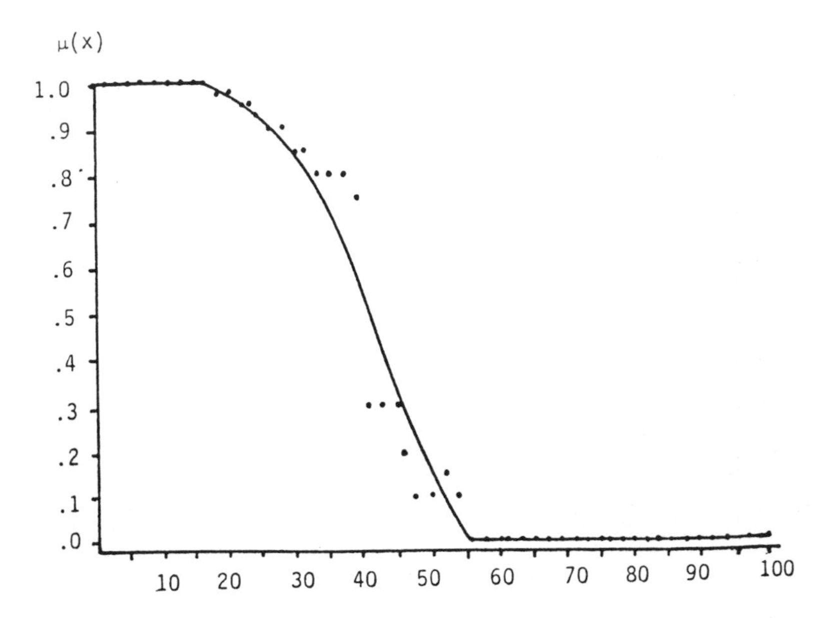

Figure 14–5. Subject 15, "Very Young Man."

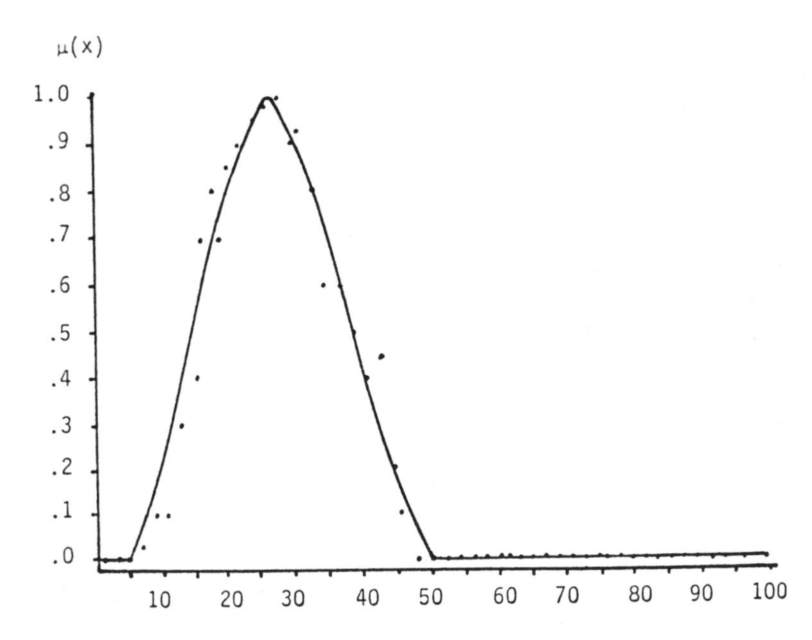

Figure 14–6. Subject 17, "Young Man."

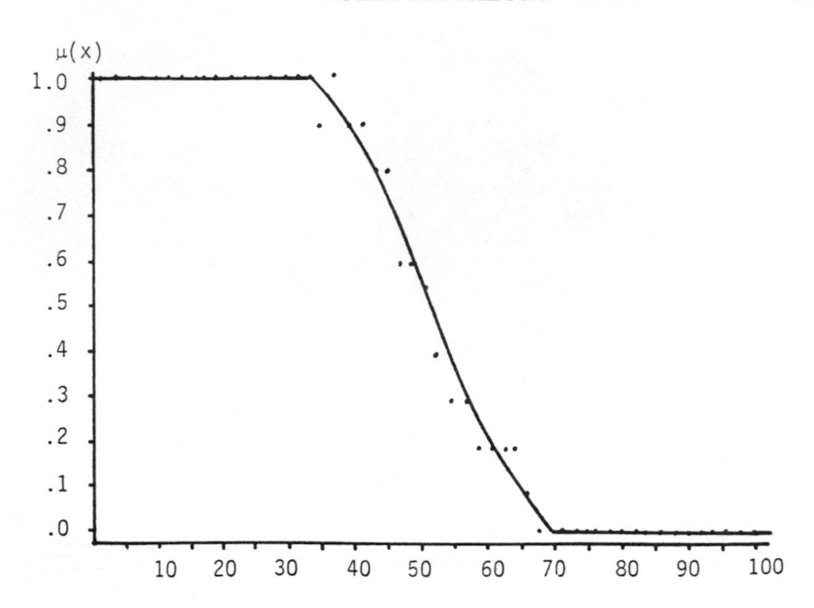

Figure 14–7. Subject 32, "Young Man."

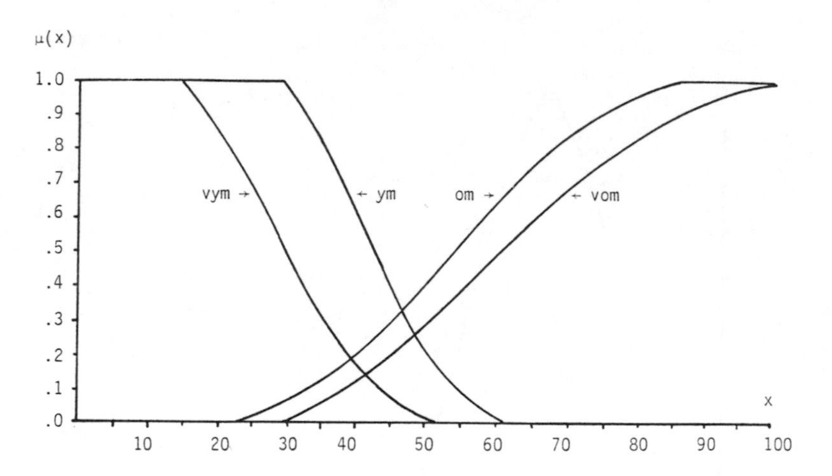

Figure 14–8. Empirical membership functions "Very Young Man," "Young Man," "Old Man," "Very Old Man."

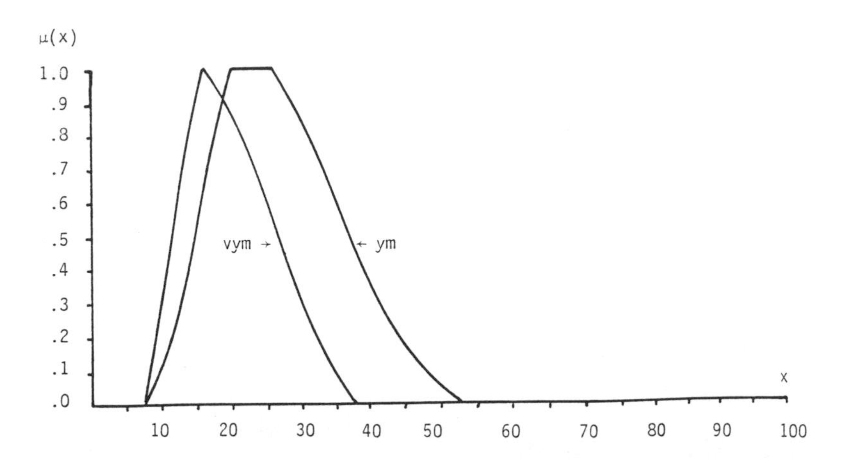

Figure 14–9. Empirical unimodal membership functions "Very Young Man," "Young Man."

indicating a difference of about five years between "old man" and "very old man." The same holds for the monotonic type of "very young man" and "young man"; their inflection points differ by nearly 15 years. It is interesting to note that the modifier "very" has a greater effect on "young" than on "old," but in both cases it can be formally represented by a constant. Several subjects provided the unimodal type in connection with "very young" and "young." Again the functions show a striking congruency.

14.3 Empirical Research on Aggregators

In section 3.2.2 a number of possible operators were mentioned. We saw that they were assigned in various ways to set theoretic operations, such as intersection, union, etc. For some of these operators axiomatic formal justifications were also given. In definition 12–1 the triple decision-intersection-min-operator was used. Some indication was given there, that from a factual point of view this triple might turn out not to be true. After what has been said in section 14.1 it should be obvious that for a factual use of fuzzy set models only empirical verification of models for the aggregators is appropriate. This can only be done in specific contexts and the results will therefore be of limited validity.

Some empirical testing of aggregators has been performed in the context of fuzzy control. We shall report on empirical research done in the context of human evaluation and decision making, that is, concerning the question "How do human beings aggregate subjective categories and which mathematical models describe this procedure adequately?"

As already mentioned, the term *decision* has been defined in many different ways. A decision also has many different aspects, for example, the logical aspect, the information-processing aspect, etc. We shall focus our attention on the last aspect: The search for and the modeling, processing, and aggregation of information. A decision in the sense of definition 12–1, rather than being some kind of optimization, is the search for an action that satisfies all constraints and all aspiration levels representing goals. "Deciding" about the creditworthiness of a person might be called an "evaluation" rather than a decision. It means, however, checking on whether a person satisfies all aspiration levels concerning security, liquidity, business behavior, and so on.

In the following we will roughly describe two experiments and their results: The first experimental design started from the triple "decision-intersection-min-operator" and tried to find out whether the min-operator was adequate for modeling the intersection. However, it did not question the pair "decision-intersection." The second experiment is no longer limited to considering a decision as the intersection; it relinquishes the set-theoretic interpretation of a decision altogether.

Test 1: Intersection-min-operator [Thole et al. 1979]

Two fuzzy sets, $\tilde{A}$ and $\tilde{B}$, were considered. It seems reasonable to demand that the following conditions concerning the judgmental "material" are satisfied:

1. The attributes characterizing the members of the sets $\tilde{A}$ and $\tilde{B}$ are independent, that is, some magnitude of $\mu_{\tilde{A}}$ is not affected by some magnitude of $\mu_{\tilde{B}}$ and vice versa. As an operational criterion for this kind of independence a correlation of zero is demanded:

$$r_{\mu_{\tilde{A}}\mu_{\tilde{B}}} = 0$$

2. If $\mu_{\tilde{A}\cap\tilde{B}}$ represents the aggregation of $\mu_{\tilde{A}}$ and $\mu_{\tilde{B}}$, modeling the intersection, and if $w_{\tilde{A}}$ and $w_{\tilde{B}}$ are weights, then $\mu_{\tilde{A}\cap\tilde{B}}$ can be described by

$$\mu_{\tilde{A}\cap\tilde{B}} = (w_{\tilde{A}}\mu_{\tilde{A}}) \circ (w_{\tilde{B}}\mu_{\tilde{B}})$$

where $\circ$ stands for some algebraic operation. But as the models proposed do not take into account different importance of the sets with respect to their intersection, equal weights are demanded:

$$w_{\tilde{A}} = w_{\tilde{B}}$$

As operational criterion for equal weights equal correlations are demanded:

$$r_{\mu_{\tilde{A}}\mu_{\tilde{A}\cap\tilde{B}}} = r_{\mu_{\tilde{B}}\mu_{\tilde{A}\cap\tilde{B}}}$$

With regard to these conditions three fuzzy sets were chosen: "metallic object" [Metallgegenstand], "container" [Behälter], and "metallic container" [Metallbehälter].[1] It has to be proved that these sets satisfy the conditions mentioned above.

Now the following hypotheses may be formulated: Let $\mu_{\tilde{M}}(x)$ be the grade of membership of some object x in the set "metallic object" and $\mu_{\tilde{C}}(x)$ be the grade of membership of x in the set "container," then the grade of membership of x in the intersection set "metallic container" can be predicted by

$$H_1: \quad \mu_{\tilde{M}\cap\tilde{C}}(x) = \min\{\mu_{\tilde{M}}(x), \mu_{\tilde{C}}(x)\}$$

$$H_2: \quad \mu_{\tilde{M}\cap\tilde{C}}(x) = \mu_{\tilde{M}}(x) \cdot \mu_{\tilde{C}}(x)$$

A pretest was carried out in order to guarantee that the assumptions were justified.

Sixty students at the RWTH Aachen from 21 to 33 years of age, all of them native speakers of the German language, served as unpaid subjects in the main experiment. Each subject was run individually through two experimental sessions, the first one taking about twenty minutes, the second one about forty minutes. In order to eliminate influences of memory as far as possible, the interviews were performed at an interval of approximately three days.

Each subject was asked to evaluate each of the objects with respect to being a member of $\tilde{A}$ (metallic object), $\tilde{B}$ (container), and $\tilde{A} \cap \tilde{B}$ (metallic container). The three resulting membership scales are shown in table 14–2.

Now, what about the prediction of the empirical data for "metallic container" by the two candidate rules? Table 14–3 shows the empirical results

[1] This investigation has been carried out in Western Germany. In brackets you find the corresponding German word. It should be realized that the German language allows the forming of compound words. Hence the intersection is labeled by one word.

Table 14–3. Empirical vs. predicted grades of membership.

Stimulus x	$\mu_{\tilde{M}\cap\tilde{C}}(x)$	$\mu_{\tilde{M}\cap\tilde{C}}(x)\mid\text{min}$	$\mu_{\tilde{M}\cap\tilde{C}}(x)\mid\text{prod.}$
1. bag	0.007	0.000	0.000
2. baking-tin	0.517	0.419	0.380
3. ball-point-pen	0.170	0.149	0.032
4. bathing-tub	0.674	0.552	0.444
5. book wrapper	0.007	0.023	0.010
6. car	0.493	0.437	0.219
7. cash register	0.537	0.400	0.252
8. container	1.000	0.847	0.847
9. fridge	0.460	0.424	0.264
10. Hollywood-swing	0.142	0.212	0.067
11. kerosene lamp	0.401	0.310	0.149
12. nail	0.000	0.000	0.000
13. parkometer	0.437	0.335	0.222
14. pram	0.239	0.283	0.127
15. press	0.101	0.130	0.067
16. shovel	0.301	0.293	0.078
17. silver-spoon	0.330	0.256	0.248
18. sledge-hammer	0.023	0.012	0.006
19. water-bottle	0.714	0.546	0.525
20. wine-barrel	0.185	0.127	0.124

together with the grades of membership computed by using the min-operator and the product-operator, respectively.

Figures 14–10 and 14–11 show graphically the relationship between empirical and theoretical grades of membership. The straight line indicates locations of perfect prediction, that is, if the operator makes perfect predictions and the data are free of error, then all points lie on the straight line.

The question arises: Are the observed deviations small enough to be tolerable? To answer this question we chose two criteria. They are:

1. if the mean difference between observed and predicted values is not different from zero ($\alpha = 0.25$; two-tailed) and
2. if the correlation between observed and predicted values is higher than 0.95, the connective operator in question should be accepted.

As the observed differences are normally distributed we used the student t as test statistic. It is entered by the mean of the population (in this case:

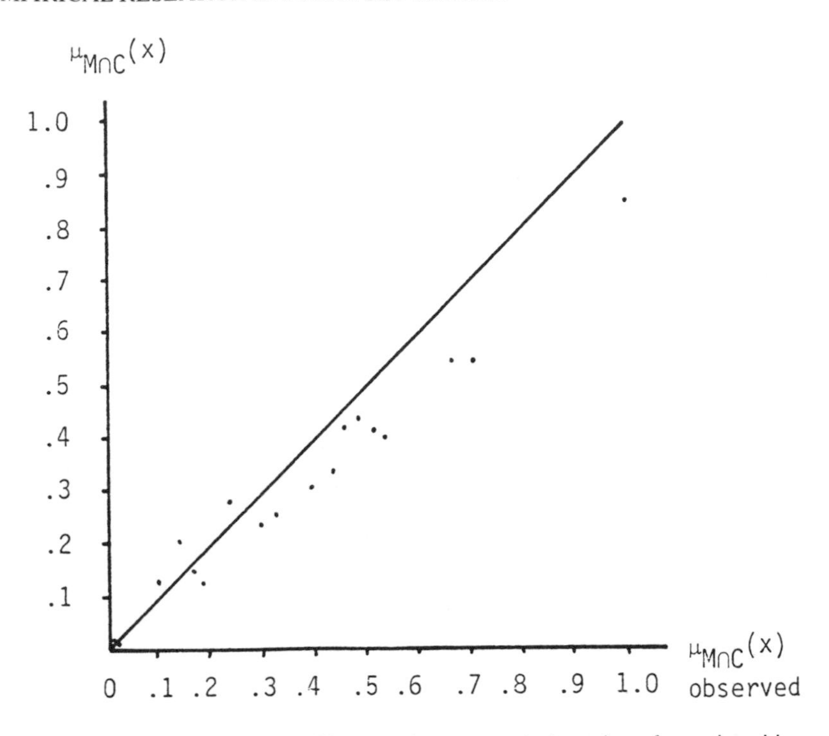

Figure 14–10. Min-operator: Observed vs. expected grades of membership.

0), the mean of the sample (0.052 for min-operator, 0.134 for product-operator), the observed standard deviation (0.067 for minimum, 0.096 for product), and the sample size (20). For the min-rules the result is $t = 3.471$, which is significant ($df = 19$; p, the probability of transition, is less than 0.01). For the product-rule the result is $t = 6.242$, which is also significant ($df = 19$; p is less than 0.001). Thus, both hypotheses H_1 and H_2 have to be rejected.

Despite the fact that none of the connective operators tested seems to be a really suitable model for the intersection of subjective categories there is a slight superiority of the min-rule, as can be seen from the figures. If one were forced to use one of these aggregation rules, then the minimum certainly would be the better choice.

The results of this experiment indicate that both product and minimum fail to be perfect models for the intersection operation in human categorizing processes.

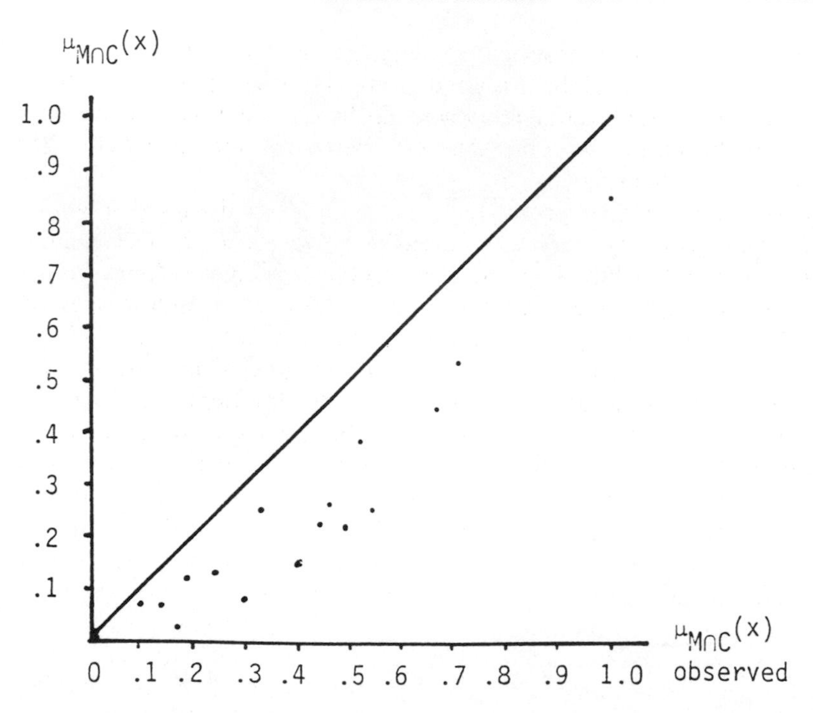

Figure 14–11. Product-operator: Observed vs. expected grades of membership.

Test 2 *[Zimmermann and Zysno 1980]*

The interpretation of a decision as the intersection of fuzzy sets implies no positive compensation (trade-off) between the degrees of membership of the fuzzy sets in question if either the minimum or the product is used as an operator. Each of them yields degrees of membership of the resulting fuzzy set (decision) that are on or below the lowest degree of membership of all intersecting fuzzy sets (see test).

The interpretation of a decision as the union of fuzzy sets, using the max-operator, leads to the maximum degree of membership achieved by any of the fuzzy sets representing objectives or constraints. This amounts to a full compensation of lower degrees of membership by the maximum degree of membership (see example 12–4).

Observing managerial decisions one finds that there are hardly any decisions with no compensation between either different degrees of goal achievement or the degrees to which restrictions are limiting the scope of decisions. The compensation, however, rarely ever seems to be "complete" such as would be assumed using the max-operator. It may be argued that compensatory tendencies in human aggregation are responsible for the failure of some classical operators (min, product, max) in empirical investigations.

Two conclusions can probably be drawn: Neither the noncompensatory 'and' represented by operators that map between zero and the minimum degree of membership (min-operator, product-operator, Hamacher's conjunction operator—see definition 3–15—Yager's conjunction operator—see definition 3–16) nor the fuzzy compensatory "or" represented by operators that map between the maximum degree of membership and 1 (maximum, algebraic sum, Hamacher's disjunction operator, Yager's disjunction operator) are appropriate to model the aggregation of fuzzy sets representing managerial decisions. It is necessary to define new additional operators which imply some degree of compensation, that is, which map also between the minimum degree of membership and the maximum degree of membership of the aggregated sets. By contrast to modeling the noncompensatory "and" or the fully compensatory "or," they should represent types of aggregation which we shall call "compensatory and."

It is possible that human beings use many nonverbal connectives in their thinking and reasoning. One type of these connectives may be called "merging connectives," which may be represented by the "compensatory and." Being forced to verbalize them men possibly map the set of "merging connectives" into the set of the corresponding language connectives ("and," "or"). Hence, when talking, they use the verbal connective they feel to be closest to their "real" nonverbal connective.

In analogy to the verbal connectives, the logicians defined the connectives $\wedge$ and $\vee$, assigning certain properties to each of them. By this, compound sentences can be examined for their truth values. In contrast to this constructive process, the empirical researcher has to analyze a given structure. Therefore, in order to induce subjects to use their own connectives, we avoided the verbal connectives "and" and "or" in our experiment, but tried to ask for combined membership values implicitly presenting a suitable experimental design and instruction, respectively.

We shall not describe in detail the experimental work in which different compensatory operators were tested and in which the γ-operator (see definition 3–19) turned out to perform best. The reader is referred to Zimmermann and Zysno [1980] for details. We shall return to figure 1–1 and explain how credit clerks arrive at a decision concerning the creditworthiness of customers by aggregating their judgments concerning the determinants of creditworthiness. For details see [Zimmermann and Zysno

1983]. A number of possible compensatory and noncompensatory models were tested.

Searching for an appropriate decision situation our choice fell on the rating of creditworthiness for the following reasons:

1. This is a decision problem that is complex enough though it is still relatively transparent and definable. In addition, this situation is highly standardized. Even though test subjects come from different organizations similar evaluation schemes can be assessed.
2. A sufficiently large number of decision makers is available with about the same training background and similar levels of competence.
3. The decision problem to be solved can be formulated and presented in a realistic manner with respect to contents and appearance.

First of all the creditworthiness hierarchy shown in figure 1–1 was developed together with 18 credit clerks.

Testing the predictive quality of the proposed models required a suitable basis of stimuli which were to rate with respect to the creditworthiness criteria and a weighting system which allowed a differentiated aggregation of these criteria.

The natural basis of information for evaluating creditworthiness is the credit file. Therefore, we would have liked to analyze original bank files. However, a selection of finished cases is always a biased sample since the initially rejected applicants are missing. Moreover, we wanted to avoid unnecessary troubles with banking secrecy. Therefore, it was decided to prepare fifty fictitious applicants for credit.

A credit application form usually contains about thirty continuous or discrete attributes of applicants. If each variable were dichotomized, 2^{30} different borrowers could be produced. Clearly, one cannot realize all possible variations. Therefore, a sample was drawn which should satisfy the following two conditions: The fifty applicants (stimuli) should

1. be distributed as evenly as possible along the continuum of each aspect, and
2. be typical for consumer credits.

The files were produced in three stages:

1. 120 applications were completed randomly with respect to the grade of extension of the thirty attributes.
2. The resulting 30×120 data matrix was purged of 40 cases most unlikely and least typical. The remaining 80 files were completed using in-

formation of an inquiry agency (Schufa) and a short record of a conversation between the client concerned and a credit clerk.

3. The applicants should represent the variability of the eight concepts. If each aspect is dichotomized into two classes ($\mu \leqslant 0.5 \rightarrow 0$, $\mu > 0.5 \rightarrow 1$) then the resulting $2^8 = 256$ patterns of evaluation can be put in a 16×16 matrix. With the assistance of two credit experts the 80 credit files were placed into this tableau. Finally 30 files were eliminated in order to obtain equal frequencies in rows and columns.

We could now expect that the 50 applicants varied evenly along each attribute and each criterion. Only one attribute was constant: the credit amount was fixed at DM 8,000. This because the judgment "creditworthy" is only meaningful with respect to a certain amount. A borrower might be good for DM 8,000, but not for DM 15,000.

Surely it would be interesting to include the credit amount as a variable into this investigation. But in order to receive a stable basis for scaling and interpretation a serious enlargement of the sample of credit experts would be necessary. This, however, would have considerably exceeded our budget.

The predictive quality of each model can be evaluated by comparing observed μ-grades with theoretical μ-grades. The latter can be computed for higher-level concepts by aggregation of the lower-level concepts using the candidate formula. The membership values for higher-level concepts should be predicted sufficiently well by any lower level of the corresponding branch. The quality of a model can be illustrated by a two-dimensional system, the axes of which represent the observed versus theoretical μ-values. Each applicant is represented by a point. In the case of exact prognosis all points must be located on a straight diagonal line. As our data are collected empirically, there will be deviations from this ideal. Figures 14–12 to 14–15 depict some of the typical results of the tests for security as being determined by 4th-level determinants.

Unfortunately, the weighted geometric mean fails drastically in predicting security by unmortgaged real estate and other net properties. In our view, this is due to the fact that the model does not regard different grades of compensation. The inclusion of different weights for the concepts does not seem to be sufficient for describing the human aggregation process adequately. Consequently it does not surprise that the γ-model comprising different weights as well as different grades of compensation yields the best results.

It should be kept in mind, however, that γ has not been determined empirically. This would have afforded a further experimental study, based

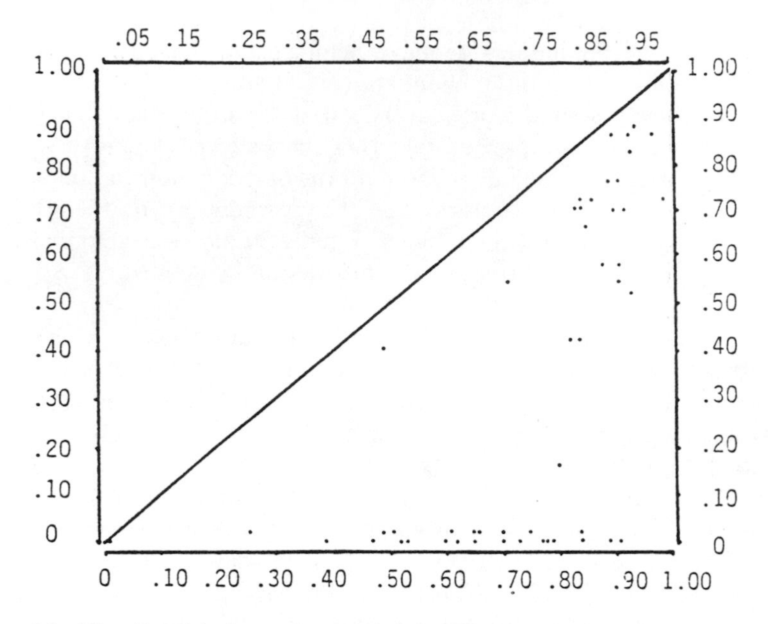

Figure 14–12. Predicted vs. observed data: Min-operator.

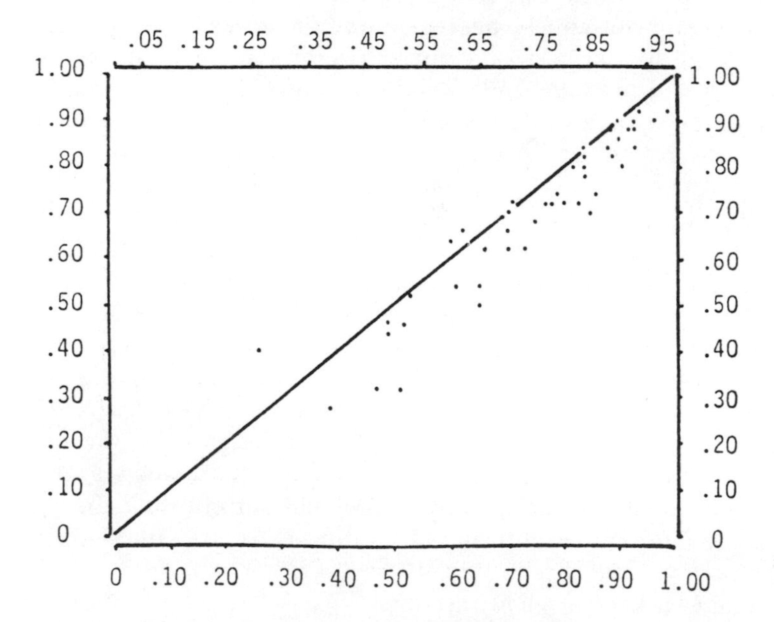

Figure 14–13. Predicted vs. observed data: Max-operator.

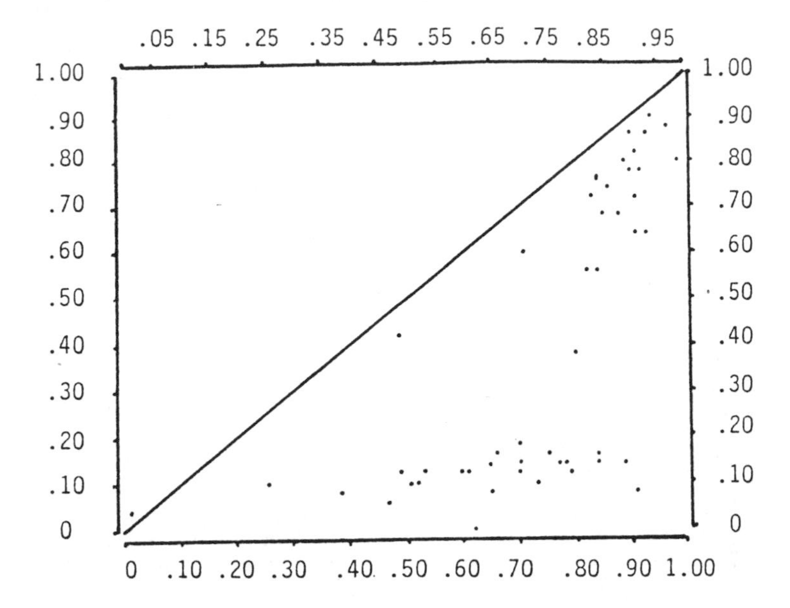

Figure 14–14. Predicted vs. observed data: Geometric mean operator.

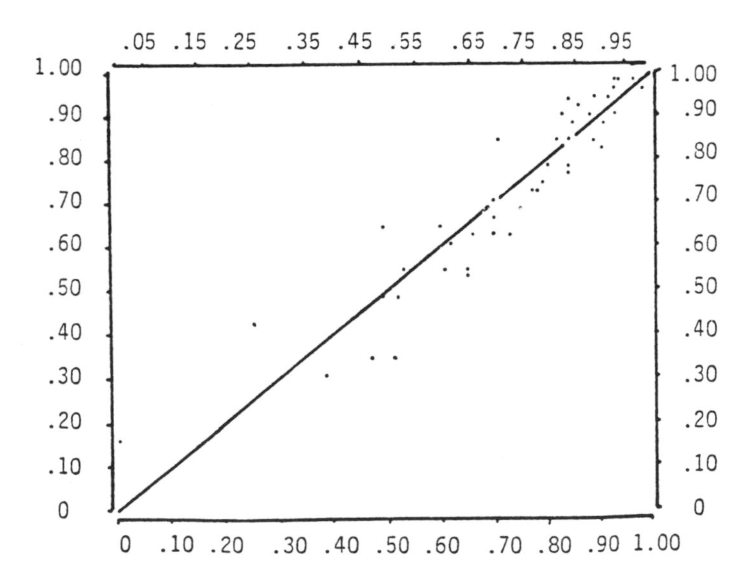

Figure 14–15. Predicted vs. observed data: γ-operator.

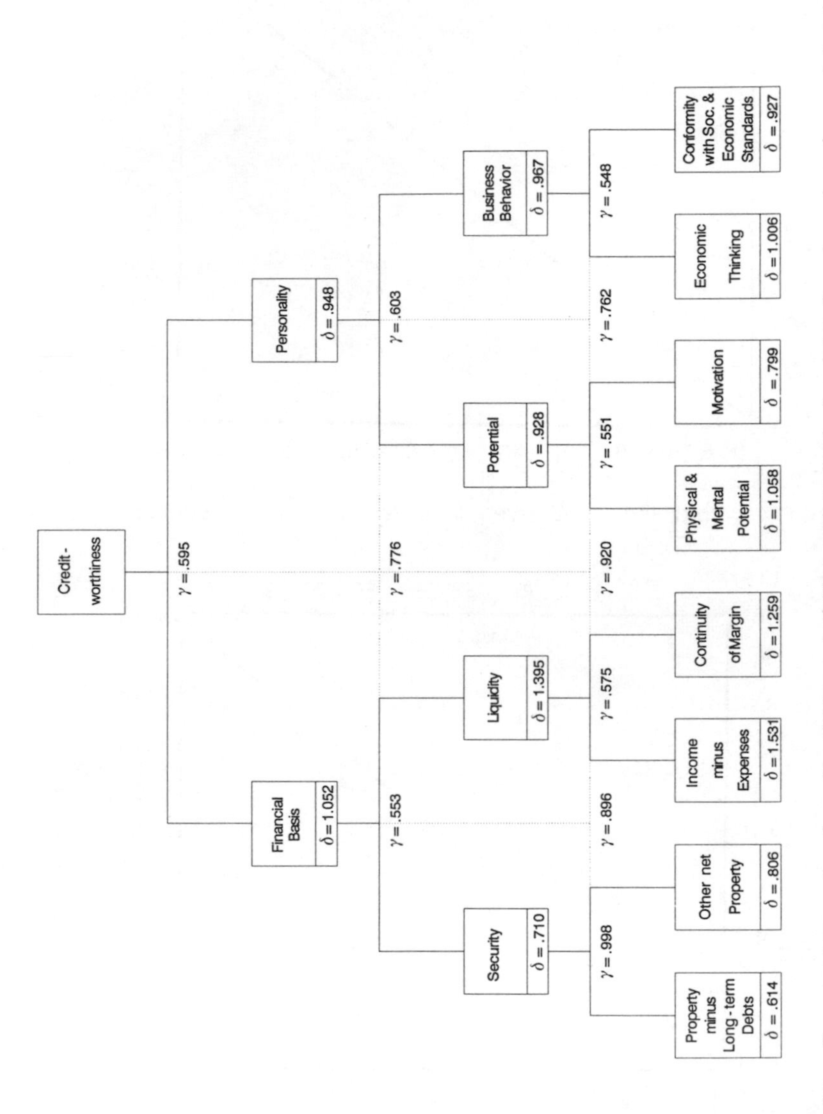

Figure 14–16. Concept hierarchy of creditworthiness together with individual weights δ and γ-values for each level of aggregation.

on a theory describing the dependence of γ-values between higher and lower levels. For the present we are content with estimations derived from the data. At least it has been shown that the judgmental behavior of credit clerks can be described quite well if this parameter is taken into account.

Finally, the complete hierarchy of creditworthiness is presented together with the elaborated weighting system and the γ values for each level of aggregation (figure 14–16).

14.4 Conclusions

Our exemplary analysis of the process of rating creditworthiness yields a criteria structure that is concept oriented and self-explanatory. The γ-model, which was from the beginning designed to satisfy mathematical requirements as well as to describe human aggregation behavior, proved most adequate with respect to prognostic power. This class of operators is continuous, monotonic, injective, commutative, and in accordance with classical truth tables, which manifests their relationship to formal logic and set theory. They aggregate partial judgments such that the formal result of the aggregation ought to make them attractive for empirically working scientists and useful for the practitioner.

Banking managers not only evaluate, they also decide. In order to complete the description of a decision process we had therefore asked to arrive at a decision for each fictitious credit application. If the creditworthiness were an attribute of the all-or-none type and all credit managers followed the same decision-making process then two homogenious blocks of credit decisions (one block with 100% yes decisions and one block with 100% no decisions) would result. The number of positive decisions, however, varied over the entire range from 45 to 0. Obviously there existed a considerable individual decision space.

15 FUTURE PERSPECTIVES

In the first 9 chapters of this book we covered the basic foundations of the theory of fuzzy sets, as they can be considered undisputed as of today. Many more concepts and theories could not be discussed either because of space limitations or because they cannot yet be considered ready for a textbook. In a recent book by Kandel [1982], 3064 references are listed which supposedly are "Key references in fuzzy pattern recognition" [Kandel 1982, p. 209]. In a recent bibliography, Ma Jiliang [1989] lists approximately 2800 references in the area of fuzzy sets, including 50 books. In the *Journal for Fuzzy Sets and Systems*, alone, almost 1000 articles have been published so far. Even though this might overestimate somewhat the total of all knowledge available in the area of fuzzy set theory today, it is indication of rather vivid research and particular publication activities.

Fuzzy set theory is certainly not a philosopher's stone which solves all the problems that confronted us today. But it has a considerable potential for practical as well as for mathematical applications, the latter of which have not been discussed at all in this book.

To indicate the scope of future applications of fuzzy set theory we shall point to some of the most relevant subject areas: Researchers have become more and more conscious that we should be less certain about uncertainty than we have been in the past. The management of uncertainty—that is, uncertainty due to lack of knowledge or evidence, due to an abundance of complexity and information, or due to the fast and unpredictable development of scientific, political, social, and other structures nowadays—will be of growing importance in the future.

In fact, in practice the "fuzzy epoch" has already begun. There already exist quite a number of expert systems and expert-system shells, which use fuzzy sets either in the form of linguistic variables or in the inference process (see chapter 10 and Gupta and Yamakawa [1988b]). Fuzzy computers were already exhibited 1987 in Tokyo. Gupta and Yamakawa [1988a] provide a very good description of the present state of development.

One of the advantages of fuzzy set theory is its extreme generality, which will enable it to accommodate quite a number of the new developments necessary to coping with existing and emerging problems and challenges. Some areas are already well developed. Possibility theory [Dubois and Prade 1988a], fuzzy clustering, fuzzy control, fuzzy mathematical programming, etc. Other areas, however, have still ample space for further development.

One of the most thriving areas of fuzzy set applications is certainly that of fuzzy control. Some indications are given in chapter 10. The state of affairs around 1985 is well described by Sugeno [1985a]. The fact that since then not too many new publications have appeared is not an indication of a slowing down of the development. On the contrary: Most of the successful commercial applications do not—for competitive reasons—lead to publications.

Considerably more research—formal as well as empirical—will be necessary in order to cope with those challenges. Much of this research will only be possible through interdisciplinary team efforts. Let us indicate some of the research that is needed. Fuzzy set theory can be considered as a modeling language for vague and complex formal and factual structures. So far, mainly the min-max version of fuzzy set theory has been used and applied, even though many other connectives, concepts, and operations have been suggested in the literature. Membership functions generally are supposed "to be given." Therefore, much empirical research and good modeling effort is needed on connectives and on the measurement of membership functions to be able to use fuzzy set theory adequately as a modeling language. Strong chances, not yet been exploited, exist in the field of artificial intelligence. Most of the approaches and methods offered there so far are dichotomous. If artificial intelligence really wants to be useful in capturing human thinking and perception the phenomenon of uncertainty will have to be modeled much more adequately than is being done so far. Here, of course fuzzy set theory offers many different opportunities.

Another (at least potential) strength of fuzzy set theory is its algorithmic, computational promise. The more we realize that there are problems—the reader might, for instance, think of NP-complete problem structures, which are far too complex for existing traditional approaches (combinatorial programming, etc.) to cope with—the more the need for new computational avenues becomes apparent. So far, fuzzy set theory has not yet proved to be computationally able to solve large and complex problems efficiently. Reasons for this are that for computation we either still have to resort to traditional techniques (linear programming, branch and bound, traditional inference) or that the additional information contained in fuzzy set models makes computations excessively voluminous. Here prudent standardization (support fuzzy logic, etc.) as well as good algorithmic combinations of heuristics and fuzzy set theory might offer some real promise. In other words, research in the direction of fuzzy algorithms is also urgently needed.

The second volume of this book, which appeared in 1987, presents and discusses some of that type of development in one specific area—namely, decision analysis. It can only be hoped that other efforts in this direction will follow to help fuzzy set theory progress and mature in a number of ways for the sake of further remarkable contributions to applications in many areas.

Abbreviations of Frequently Cited Journals

ECECSR	Economic Computation and Economic Cybernetics Studies and Research
EIK	Elektronische Informationsverarbeitung und Kybernetik
EJOR	European Journal of Operational Research
FSS	Fuzzy Sets and Systems
JMAA	Journal of Mathematics, Analysis and Applications
J.Op.Res.Soc.	Journal of the Operational Research Society

Bibliography

Ackoff, R. [1962]. *Scientific method: Optimising applied research decisions.* New York.

Adamo, J.M. [1980]. L.P.L. A fuzzy programming language: 1. Syntactic aspects. *FSS* 3, 151–179.

Adamo, J.M. [1980]. L.P.L. A fuzzy programming language: 2. Semantic aspects. *FSS* 3, 261–289.

Adamo, J.M. [1980]. Fuzzy decision trees. *FSS* 4, 207–219.

Adamo, J.M. [1981]. Some applications of the L.P.L. language to combinatorial programming. *FSS* 6, 43–59.

Adlassnig, K.-P. [1980]. A fuzzy logical model of computer-assisted medical diagnosis. *Models of Inform. in Sci.* 19, 141–148.

Adlassnig, K.-P. [1982]. A survey on medical diagnosis and fuzzy subsets. In Gupta and Sanchez, 203–217.

Adlassnig, K.-P., and Kolarz, G. [1982] CADIAG-2: Computer assisted medical diagnosis using fuzzy subsets. In: Gupta and Sanchez, 141–148.

Adlassnig, K.-P., Kolarz, G., and Scheithauer, W. [1985]. Present state of the medical expert system CADIAG-2. *Medical Information* 24, 13–20.

Albert, P. [1978]. The algebra of fuzzy logic. *FSS* 1, 203–230.

Albrycht, J., Wiesniewski, H., (eds.) [1983]. *Polish Symposium on Interval and Fuzzy Mathematics*. Poznan.

Alsina, C. [1985] On a family of connectives for fuzzy sets. *FSS* 16, 231–235.

von Altrock, C. [1990]. Konzipierung eines Lösungsverfahrens zur Produktionsplanung und -steuerung in der chemischen Industrie. Master Thesis, Institute for OR, RWTH University of Aachen, Germany.

Apostel, L. [1961]. Formal study of models. In Freudenthal, H. (ed.). *The Concept and the Role of the Model in Mathematics and Natural and Social Science*. Dordrecht.

Arrow, K. J. [1951]. Social Choice and Individual Values. New York.

Asai, K., Tanaka, H., and Okuda, T. [1975]. Decision making and its goal in a fuzzy environment. In Zadeh et al., 257–277.

Atanassov, K. T. [1986]. Intuitonistic fuzzy sets. *FSS* 20, 87–96.

Atanassov, K. T, and Stoeva, S. [1983]. Intuitonistic fuzzy sets. In Albrycht and Wiesniewski, 23–26.

Atanassov, K. T, and Stoeva, S. [1984]. Intuitonistic L-fuzzy Sets. In Trappl, 539–540.

Aumann, R.J. [1965]. Integrals of Set-Valued Functions. J. Math. Anal. Appl. 12, 1–12.

Baas, M.S., and Kwakernaak, H., [1977]. Rating and ranking of multiple-aspect alternatives using fuzzy sets. *Automatica* 13, 47–58.

Backer, E. [1978]. Cluster analysis formalized as a process of fuzzy identification based on fuzzy relations. Rep. IT-78-15, Delft University.

Backer, E. [1978]. Cluster analysis by optimal decomposition of induced fuzzy sets. *Diss. Delft.*

Baldwin, J.F. [1979]. A new approach to approximate reasoning using a fuzzy logic. *FSS* 2, 309–325.

Baldwin, J.F. [1981]. Fuzzy logic and fuzzy reasoning. In Mamdani and Gaines, 133–148.

Baldwin, J.F. [1981]. Fuzzy logic knowledge basis and automated fuzzy reasoning. In Lasker, 2959–2965.

Baldwin, J.F. [1982]. Pi-fuzzy logic. *International Report*, Univ. of Bristol.

Baldwin, J.F. [1986]. Support Logic Programming. Int. J. of Intelligent Systems, Vol. I, 73–104.

Baldwin, J.F. [1987]. Evidential support logic programming. *FSS* 24, 1–26.

Baldwin, J.F. [1989]. Combining Evidences for Evidential Reasoning. ITRC—Report, University of Bristol.

Baldwin, J.F., and Guild, N.C.F. [1979]. FUZLOG: A computer program for fuzzy reasoning. *Proc. 9th Inter. Symp. MVL*, Bath 1979, pp. 38–45.

Baldwin, J.F., and Guild, N.C.F. [1980a]. Feasible algorithms for approximate reasoning using fuzzy logic. *FSS* 3, 225–251.

Baldwin, J.F., and Guild, N.C.F. [1980b]. Modelling controllers using fuzzy relations. *Kybernetes* 9, 223–229.

Baldwin, J.F., and Pilsworth, B.W. [1979]. Fuzzy truth definition of possibility measure for decision classification. *Inter. J. Man-Machine Studies* 2, 447–463.

Baldwin, J.F., and Pilsworth, B.W. [1980]. Axiomatic approach to implication for approximate reasoning with fuzzy logic. *FSS* 3, 193–219.

Baldwin, J.F., and Pilsworth, B.W. [1982]. Dynamic programming for fuzzy systems with fuzzy environment. *J. Math. Anal. Applic.* 85, 1–23.

Banon, G. [1981]. Distinction between several subsets of fuzzy measures. *FSS* 5, 291–305.

Barr, A. and Feigenbaum, E.A., eds. [1981, 1982]. *The Handbook of Artificial Intelligence*. Los Altos, CA, 1981, vol. 1; 1982, vols. 2–3.

Bellman, R. [1957]. *Dynamic Programming*. Princeton.

Bellman, R., and Giertz, M. [1973]. On the analytic formalism of the theory of fuzzy sets. *Information Sciences* 5, 149–156.

Bellman, R., Kalaba, R., and Zadeh, L.A. [1966]. Abstraction and pattern classification. *J. Math. Anal. Applic.* 13, 1–7.

Bellman, R., and Zadeh, L.A. [1970]. Decision-making in a fuzzy environment. *Management Science* 17, B-141–164.

Bensana, E., Bel, G., and Dubois, D. [1988]. Opal: A multi-knowledge-based system for industrial job-shop scheduling. *Inter. J. Product. Res.* 26, 795–819.

Benson, I. [1986]. Prospector: An expert system for mineral exploration. In Mitra, 17–26.

Bezdek, J.C. [1981]. *Pattern Recognition with Fuzzy Objective Function Algorithms*. New York, London.

Bezdek, J.C., ed. [1987a]. *Analysis of Fuzzy Information, vol. II Artificial Intelligence and Decision Systems*. Boca Raton.

Bezdek, J.C., ed. [1987b]. *Analysis of Fuzzy Information, vol. III Applications in Engineering and Science*. Boca Raton.

Bezdek, J.C. [1987c]. Partition structures: A tutorial. In Bezdek [1987b], 81–107.

Bezdek, J.C., ed. [1989]. The coming of age of fuzzy logic. *Proc. 3rd IFSA Congress*, Seattle, 1989.

Bezdek, J.C., and Harris, J.D. [1978]. Fuzzy partitions and relations. *FSS* 1, 111–127.

Blockley, D.I. [1979]. The role of fuzzy sets in civil engineering. *FSS* 2, 267–278.

Blockley, D.I. [1980]. *The Nature of Structural Design and Safety*. Chichester.

Bock, H.-H., ed. [1979a]. *Klassifikation und Erkenntnis III*. Meisenheim/Glan.

Bock, H.-H. [1979b]. Clusteranalyse mit unscharfen Partitionen. In Bock [1979a], 137–163.

Boender, C.G.E., Graan, de J.G., and Lootsma, F.A. [1989]. Multi-criteria decision analysis with fuzzy pairwise comparisons. *FSS* 29, 133–143.

Bonissone, P.P., and Decker, K.S. [1986]. Selecting uncertainty calculi and granularity: An experiment in trading-off precision and complexity. In Kanal and Lemmer [1986], 217–247.

Boose, J.H. [1989]. A survey of knowledge acquisition techniques and tools. *Knowledge Acquisition* 1, 3–37.

Bossel, H., Klaczko, S., and Mueller, N., eds. [1976]. *Systems Theory in the Social Sciences*. Basel, Stgt.

Bowman, E.H. [1963]. Consistency and optimality in managerial decision making. In Muth, J.F., and Thompson, G.L. eds., *Industrial Scheduling*. Englewood Cliffs, NJ.

Braae, M., and Rutherford, D.A. [1979]. Selection of parameters for a fuzzy logic controller. *FSS* 2, 185–199.

Brand, H.W. [1961]. *The Fecundity of Mathematical Methods*. Dordrecht.

Buchanan, B.G., et al. [1983]. Constructing an expert system. In Hayes-Roth et al., 127–168.

Buchanan, B., Shortliffe, E. [1984]. *Rule-Based Expert Systems*. Reading, MA.

Buckley, J.J. [1984]. The multiple judge, multiple criteria ranking problem: A fuzzy set approach. *FSS* 13, 25–38.

Buckley, J.J. [1988a]. Possibility and necessity in optimization. *FSS* 25, 1–13.
Buckley, J.J. [1988b]. Possibilistic linear programming with triangular fuzzy numbers. *FSS* 26, 135–138.
Campos, L., and Verdegay, J.L. [1989]. Linear programming problems and ranking of fuzzy numbers. *FSS* 32, 1–11.
Cao, H., and Chen, G. [1983]. Some applications of fuzzy sets of meteorological forecasting. *FSS* 9, 1–12.
Capocelli, R.M., and de Luca, A. [1973]. Fuzzy sets and decision theory. *Information and Control* 23, 446–473.
Carlsson, C., and Korhonen, P. [1986]. A parametric approach to fuzzy linear programming. *FSS* 20, 17–30.
Carlucci, D., and Donati, F. [1977]. Fuzzy cluster of demand within a regional service system. In Gupta, Saridis, and Gaines, 379–385.
Carnap, R. [1946]. *Introduction to Semantics*. Cambridge.
Carnap, R. and Stegmueller, W. [1959]. *Induktive Logik und Wahrscheinlichkeit*. Wien.
Chanas, S. [1982]. Fuzzy sets in few classical operational research problems. In Gupta and Sanchez, 351–363.
Chanas, S. [1983]. Parametric programming in fuzzy linear programming. *FSS* 11, 243–251.
Chanas, S. [1989]. Parametric techniques in fuzzy linear programming. In Verdegay and Delgado 105–116.
Chanas, S., and Kamburowski, J. [1981]. The use of fuzzy variables in PERT. *FSS* 5, 11–19.
Chanas, S., Kolodziejczyk, W., and Machaj, A. [1984]. A fuzzy approach to the transportation problem. *FSS*, 13, 211–221.
Chandhuri, B.B., and Magumder, D.D. [1982]. On membership evaluation in fuzzy sets. In Gupta and Sanchez, 3–12.
Chang, S. and Zadeh, L. [1972]. On fuzzy mapping and control. *IEEE Trans. Syst. Man Cyber*. 2, 30–34.
Charnes, A., and Cooper, W.W. [1961]. *Management Models and Industrial Applications of Linear Programming*. New York.
Chatterji, B.N. [1982]. Character recognition using fuzzy similarity relations. In Gupta and Sanchez, 131–137.
Ciobanu, Y., and Stoica, M. [1981]. Production scheduling in fuzzy conditions. *Econ. Comput. Econ. Cyber. Stud. Res*. 15, 67–79.
Correa-Guzman, E.Y. [1984]. Erweiterung des unscharfen linearen Programmierens bei Mehrfachzielsetzungen Anwendung auf Energieanalysen (M. Thesis Aachen 1984) Juelich 1984.
Cronbach, L.J. [1950]. Further evidence on response sets and test design, *Educ. Psychol. Measmt*. 10, 3–31.
Czogala, E., and Pedrycz, W. [1981]. On identification in fuzzy systems and its applications in control problems. *FSS* 6, 73–83.
Czogala, E., and Pedrycz, W. [1982]. Control problems in fuzzy systems. *FSS*

7, 257–273.

Czogala, E., and Zimmermann, H.-J. [1984a]. Some aspects of synthesis of probabilistic controllers. *FSS* 13, 169–177.

Darzentas, J. [1987]. On fuzzy location models. In Kacprzyk and Orlovski, 328–341.

Delgado, M., Verdegay, J.L., and Vila, M.A. [1989]. A general model for fuzzy linear programming. *FSS* 29, 21–29.

Dempster, A.P. [1967]. Upper and lower probabilities induced by a multivalued mapping. *Ann. Math. Statistics* 38, 325–339.

Diederich, G.W., Messick, S.J., and Tucker, L.R. [1957] A general least squares solution for successive intervals. *Psychometrika* 22, 159–173.

Dijkman, J.G., van Haeringen, I., and de Lange, S.J. [1981]. Fuzzy numbers. In Lasker, 2753–2756.

Dimitrescu, D. [1988]. Hierarchical Pattern Classification. *FSS* 28, 145–162.

Dishkant, H. [1981]. About membership function estimation. *FSS* 5, 141–147.

Dodson, C.T.J. [1981]. A new generalisation of graph theory. *FSS* 6, 293–308.

Dombi, J. [1982]. A general class of fuzzy operators, the De Morgan class of fuzzy operators and fuzziness measures induced by fuzzy operators. *FSS* 8, 149–163.

Dubois, D. [1989]. Fuzzy knowledge in an artificial intelligence system for job-shop scheduling. In Evans et al., 73–79.

Dubois, D., and Prade, H. [1979]. Fuzzy real algebra: Some results. *FSS* 2, 327–348.

Dubois, D., and Prade, H. [1980a]. *Fuzzy Sets and Systems: Theory and Applications*. New York, London, Toronto.

Dubois, D., and Prade, H. [1980b]. Systems of linear fuzzy constraints. *FSS* 3, 37–48.

Dubois, D., and Prade, H. [1980c]. New results about properties and semantics of fuzzy set-theoretic operators. In Wang and Chang, 59–75.

Dubois, D., and Prade, H. [1982a]. A class of fuzzy measures based on triangular norms. *Inter. J. Gen. Syst.* 8, 43–61.

Dubois, D., and Prade, H. [1982b]. Towards fuzzy differential calculus: Part 1, Integration of fuzzy mappings: Part 2, Integration of fuzzy intervals: Part 3, Differentiation. *FSS* 8, 1–17, 105–116, 225–233.

Dubois, D., and Prade, H. [1984]. Criteria aggregation and ranking of alternatives in the framework of fuzzy set theory. In Zimmermann, Zadeh, and Gaines [1984], 209–240.

Dubois, D., and Prade, H. [1985]. A review of fuzzy set aggregation connectives. *Information Science* 36, 85–121.

Dubois, D., and Prade, H. [1988a]. *Possibility Theory*. New York, London.

Dubois, D., and Prade, H. [1988b]. Processing of imprecision and uncertainty in expert system reasoning models. In C.J. Ernst, 67–88.

Dubois, D., and Prade, H. [1989]. Fuzzy sets, probability and measurement. *EJOR*. 40, 135–154.

Dumitru, V., and Luban, F. [1982]. Membership functions, some mathematical programming models and production scheduling. *FSS* 8, 19–33.

Dunn, J.C. [1974]. A fuzzy relative to the ISODATA process and its use in detecting compact well-separated clusters. *J. Cyber.* 3, 310–313.

Dunn, J.C. [1977]. Indices of partition fuzziness and detection of clusters in large data sets. In Gupta, Saridis, and Gaines, 271–284.

Dyer, J.S. [1972/73]. Interactive goal programming. *Management Science* 19, 62–70.

Ernst, C.J. [1981]. An approach to management expert systems using fuzzy logic. In Lasker, 2898–2905.

Ernst, C.J. ed. [1988]. *Management Expert Systems*. Wokingham, Reading, Menlo Park, New York.

Ernst, E. [1982]. Fahrplanerstellung und Umlaufdisposition im Containerschiffsverkehr (Diss. Aachen) Frankfurt/M., Bern.

Esogbue, A.O., and Bellman, R.E. [1984]. Fuzzy dynamic programming and its extensions. In Zimmermann et al., 147–167.

Esogbue, A.O., and Elder, R.C. [1979]. Fuzzy sets and the modelling of physician decision processes, Part I: The initial interview—Information gathering sessions. *FSS* 2, 279–291.

Esogbue, A.O., and Elder, R.C. [1983]. Measurement and valuation of a fuzzy mathematical model for medical diagnosis. *FSS* 10, 223–242.

Evans, G. W., Karwowsky, W., and Wilhelm, M.R., eds. [1989]. *Applications of Fuzzy Set Methodologies in Industrial Engineering*. Elsevier.

Fabian, C., and Stoica, M. [1984]. Fuzzy integer programming. In Zimmermann et al. [1984], 123–132.

Fieschi, M., Joubert, M., Fieschi, D., Soula, G., and Roux, M. [1982]. SPHINX, An interactive system for medical diagnosis aids. In Gupta and Sanchez, 269–275.

Fiksel, J. [1981]. Applications of fuzzy set and possibility theory to systems management. In Lasker, 2966–2973.

Ford, N. [1987]. *How Machines Think*. Chichester, New York, Brisbane, Toronto, Singapore.

Fordon, W.A., and Bezdek, J.C. [1979]. The application of fuzzy set theory to medical diagnosis. In Gupta, Ragade, and Yager, 445–461.

Fordyce, K., Norden, P., and Sullivan, G. [1989]. Artificial intelligence and the management science practitioner: One definition of knowledge-based expert systems. *Interfaces* 19, 66–70.

Freeling, A.N.S. [1984]. Possibilities versus fuzzy probabilities—two alternative decision aids. In Zimmermann et al., 67–82.

Freksa, Ch. [1982]. Linguistic description of human judgments in experts systems and in the 'soft' sciences. In Gupta and Sanchez, 297–306.

Fu, K.S., Ishizuka, M., and Yao, J.T.P. [1982]. Application of fuzzy sets in earthquake engineering. In Yager, 504–518.

Gaines, B.R. [1975]. Multivalued logics and fuzzy reasoning. BCS AISB Summer School Cambridge.

Giles, R. [1976]. Lukasiewicz logic and fuzzy theory. *Inter. J. Man-Mach. Stud.* 8, 313–327.

Giles, R. [1979]. A formal system for fuzzy reasoning. *FSS* 2, 233–257.

Giles, R. [1980]. A computer program for fuzzy reasoning. *FSS* 4, 221–234.

Giles, R. [1981]. Lukasiewicz logic and fuzzy set theory. In Mamdani and Gaines, 117–131.

Glover, F., and Greenberg, H.J. [1989]. New approaches for heuristic search: A bilateral linkage with artificial intelligence. *EJOR* 39, 119–130.

Goguen, J.A. [1967]. L-fuzzy sets. *JMAA* 18, 145–174.

Goguen, J.A. [1969]. The logic of inexact concepts. *Synthese* 19, 325–373.

Goguen, J.A., Jr. [1981]. Concept representation in natural and artificial languages: Axioms, extensions and applications for fuzzy sets. In: Mamdani and Gaines, 67–115.

Gordon, J., and Shortliffe, E.H. [1984]. The Dempster-Shafer theory of evidence. In Buchanan and Shortliffe Reading, MA, 272–292.

Gottwald, S. [1979a]. Set theory for fuzzy sets of higher level. *FSS* 2, 125–151.

Gottwald, S. [1979b]. A note on measures of fuzziness. *EIK* 15, 221–223.

Gottwald, S. [1980]. Fuzzy propositional logics. *FSS* 3, 181–192.

Graham, I., and Jones, P.L. [1988]. *Expert Systems: Knowledge, Uncertainty and Decision*. London, New York.

Gu, T., and Dubuisson, B. [1990]. Similarity of classes and fuzzy clustering. *FSS* 34, 213–221.

Gupta, M.M., and Sanchez, E., eds. [1982]. *Approximate Reasoning in Decision Analysis*. Amsterdam, New York, Oxford.

Gupta, M.M., Ragade, R.K., Yager, R.R., eds. [1979]. *Advances in Fuzzy Set Theory and Applications*. Amsterdam. New York, Oxford.

Gupta, M.M., Saridis, G.N., and Gaines, BR., eds. [1977]. *Fuzzy Automata and Decision Processes*. Amsterdam. New York.

Gupta, M.M., and Yamakawa, T., eds. [1988a]. *Fuzzy Computing Theory, Hardware, and Applications*. Amsterdam, New York, Oxford, Tokyo.

Gupta, M.M., and Yamakawa, T., eds. [1988b]. *Fuzzy Logic in Knowledge-Based Systems, Decision and Control*. Amsterdam, New York.

Hajnal, M., and Koczy, L.T. [1982]. Classification of textures by vectorial fuzzy sets. In Gupta and Sanchez, 157–164.

Hamacher, H. [1978]. Über logische Aggregationen nicht-binär expliziter Entscheidungskriterien. Frankfurt/Main.

Hamacher, H., Leberling, H., and Zimmerman, H.-J. [1978]. Sensitivity analysis in fuzzy linear programming. *FSS* 1, 269–281.

Hammerbacher, J.M., and Yager, R.R. [1981]. The personalization of security: An application of fuzzy set theory. *FSS* 5, 1–9.

Harmon, P., and King, D. [1985]. *Expert Systems*. New York, Chichester, Toronto.

Harré, R. [1967]. *An Introduction to the Logic of Sciences*. London, Melbourne, Toronto.

Harré, R. [1972]. *The Philosophies of Science*. London, Oxford, New York.

Hartley, J. [1984]. *FMS at Work*. Bedford (UK).

Hax, A.C., and Majluf, N.S. [1984]. *Strategic Management: An Integrative Perspective*. Englewood Cliffs.

Hayes-Roth, F., Waterman, D.A., and Lenat, D.B., eds. [1983]. *Building Expert Systems*. London, Amsterdam.

Helson, H. [1964]. *Adaption-level Theory*. New York.

Hersh, H.M., Caramazza, A., and Brownell, H.H. [1979]. Effects of context on fuzzy membership functions. In Gupta et al. 389–408.

Higashi, M., and Klir, G.J. [1982]. Measures of uncertainty and information based on possibility distributions. *Inter. J. Gen. Syst.* 9, 43–58.

Hintz, G.W., and Zimmermann, H.-J. [1989]. A method to control flexible manufacturing systems. *EJOR* 321–334.

Hirota, K. [1981]. Concepts of probabilistic Sets. *FSS* 5, 31–46.

Hisdal, E. [1978]. Conditional possibilities independence and noninteraction. *FSS* 1, 283–297.

Holmblad, L.P., and Ostergaard, J.J. [1982]. Control of cement kiln by fuzzy logic. In Gupta and Sanchez, 389–400.

Holt, C.C., Modigliani, F., Muth, J.F., and Simon, H. [1960]. *Planning Production, Inventories and Workforce*. New York.

Holtz, M., and Desonki, Dr. [1981]. Fuzzy-Model für Instandhaltung. *Unscharfe Modellbildung und Steuerung* IV, 54–62, Karl-Marx-Stadt.

Hudson, D.L., and Cohen, M.E. [1988]. Fuzzy logic in a medical expert system. In Gupta and Yamakawa [1988a], 273–284.

Hughes, G.E., and Cresswell, M.J. [1968] *An Introduction to Modal Logic*. London.

Hwang, Ch.-L., and Masud, A.S.M. [1979]. *Multiple Objective Decision Making-Methods and Applications*. Berlin, Heidelberg, New York.

Hwang, Ch.-L., and Yoon, K. [1981]. *Multiple Attribute Decision Making*. Berlin, Heidelberg, New York.

Ishizuka, M., Fu, K.S., and Yao, J.T.P. [1982]. A rule-based inference with fuzzy set for structural damage assessment. In Gupta and Sanchez, 261–275.

Ismail, M.A. [1988]. Soft clustering: Algorithms and validity of solutions. In Gupta and Yamakawa [1988a], 445–471.

Jain, R. [1976]. Tolerance analysis using fuzzy sets. *Inter. J. Syst. Sci.* 7(12), 1393–1401.

Jain, R., and Nagel, H.H. [1977]. Analysing a real world scene sequence using fuzziness. *Proc. IEEE Conf. Dec. Control*, 1367–1372.

Jardine, N., and Sibson, R. [1971]. *Mathematical Taxonomy*. New York.

Jensen, J.H. [1976]. *Application for Fuzzy Logic Control*. Techn. Univ. of Denmark, Lyngby.

Jones, C.H. [1967]. Parametric production planning. *Management Science* 13, 843–866.

Kacprzyk, J. [1983]. *Multistage Decision Making and Fuzziness*. Köln.

Kacprzyk, J., and Orlovski, S.A., eds. [1987]. *Optimization Models Using Fuzzy*

Sets and Possibility Theory. Dodrecht, Boston.

Kacprzyk, J., and Staniewski, P. [1982]. Long-term inventory policy-making through fuzzy decision-making models. *FSS* 8, 117–132.

Kacprzyk, J., and Yager, R.R., eds. [1985]. *Management Decision Support Systems Using Fuzzy Sets and Possibility Theory*. Köln.

Kanal, L.N., and Lemmer, J.F., eds. [1986]. *Uncertainty in Artificial Intelligence*. Amsterdam.

Kandel, A., Lee, S.C. [1979]. Fuzzy Switching and Automata. New York.

Kandel, A. [1982]. *Fuzzy Techniques in Pattern Recognition*. New York, Toronto, Singapore.

Kandel, A. [1986]. *Fuzzy Mathematical Techniques with Applications*. Reading.

Kandel, A., and Byatt, W.J. [1978]. Fuzzy sets, fuzzy algebra, and fuzzy statistics. *Proc. IEEE 66*, 1619–1637.

Kastner, J.K., and Hong, S.J. [1984]. A review of expert systems. *EJOR* 18 285–292.

Kaufmann, A. [1975]. *Introduction to the Theory of Fuzzy Subsets*, vol. I. New York, San Francisco, London.

Kaufmann, A., and Gupta, M.M. [1988]. *Fuzzy Mathematical Models in Engineering and Management Science*. Amsterdam, New York.

Keeney, R.L., and Raiffa, H. [1976]. *Decisions with Multiple Objectives*. New York, Santa Barbara, London.

Kickert, W.J.M. [1978]. *Fuzzy Theories on Decision-making*. Leiden, Boston, London.

Kickert, W.J.M. [1979]. Towards analysis of linguistic modelling. *FSS* 2, 293–307.

Kickert, W.J.M. and Mamdani, E.H. [1978]. Analysis of a fuzzy logic controller. *FSS* 1, 29–44.

Kickert, W.J.M., and Van Nauta Lemke, H.R. [1976]. Application of a fuzzy controller in a warm water plant. *Automatica* 12, 301–308.

Kim, K.H., and Roush, F.W. [1982]. Fuzzy flows in networks. *FSS* 8, 35–38.

King, P.Y., and Mamdani, E.J. [1977]. The application of fuzzy control systems to industrial processes. In Gupta et al., 321–330.

King, R.E., and Karonis, F.C. [1988]. Multi-level expert control of a large-scale industrial process. In Gupta and Yamakawa, [1988a] 323–340.

Klement, E.P. [1981]. On the relationship between different notions of fuzzy measures. In Lasker, 2837–2842.

Klement, E.P., and Schwyhla, W. [1982]. Correspondence between fuzzy measures and classical measures. *FSS* 7, 57–70.

Kling, R. [1973]. Fuzzy planner: Reasoning with inexact concepts in a procedural problem-solving language. *J. Cybernetics* 3, 1–16.

Klir, G.J. [1987]. Where do we stand on measures of uncertainty, ambiguity, fuzziness, and the like? *FSS* 24, 141–160.

Klir, G.J., and Folger, T.A. [1988]. *Fuzzy Sets, Uncertainty, and Information*. Englewood Cliffs.

Knopfmacher, J. [1975]. On measures of fuzziness. *JMAA* 49, 529–534.

Kokawa, M. [1982]. Heuristic approach to pump operations using multi-valued logic. In Gupta and Sanchez, 415–422.

Kolmogoroff, A. [1950]. Foundation of Probability. New York.

Konopasek, M., and Jayaraman, S. [1984]. Expert systems for personal computers. *Byte*, 1984, 137–154.

Koonth, W.G.L., Narenda, P.M., and Fukunaga, K. [1975]. A branch and bound clustering algorithm. *IEEE Trans. Comput.* C24, 908–915.

Koopman, B.O. [1940]. The axioms and algebra of intuitive probability. *Ann. Math.* 41, 269–292.

Krarup, J., and Pruzan, P.M. [1983]. The simple plant location problem: Survey and synthesis. *EJOR* 12, 36–81.

Kuhn, H.W. and Tucker, A.W. [1951]. Nonlinear programming. *Proc. 2nd Berkeley Symp. Math. Stat. Prob.*, 481–492.

Kuz'min, V.B. [1981a]. A parametric approach to description of linguistic values of variables and hedges. *FSS* 6, 27–41.

Kuz'min, V.B. [1981b]. Corrections to "A parametric approach to description of linguistic values of variables and hedges" (erratum). *FSS* 6, 205.

Laarhoven, P.J.M., von, Pedrycz, W. [1983] A Fuzzy extension of Saaty's priority theory. *FSS* 11, 229–241.

Larsen, P.M. [1981]. Industrial applications of fuzzy logic control. In Mamdani and Gaines, 335–343.

Lasker, G.E. ed. [1981]. *Applied Systems and Cybernetics*, vol. VI. New York, Oxford, Toronto.

Lesmo, L., Saitta, L., and Torassa, P. [1982]. Learning of fuzzy production rules for medical diagnosis. In Gupta and Sanchez [1982], 249–60.

Lindsay, R.K., Buchanan, B.G., Feigenbaum, E.A., and Lederberg J. [1980]. *Applications of Artificial Intelligence for Organic Chemistry*: The DENDRAL project. New York.

Lipp, H.P. [1981]. Anwendung eines Fuzzy Petri Netzes zur Bestimmung instationärer Steuervorgänge in komplexen Produktionssystemen. *Unscharfe Modelllbildung und Steuerung* IV, 63–81 Karl-Marx-Stadt.

Little, J.D.C. [1970]. Models and managers: The concept of a decision calculus in: *Management Science* 16, B 446–458.

Loo, S.G. [1977]. Measures of fuzziness. *Cybernetica* 20, 201–210.

Lowen, R. [1978]. On fuzzy complements. *Information Science* 14, 107–113.

de Luca, A., and Termini, S. [1972]. A definition of a nonprobabilistic entropy in the setting of fuzzy sets theory. *Information and Control* 20, 301–312.

Luhandjula, M.K. [1982]. Compensatory operators in fuzzy linear programming with multiple objectives. *FSS* 8, 245–252.

Luhandjula, M.K. [1984]. Fuzzy approaches for multiple objective linear fractional optimization. *FSS* 13, 11–24.

Luhandjula, M.K. [1986]. On possibilistic linear programming. *FSS* 18, 15–30.

Luhandjula, M.K. [1987]. Multiple objective programming problems with possibilistic coefficients. *FSS* 21, 135–145.

Ma Jiliang [1989]. *A Bibliography of Fuzzy Systems*. Beijing.

Mamdani, E.H. [1977a]. Application of fuzzy logic to approximate reasoning. *IEEE Trans. Comput.* 26, 1182–1191.

Mamdani, E.H. [1977b]. Applications of fuzzy set theory to control systems. In Gupta et al., 77–88.

Mamdani, E.H. [1981]. Advances in the linguistic synthesis of fuzzy controllers. In Mamdani and Gaines, 325–334.

Mamdani, E.H., and Assilian, S. [1975]. An experiment in linguistic synthesis with a fuzzy logic controller. *Inter. J. Man-Mach. Studies* 7, 1–13.

Mamdani, E.H., and Assilian, S. [1981]. An experiment in linguistic synthesis with a fuzzy logic controller. Mamdani and Gaines [1981], 311–323.

Mamdani, E.H., and Gaines, G.R., eds. [1981]. *Fuzzy Reasoning and Its Applications*. London, New York, Toronto.

Mamdani, E.H., Ostergaard, J.J., and Lembessis, E. [1984]. Use of fuzzy logic for implementing rule-based control of industrial processes. In Zimmermann et al., 429–445.

Manes, E.G. [1982]. A class of fuzzy theories. *JMAA* 85, 409–451.

McDermott, J. [1982] R1: A rule-based configurer of computer systems. *Artificial Intelligence* 19, 39–88.

Michalski, R.S., Carbonell, J.G., and Mitchell, T.M. eds. [1986]. *Machine Learning: An Artificial Intelligence Approach* vol. II. Los Altos, California.

Michalski, R.S., and Chilausky, R.L. [1981]. Knowledge acquisition by encoding expert rules versus computer induction from examples: A case study involving soybean pathology. In Mamdani and Gaines, 247–271.

Minsky, M.A. [1975]. A framework for representing knowledge. In Winston 221–242.

Mitra, G., ed. [1986]. *Computer Assisted Decision Making*. Amsterdam, New York, Oxford.

Mitra, G., ed. [1988]. *Mathematical Models for Decision Support*. Berlin, Heidelberg, New York, London, Paris, Tokyo.

Miyamoto, S., Yasunobu, S., and Ihara, H. [1987]. Predictive fuzzy control and its application to automatic train operation systems. In Bezdek [1987a], 59–68.

Mizumoto, M. [1981, 1982] Fuzzy sets and their operations. *Information and Control* 48, 30–48, 50, 160–174.

Mizumoto, M. [1989]. Pictorial representations of fuzzy connectives, Part I: cases of *T*-norms, *T*-conorms and averaging operators. *FSS* 31, 217–242.

Mizumoto, M., Fukami, S., and Tanaka, K. [1979]. Some methods of fuzzy reasoning. In Gupta et al., 117–136.

Mizumoto, M., and Tanaka, K. [1976]. Some properties of fuzzy sets of type 2. *Information and Control* 31, 312–340.

Mizumoto, M., and Zimmermann, H.-J. [1982]. Comparison of fuzzy reasoning methods. *FSS* 8, 253–283.

Moon, R.E., Jordanow, S., Perez, A., and Turksen, I.B. [1977]. Medical diagnostic system with human-like reasoning capability. In Shires, H.B., Wolf, H.

eds. *Medinfo 77*, Amsterdam-New York [1977], pp. 115–119.
Morik, K., ed. [1989]. *Knowledge Representation and Organization in Machine Learning*. Berlin, Heidelberg 1989.
Muñoz-Rodriguez, D, Cattermole, K.W. [1987] Multiple criteria for hand-off in cellular mobile radio. *IEE Proc.* 134, 85–88.
Murakami, S., and Maeda, H. [1983]. Fuzzy decision analysis on the development of centralized regional energy control systems. In Sanchez and Gupta, 353–358.
Murofushi, T., and Sugeno, M. [1989]. An interpretation of fuzzy measures and the choquet integral as an integral with respect to a fuzzy measure. *FSS* 29, 201–227.
Nagel, E. [1969]. *The Structure of Science*. London.
Nahmias, S. [1979]. Fuzzy variables. *FSS* 1, 97–110.
Negoita, C.V. [1985]. *Expert Systems and Fuzzy Systems*. Menlo Park, Reading, London, Amsterdam.
Negoita, C.V., Minoiu, S., and Stan, E. [1976]. On considering imprecision in dynamic linear programming. *ECECSR* 3, 83–95.
Negoita, C.V., and Ralescu, D.A. [1975]. *Application of Fuzzy Sets to Systems Analysis*. Basel, Stuttgart.
Nguyen, H.T. [1978]. On conditional possibility distributions. *FSS* 1, 299–309.
Nguyen, H.T. [1979]. Some mathematical tools for linguistic probabilities. *FSS* 2, 53–65.
Nguyen, H.T. [1981] On the possibilistic approach to the analysis of evidence. In Lasker, 2959–2965.
Nilsson, N.J. [1980]. *Principles of Artificial Intelligence*. Palo Alto.
Norwich, A.M., and Turksen, I.N. [1981]. Measurement and scaling of membership functions. In Lasker, 2851–2858.
Norwich, A.M., and Turksen, I.B. [1984]. A model for the measurement of membership and consequences of its empirical implementation. *FSS* 12, 1–25.
Oh Eigeartaigh, M. [1982]. A fuzzy transportation algorithm. *FSS* 8, 235–243.
Ono, H., Ohnishi, T., and Terada, Y. [1989] Combustion control of refuse incineration plant by fuzzy logic. *FSS* 32, 193–206.
Orlovsky, S.A. [1977]. On programming with fuzzy constraint sets. *Kybernetes* 6, 197–201.
Orlovsky, S.A. [1980]. On formalization of a general fuzzy mathematical problem. *FSS* 3, 311–321.
Orlovsky, S.A. [1985]. Mathematical programming problems with fuzzy parameters. In Kacprzyk and Yager [1985], 136–145.
Ostergaard, J.J. [1977]. Fuzzy logic control of a heat exchanger process. In Gupta, et al. [1977], 285–320.
Ovchinnikov, S.V., and Ozernoy, V.M. [1988]. Using fuzzy binary relations for identifying noninferior decision alternatives. *FSS* 25, 21–32.
Pawlak, Z. [1982]. Rough sets. *Inter. J. Inform. Comput. Sci.* 11(5), 341–356.
Pawlak, Z. [1985]. Rough sets. *FSS* 17, 99–102.
Pawlak, Z., Wong, S.K.M., and Ziarko, W. [1988]. Rough sets: Probabilistic

versus deterministic approach. *Inter J. Man-Mach. Stud.* 29, 81–95.
Pedrycz, W. [1982]. *Fuzzy Control and Fuzzy Systems.* Delft.
Pedrycz, W. [1983]. Some applicational aspects of fuzzy relational equations in systems analysis. In Gupta and Sanchez, 125–132.
Pedrycz, W. [1989]. *Fuzzy Control and Fuzzy Systems.* New York, Chichester, Toronto.
Peng, X.-T., Liu, S.-M., Yamakawa, T., Wang, P., and Liu, X. [1988]. Self-regulating PID controllers and its applications to a temperature controlling process. In Gupta and Yamakawa, [1988a] 355–364.
Pfeilsticker, A. [1981]. The systems approach and fuzzy set theory bridging the gap between mathematical and language-oriented economists. *FSS* 6, 209–233.
Popper, K. *The Logic of Scientific Discovery.* London 1959.
Prade, H.M. [1977]. Ordonnancement et temps Reel. *Diss. Toulouse.*
Prade, H.M. [1979]. Using fuzzy set theory in a scheduling problem: A case study. *FSS* 2, 153–165.
Prade, H.M. [1980a]. An outline of fuzzy or possibilistic models for queuing systems. In Wang and Chang, 47–154.
Prade, H.M. [1980b]. Operations research with fuzzy data. In Wang and Chang, 155–169.
Pun, L. [1977]. Use of fuzzy formalism in problems with various degrees of subjectivity. In Gupta et al., 357–378.
Puri, M.L. and Ralescu, D. [1982]. A possibility measure is not a fuzzy measure (Short Communication). *FSS* 7, 311–313.
Ralston, P.A.S., and Ward, T.L. [1989]. Fuzzy control of industrial processes. In Evans et al., 29–45.
Ramik, J., and Rimanek, J. [1985] Inequality relation between fuzzy numbers and its use in fuzzy optimization. *FSS* 16, 123–138.
Ramik, J., and Rimanek, J. [1989]. The linear programming problem with vaguely formulated relations of coefficients. In Verdegay and Delgado, 177–194.
Rao, J.R., Tiwari, R.N., and Mohanty, B.K. [1988]. A preference structure on aspiration levels in a goal programming problem—A fuzzy approach. *FSS* 25, 175–182.
Rijckaert, M.J., Debroey, V., and Bogaerts, W. [1988]. Expert systems: The state of the art. In Mitra , 487–517.
Rinks, D.B. [1981]. A Heuristic approach to aggregate production scheduling using linguistic variables. In Lasker, 2877–2883.
Rinks, D.B, [1982a]. The performance of fuzzy algorithm models for aggregate planning and differing cost structures. In Gupta and Sanchez, 267–278.
Rinks, D.B. [1982b]. A heuristic approach to aggregate production scheduling using linguistic variables: Methodology and application. In Yager, 562–581.
Rivett, H.P. [1972]. *Principles of Model Building.* London, New York, Toronto.
Rodabaugh, S.E. [1981]. Fuzzy arithmetic and fuzzy topology. In Lasker, 2803–2807.
Rodabaugh, S.E. [1982]. Fuzzy addition in the L-fuzzy real line. *FSS* 8, 39–52.

Rödder, W., and Zimmermann, H.-J. [1980]. Duality in fuzzy linear programming. In A.V. Fiacco and K.O. Kortanek eds. *Extremal Methods and Systems Analysis*. Berlin, Heidelberg, New York, pp. 415–429.

Rommelfanger, H., Hanuschek, R., and Wolf, J. [1989]. Linear programming with fuzzy objectives. *FSS* 29, 31–48.

Rosenfeld, [1975]. A fuzzy graph. In Zadeh et al., 77–96.

Roubens, M. [1978]. Pattern classification problems and fuzzy sets. *FSS* 1, 239–253.

Ruspini, E. [1969]. A new approach to fuzzy clustering. *Information and Control* 15, 22–32.

Ruspini, E. [1973]. New experimental results in fuzzy clustering. *Information.* Science 6, 273–284.

Ruspini, E. [1982]. Recent develpments in fuzzy clustering. In Yager, 133–147.

Russell, B. [1923]. Vagueness. *Australasian J. Psychol. Philos.* 1, 84–92.

Saaty, Th.L. [1978]. Exploring the interface between hierarchies, multiple objectives and fuzzy sets. *FSS* 1, 57–68.

Sakawa, M., and Yano, H. [1987]. An interactive satisficing method for multiobjective nonlinear programming problems with fuzzy parameters. In Kacprzyk and Orlovski, 258–271.

Sanchez, E. [1979]. Medical diagnosis and composite fuzzy relations. In Gupta, Ragada, and Yager, 437–444.

Sanchez, E., Gouvernet, J., Bartolin, R., and Voran, L. [1982]. Linguistic approach in fuzzy logic of W.H.O. classification of dyslipoproteinemias. In Yager, 522–588.

Sanchez, E., Gupta, M.M. [1983]. Fuzzy Information, Knowledge Representation, and Decision Analysis. New York.

Schefe, P. [1981]. On foundations of reasoning with uncertain facts and vague concepts. In Mamdani and Gaines, 189–216.

Schwartz, J. [1962]. The pernicious influence of mathematics in science. In Nagel, Suppes, and Tarski *Logic Methodology and Philosophy of Science*. Standford.

Scott, L.L. [1980]. Necessary and sufficient conditions for the values of a function of fuzzy variables to lie in a specified subinterval of [0, 1]. In Wang and Chang, 35–47.

Shafer, G.A. [1976]. *A Mathematical Theory of Evidence*. Princeton.

Silverman, B.G. ed. [1987]. *Expert Systems for Business*. Reading, MA.

Silvert, W. [1979]. Symmetric summation: A class of operations on fuzzy sets. *IEEE Trans. Syst., Man Cyb.* 9, 657–659.

Skala, H.J. [1978]. On many-valued logics, fuzzy sets, fuzzy logics and their applications. *FSS*, 1, 129–149.

Smets, P. [1982]. Probability of a fuzzy event: An axiomatic approach. *FSS* 7, 153–164.

Smithson, M. [1987]. *Fuzzy Set Analysis for Behavioral and Social Sciences*. New York.

Sneath, P.H.A., and Sokal, R. [1973]. *Numerical Taxonomy*. San Francisco.

Sommer, G. [1981]. Fuzzy inventory scheduling. In Lasker, 3052–3060.

Stamp, J. [1975]. Quoted by White [1975].

Starr, M.K., ed. [1965]. *Executive Readings in Management Science*. New York, London.

Stein, W.E. [1980]. Optimal stopping in a fuzzy environment. *FSS* 3, 253–259.

Sticha, P.J., Weiss, J.J., and Donnell, M.L. [1979]. Evaluation and integration of imprecise information. Final Technical Report PR 79-21-90, Decisions & Designs, Inc., Suite 600, 8400 Westpark Drive, P.O. Box 907, McLean, VA 22101, 1979.

Stoica, M., and Serban, R. [1983]. Fuzzy algorithms for production programming. ECECSR 18, 55–63.

Sugeno, M. [1972]. Fuzzy measures and fuzzy integrals. *Trans. S.I.C.E.* 8(2).

Sugeno, M. [1977]. Fuzzy measures and fuzzy integrals—A survey. In Gupta, Saridis, and Gaines, 89–102.

Sugeno, M., ed. [1985a]. *Industrial Applications of Fuzzy Control*. Amsterdam, New York.

Sugeno, M. [1985b]. An introductory survey of fuzzy control. *Inform. Sci.* 36, 59–83.

Sugeno, M., Murofushi, T., Mori, T., Tatematsu, T., and Tanaka, J. [1989]. Fuzzy algorithmic control of a model car by oral instructions. *FSS* 32, 207–219.

Suppes, P. [1969]. Meaning and uses of models. In H. Freudenthal, ed. [1969]. *The Concept and Role of the Model in Mathematics and Natural and Social Sciences*. London.

Symonds, G.H. [1965]. The Institute of Management Sciences: Progress Report. In Starr, 376–389.

Tanaka, H., and Asai, K. [1984]. Fuzzy linear programming problems with fuzzy numbers. *FSS* 13, 1–10.

Tanaka, H., Ishihashi, H., and Asai, K. [1985]. Fuzzy decision in linear programming problems with trapezoid fuzzy parameters. In Kacprzyk and Yager, 146–154.

Tanaka, H., Ishihashi, H., and Asai, K. [1986]. A value of information in FLP problems via sensitivity analysis. *FSS* 18, 119–129.

Tatzagi, T., and Sugeno, M. [1983]. Derivation of fuzzy control rules from human operator's control actions. *IFAC Symposium of Fuzzy Information, Knowledge Representation and Decision Analysis*, Marseille.

Taubert, W.H. [1967]. A search decision rule for the aggregate scheduling problem *Management Science* 13, 343–359.

Teichrow, J., Horstkotte, E., and Togai, M. [1989]. The fuzzy-C compiler: A software tool for producing portable fuzzy expert systems. In Bezdek, 708–711.

Thole, U., Zimmermann, H.-J., and Zysno, P. [1979]. On the suitability of minimum and product operators for the intersection of fuzzy sets. *FSS* 2, 167–180.

Thurstone, L.L. [1927]. A law of comparative judgmenet. *Psychol. Rev.* 34, 273–286.

Tiwari, R.N., Dharmar, J.R., and Rao, J.R. [1987]. Fuzzy goal programming—

An additive model. *FSS* 24, 27–34.

Tobi, T., Hanafusa, T., Ito, S., and Kashiwagi, N. [1989]. Application of fuzzy control system to coke oven gas cooling plant. In Bezdek, 16–22.

Togai InfraLogic Inc. [1989]. Fuzzy-C Development System User's Manual Release 2.0. Irvine, CA.

Tong, R.M. [1977]. A control engineering review of fuzzy systems. *Automatica* 13, 559–569.

Tong, R.M. [1978]. Synthesis of models for industrial processes—some recent results. *Inter. J. Gen. Syst.* 4, 143–162.

Tong, R.M. [1984]. A retrospective view of fuzzy control systems. *FSS* 14, 199–210.

Tong, R.M., and Bonissone, P.P. [1979]. Linguistic decision analysis using fuzzy sets. *Memo UCB/ERL M* 79/72 Berkeley.

Trappl, R., ed. [1984]. *Cybernetics and Systems Research*. Amsterdam, 539–540.

Tsukamoto, Y. [1979]. An approach to fuzzy reasoning method. In Gupta et al., 137–149.

Umbers, I.G., and King, P.Y. [1981]. An analysis of human decision making in cement kiln control and the implementations for automation. In Mamdani and Gaines, 369–380.

Verdegay, J.-L., and Delgado, M., eds. [1989]. *The Interface between Artificial Intelligence and Operations Research in Fuzzy Environment*. Köln.

Verhagen, C.J.D.M. [1975]. Some general remarks about pattern recognition; its definition; its relationship with other disciplines. *J. Pattern Recogn.* 8(3), 109–116.

Vila, M.A., and Delgado, M. [1983]. On medical diagnosis using possibility measures. *FSS* 10, 211–222.

Wang, P.-Z. [1982]. Fuzzy contactability and fuzzy variables. *FSS* 8, 81–92.

Wang, P.P., and Chang, S.K., eds. [1980]. *Fuzzy Sets—Theory and Applications to Policy Analysis and Information Systems*. New York, London, 59–75.

Watada, J., Fu, K.S., and Yao, J.T.P. [1984]. Linguistic assessment of structural damage. *Rcp. CE-STR-84-30*, Purdue.

Watada, J., Tanaka, H., and Asai, K. [1982]. A heuristic method of hierarchical clustering for fuzzy intrasitive relations. In Yager, 148–166.

Waterman, D.A. [1986]. *A Guide to Expert Systems*. Reading, Menlo Park, Wokingham, Amsterdam.

Weber, R., Werners, B., and Zimmermann, H.-J. [1990]. Planning models for research and development. *EJOR* 48. (2) Forthcoming.

Weiss, S.M., and Kulikowski, C.A. [1981]. Expert consultation systems: The EXPERT and CASNET projects. In *Machine Intelligence*, Infotech State of the Art Report 9, No. 3.

Wenstop, F. [1980]. Quantitative analysis with linguistic values. *FSS* 4, 99–115.

Werners, B. [1984]. *Interaktive Entscheidungsunterstützung durch ein flexibles mathematisches Programmierungssystem*. München.

Werners, B. [1987a] An interactive fuzzy programming system. *FSS* 23, 131–147.

Werners, B. [1987b]. Interactive multiple objective programming subject to flexible constraints. *EJOR* 342–349.

Werners, B. [1988] Aggregation models in mathematical programming. In Mitra, 295–319.

Whalen, T., Schott, B., Hall, N.G., and Ganoe, F. [1987]. Fuzzy knowledge in rule-based systems. In Silverman, 99–119.

White, D.J. [1975] *Decision Methodology*. London, New York.

Whiter, A.M. [1983]. PFL: Pi-fuzzy logic. A practical fuzzy logic. In *Shortdeskription Systems* Des. Ltd., Hampshire.

Wiedey, G., and Zimmermann, H.-J. [1978]. Media selection and fuzzy linear programming. *J. Operat. Res. Soc.* 29, 1071–1084.

Wierzchon, S.T. [1982]. Applications of fuzzy decision-making theory to coping with ill-defined problems. *FSS* 7, 1–18.

Windham, M.P. [1981]. Cluster validity for fuzzy clustering algorithms. *FSS* 5, 177–185.

Windham, M.P. [1982]. Cluster validity for the fuzzy C-means clustering algorithm. *IEEE Trans. PAMI* 4, 358.

Windham, M.P. [1983]. Geometrical fuzzy clustering algorithms. *FSS* 10, 271–279.

Winston, P.H., ed. [1975]. *The Psychology of Computer Vision*. New York.

Witte, E. [1968]. Phasen-Theorem und Organisation komplexer Entscheidungen. *Zeitschr. f. betriebsw. Forschung* p. 625.

Yager, R.R. [1978]. Fuzzy decision making including unequal objectives. *FSS* 1, 87–95.

Yager, R.R. [1979]. On the measure of fuzziness and negation part I: Membership in the unit interval. *Inter. J. Gen. Syst.* 5, 221–229.

Yager, R.R. [1980]. On a general class of fuzzy connectives. *FSS* 4, 235–242.

Yager, R.R. ed. [1982]. *Fuzzy Set and Possibility Theory*. New York, Oxford, Toronto.

Yager, R.R. [1984]. A Representation of the Probability of Fuzzy Subsets. *FSS* 13, 273–283.

Yagishita, O., Itoh, O., and Sugeno, M. [1985]. Application of fuzzy reasoning to the water purification process. In Sugeno, 19–40.

Yasunobu, S., and Miyamoto, S. [1985]. Automatic train operation system by predictive fuzzy control. In Sugeno [1985b], 1–18.

Yazenin, A.V. [1987]. Fuzzy and stochastic programming. *FSS* 22, 171–180.

Yeh, R.T., and Bang, S.Y. [1975]. Fuzzy relations, fuzzy graphs and their applications to clustering analysis. In Zadeh et al., 125–150.

Zadeh, L.A. [1965]. Fuzzy sets. *Information and Control* 8, 338–353.

Zadeh, L.A. [1968]. Probability Measures of Fuzzy Events. JMAA 23, 421–427.

Zadeh, L.A. [1969]. Fuzzy algorithms. *Information Control* 19, 94–102.

Zadeh, L.A. [1971] Similarity relations and fuzzy orderings. *Information Science* 3, 177–206.

Zadeh, L.A. [1972] On fuzzy algorithms, memo UCB/ERL M 325 Berkeley.

Zadeh, L.A. [1973a] The concept of a linguistic variable and its application to approximate reasoning. Memorandum ERL-M 411 Berkeley, October 1973.

reasoning. Memorandum ERL-M 411 Berkeley, October 1973.

Zadeh, L.A. [1973b]. Outline of a new approach to the analysis of complex systems and decision processes. *IEEE Trans. Syst. Man, Cybern.* 3, 28–44.

Zadeh, L.A. [1977]. Fuzzy sets and their application to pattern recognition and clustering analysis. In *Classification and Clustering*. New York, San Francisco, London, 251–299.

Zadeh, L.A. [1978]. Fuzzy sets as a basis for a theory of possibility. *FSS* 1, 3–28.

Zadeh, L.A. [1981a]. PRUF-A Meaning representation language for natural languages. In Mamdani and Gaines, 1–66.

Zadeh, L.A. [1981b]. Possibility theory and soft data analysis. In L. Cobb, and R.M. Thrall eds. *Mathematical Frontiers of the Social and Policy Sciences*. Boulder, CO, 69–129.

Zadeh, L.A. [1981c]. Test-score semantics for natural languages and meaning representation via PRUF. *Techn. Note* 247, SRI.

Zadeh, L.A. [1983a]. The role of fuzzy logic in the management of uncertainty in expert systems. *FSS* 11, 199–227.

Zadeh, L.A. [1983b]. A computational approach to fuzzy quantifiers in natural languages. *Comput. and Maths. with Appl.* 9, 149–184.

Zadeh, L.A. [1984]. A computational theory of dispositions ERL rep., UCB, Berkeley.

Zadeh, L.A., Fu, K.S., Tanaka, K. and Shimura, M., eds. [1975]. *Fuzzy Sets and Their Applications to Cognitive and Decision Processes*. New York, London.

Zeisig, G., Wagenknecht, M, and Hartmann, K. [1984]. Synthesis of destillation trains with heat integration by a combined fuzzy and graphical approach. *FSS* 12, 103–115.

Zemankova-Leech, M., and Kandel, A. [1984]. *Fuzzy Relational Data Bases—A Key to Expert Systems*. Köln.

Zimmermann, H.-J. [1976]. Description and optimization of fuzzy systems. *Inter. J. Gen. Syst.* 2, 209–215.

Zimmermann, H.-J. [1978]. Fuzzy programming and linear programming with several objective functions. *FSS* 1, 45–55.

Zimmermann, H.-J. [1980]. Testability and meaning of mathematical models in social sciences. In *Mathematical Modelling* 1, 123–139.

Zimmermann, H.-J. [1983a]. Fuzzy mathematical programming. *Comput. Op. Res.* 10, 291–298.

Zimmermann, H.-J. [1983b]. Using fuzzy sets in operational research. *EJOR* 13, 201–216.

Zimmermann, H.-J. [1987]. *Fuzzy Sets, Decision Making, and Expert Systems*. Boston, Dordrecht, Lancaster.

Zimmermann, H.-J. [1988]. Fuzzy sets theory—and inference mechanism. In Mitra, 727–741.

Zimmermann, H.-J. [1989]. Strategic planning, operations research and knowledge based systems. In Verdegay and Delgado, 253–274.

Zimmermann, H.-J. and Pollatschek, M.A. [1984]. Fuzzy 0-1 Programs. In Zimmermann, Zadeh, and Gaines, 133–146.

Zimmermann, H.-J., Zadeh, L.A., and Gaines, B.R., eds. [1984]. *Fuzzy Sets and Decision Analysis*. Amsterdam, New York, Oxford.

Zimmermann, H.-J., and Zysno, P. [1980]. Latent connectives in human decision making. *FSS* 4, 37–51.

Zimmermann, H.-J., and Zysno, P., [1982]. Zugehörigkeitsfunktionen unscharfer Mengen (DFG-Forschungsbericht).

Zimmermann, H.-J., and Zysno, P. [1983]. Decisions and evaluations by hierarchical aggregation of information. *FSS* 10, 243–266.

Index